CONNECTING A NATION

CONNECTING A NATION

THE STORY OF TELECOMMUNICATIONS IN IRELAND

BY DERYCK FAY

Published by
UNIVERSITY COLLEGE DUBLIN PRESS
PREAS CHOLÁISTE OLLSCOILE BHAILE ÁTHA CLIATH
2021

First published 2021
by University College Dublin Press
UCD Humanities Institute, Room H103,
Belfield,
Dublin 4
www.ucdpress.ie

ISBN 978-1-910-08208-72

CIP data available from the British Library

Design by iota (www.iota-books.com)
Typeset in Scotland by Ryan Shiels
Printed in Scotland on acid-free paper
by Bell & Bain Ltd, Burnfield Road,
Thornliebank, Glasgow, G46 7UQ, UK

CONTENTS

LIST OF ILLUSTRATIONS

FIGURES

Copyright owners are credited in the caption under the appropriate figure. I have tried and failed to find the owners of the copyright for some figures I have used. I would be pleased to insert an acknowledgement in future editions.

TABLES

ACKNOWLEDGEMENTS

First and foremost, I must thank my manuscript editor Richard Lever for his patience, attention to detail and encouragement. This book would simply not have happened without him.

A large number of other people have also been enormously helpful along the way. These include John McCourt, Doreen Finn, Lidia Lonergan, Pauline McGuirk, Margaret Hynds O'Flanagan and Vawn Corrigan for help with proposals and dealing with publishers, Ian Shuttleworth for help with GIS, and Colum Ring, Nevan Bermingham and Andrew Corcoran for their sleuth work. I must especially thank Tom Wall, former archivist with Telecom Éireann, for patiently answering my various queries with replies full of detail and charm; and Derek Cassidy, Mick Fox, Frank McCurry, Brendan Smith, Donal Fallon and Dani Colbert for help and advice on various questions from submarine cable capacity to dental care in the 1970s.

I am very grateful to everyone who provided or helped me locate the many illustrations in the book, including John Mulrane, Dennis Jennings, Paul Mulvey, Florence Bugler, Stephen Ferguson, Angela Sturdy and Padraig Mitchell. Tragically, Adrian Clayton, who supplied the beautiful photograph of a cable landing at Bull Island which closes Chapter Twelve, passed away during the course of writing.

This book is brought alive by the individuals who contributed their own stories, including Vincent Sweeney, Florence Bugler, Sarah Carey, Brian Noble, John Beckett, Dennis Jennings, Tom Allen, Mary O'Rourke, Liam Ferrie, Ben Jones, Isolde Goggin, Margaret Furnell, Joe Kearney, Nevan Bermingham and Michelle Ward. I am also indebted to my uncle, Tony Rice, for jogging my memory of the family story which forms the basis of Chapter Eight. Thank you all for taking the time to contribute.

I also want to thank Noelle Moran, Conor Graham, Ryan Shiels, Jane Rogers and all the UCD Press team, and the readers of my initial manuscript for their helpful feedback.

Before embarking on this book, I don't think I had ever spoken to an archivist and had no idea what a helpful and enthusiastic bunch they are. I have to especially thank James Elder, Anne Archer and Dave Shawyer at the BT Archives, Aisling Dunne in the Irish Architectural Archive, and Aisling O'Malley of the IET for their assistance digging out material for me. In the same vein, I must also thank staff at the various libraries I have visited including TCD, the National Library of Ireland and the Dublin City Council library network. The ability to search online for a book held in any public library in the State and pick it up a few days later from the friendly staff in my local library is a joy and a credit to the dedicated staff of our public library service.

Throughout the writing process, my wonderful husband Justin has provided endless patience and constant support, from driving me to some Napoleonic-era Martello tower on a far-flung headland to providing a live thesaurus service. I am eternally grateful for his love and help.

Deryck Fay
Dublin, October 2021

PREFACE

As you walk up Dublin's Grafton Street every second person is head-down on their mobile phone. It's the same story on Patrick Street in Cork and almost every street of every town in Ireland. Those same streets are bursting with shops selling mobile phones and unlimited broadband deals with fibre-this and wireless-that. Media constantly report on how Ireland's software industry includes nine of the world's top ten tech firms and generates €50 bn in annual exports. Sitting silently around Dublin is a necklace of unmarked data centres, storing everything from airline bookings to our personal videos of cute cats. Even more anonymous are the 17 underwater cables stretching out from the coastline, carrying our text messages, phone calls and internet data in and out of the country. Beyond Dublin, a national programme to connect over half a million rural homes to the internet by fibre is underway. We are awash, some might say swimming, in modern telecommunications.

And yet I'm old enough to remember when Ireland was a largely agrarian society with a two-year waiting list just to get a landline phone installed. How did we get from that old-world agrarian society to this modern, super-connected one?

I've always been curious about technology and can still remember the thrill I felt at age seven when I discovered an extra, fourth channel on our television, a fuzzy BBC2 irradiated by a new cable system to which we were not yet officially connected. A year or two later, on a walk up Howth Hill with my cousins, we came across a huge metal sheet standing proud of the gorse, like a tarnished mirror made for a giant. My young mind was fascinated as my uncle Jim Prendergast explained that this contraption was used to bounce telephone calls carried by radio waves. (You can read a fuller description of the Dublin Arklow microwave link in Chapter Eight).

Having lived through the massive technological changes of the last few decades and worked in telecommunications on and off for 16 years, I found myself wanting to know about the journey to this modern, connected Ireland: the technologies involved, the people who had made this progress possible, and who or what had hindered it. But when I searched for books that charted this journey, I couldn't find any. I could find several books about the history of electricity in Ireland, many about radio and television broadcasting, and innumerable books about the most obscure Irish railways. I could find little written about the history of telephones or telegraphs and even less about topics like broadband and mobile phones. In fact, the whole story of telecommunications – what the dictionary defines as 'the technology of sending signals and messages over long distances using electronic equipment' – seems to have been ignored in Ireland.

There was only one solution! Before I started writing this book, I'd envisaged a story of how Ireland, a telecoms laggard, had finally caught up with its peers, of technological advances pioneered by well-known heroes, and of politicians who failed to grasp the importance of telecommunications to the economic and social development of the State. There were plenty of such politicians, and some honourable exceptions. But as I researched the topic, I found there were also many unsung heroes and heroines from Ireland, some working to connect an empire, their contributions forgotten in an Ireland that sees itself as the colonised rather than the coloniser. And it turned out that Ireland's current role as hyper-connected home to multiple data centres had a precedent in the nineteenth century when the country was a vital hub in a global telecommunications network.

The more I researched the more I found that not only were there the big stories about technical breakthroughs, policy shifts and dodgy dealings but lesser-known ones about experimentation, social change and personal contributions. For instance, as well as Marconi's pioneering experiments at Rathlin Island and his transatlantic telegraph station illuminating a Connemara bog with its sparks, there was 16-year-old Agnes Duggan's first day as a 'telephone operator' in Dublin in 1881, and Maureen Flavin's phone calls with the weather reports from Blacksod which helped to change the course of Second World War. The stories came from all corners of the island, from the transatlantic telegraph hub at Valentia in Co Kerry to the maritime communications centre of Malin Head in Co Donegal, and everywhere in between.

With little published material about the subject of telecommunications, I sought out primary sources such as the records of telecommunications operators like BT, which holds much of the pre-1922 material for Ireland. At the start of my research, I learnt that just after its privatisation in 1999, Eircom had transferred its archive to the National Archives of Ireland and anticipated that this would be a rich source of information for me. Unfortunately, it emerged that during the intervening 20 years the National Archives hadn't managed to even catalogue the data, and so it remained inaccessible to researchers. Indeed, most of the records of the Department of Posts and Telegraphs and its successor departments appear to have vanished.

The lack of official records of the country's telecoms infrastructure, however, did not turn out to be as fatal as I'd feared. There proved to be a wealth of information out there in newspapers and journals, records of Dáil debates and, often best of all, correspondence and personal memories. Some of the material was poignant, such as the letter tucked in a Treasury file in London seeking government approval for a new telegraph line to Belmullet in Co Mayo to facilitate the arrangements 'for the emigration of poor persons'. The beautiful copperplate handwriting of the author could not disguise the desperation of the people in this remote part of the west of Ireland, still suffering 38 years after the Famine.

The Coronavirus pandemic struck as I was writing the closing chapters. With archives and libraries shuttered, I began to seek out more of these personal narratives, hunting down contributors through Facebook, LinkedIn, and sometimes by dispatching a letter with just a name and a sketchy address. The individual accounts I received provided a rich source of first-hand material to add to the other sources.

Connecting A Nation has turned out to be a fascinating and uniquely Irish story.

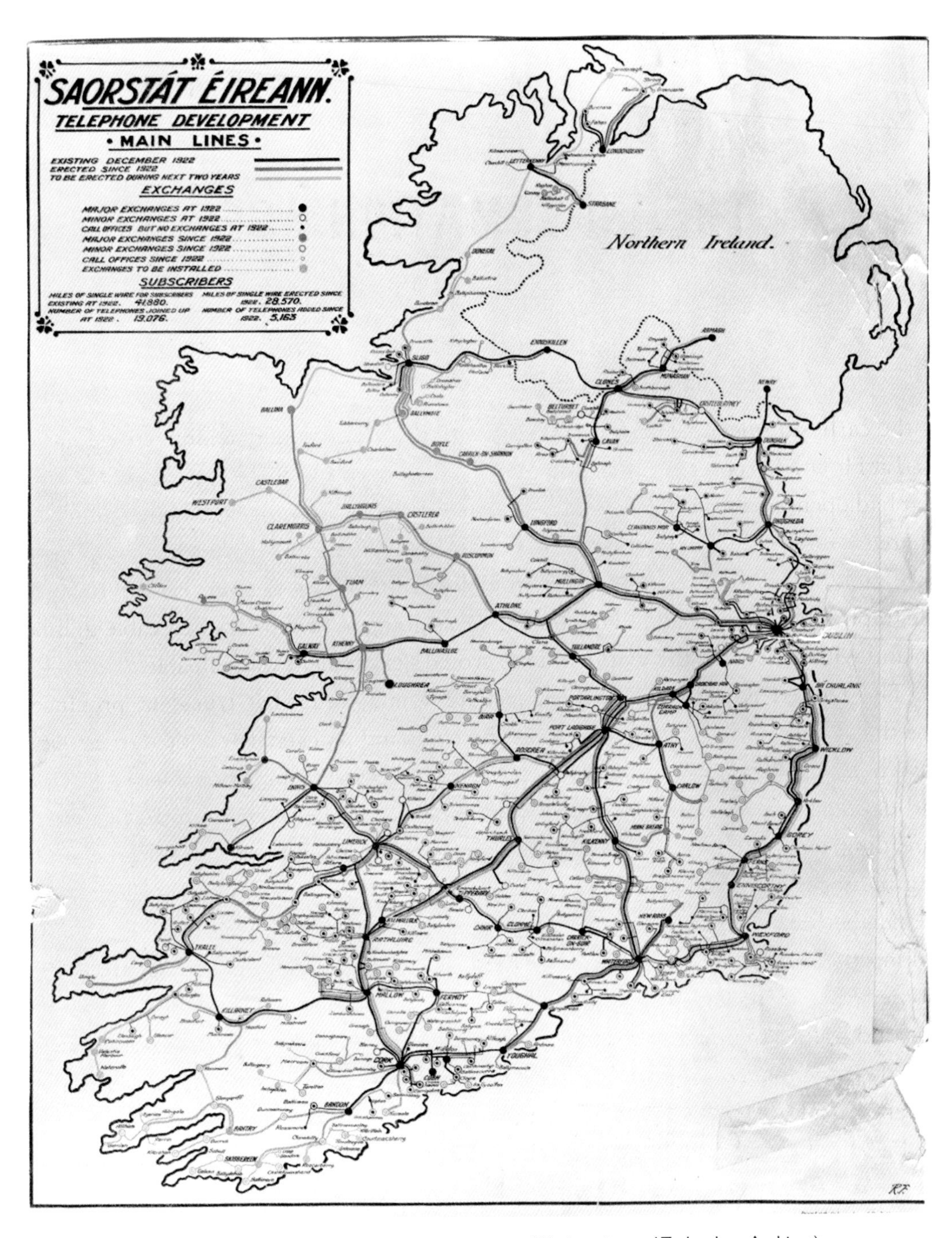

1926 Map of Telephone Development. (Courtesy of the Institution of Engineering and Technology Archives)

CHAPTER ONE

THE INFORMATION AGE

In the early summer of 1852, on a pleasant Wednesday evening, a large crowd gathered at Howth harbour, many having travelled from Dublin city by train along the line that had opened just five years before. They peered northwards towards Ireland's Eye where two ships, the *Britannia* and the *Prospero*, were coming slowly into view. These were cable ships, which had set off from Holyhead in Wales 16 hours earlier[1] to lay a telegraph cable all the way to Howth. This was going to be the very first cable to link Ireland with the rest of the world and the crowd's excitement was palpable.

The United Kingdom, of which Ireland was a somewhat truculent province, was the richest country in the world at the time and its capital was the largest city anywhere. It was an era of dizzying technological advancement. The railway network had doubled in five years and lines for the recently invented telegraph service were spreading across Britain. A number of companies realised the potential for transmitting messages across the Irish Sea by telegraph using a copper wire insulated with the newly discovered gutta-percha and a battle of competing routes ensued. A cable linking the two islands would allow Dublin Castle to receive commands from the Home Office in London or a linen mill in Belfast to accept orders from a merchant in England – all in an instant. An even greater prize was the ability to transmit a message from London to a port on the west coast such as Galway, from where it could then be put aboard a liner that would head straight across the Atlantic to America.

Back in Howth, just after eight o'clock the crowd grew animated as the two ships drew close to the shoreline. The *Britannia*, which had been transmitting regular test communications over the cable stretching behind it back to Holyhead as it crossed the Irish Sea, now sent a message to say that it had moored at Howth.[2] Thirty minutes later, the *Britannia* fired its cannon and a cheer went up from the jubilant crowd, for they knew that the signal to fire the cannon had been given all the way from Holyhead, 123 km distant, over the new telegraph cable. The cable was working!

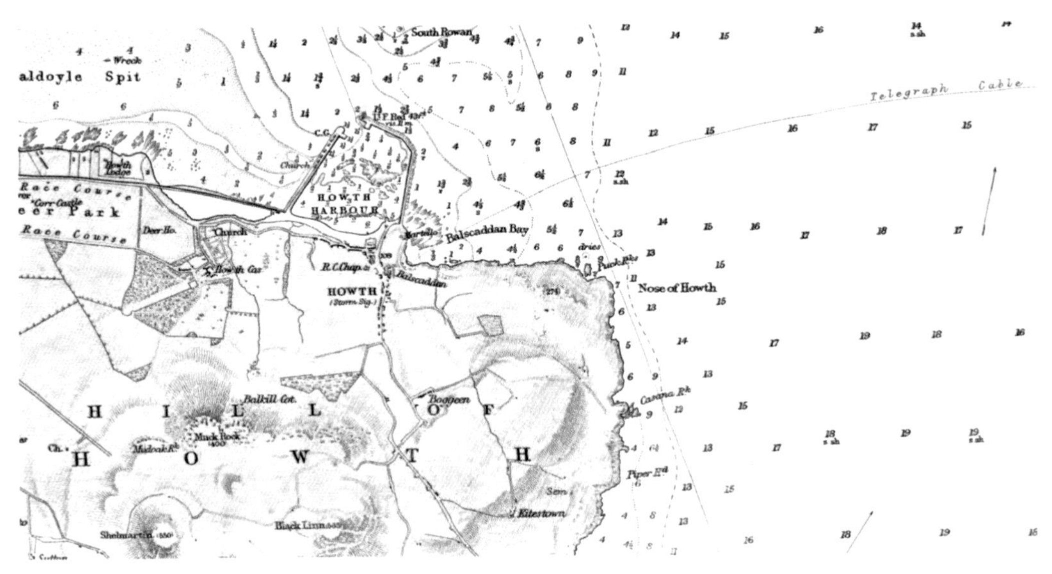

Figure 1.1: Howth, as shown on an Admiralty chart of Dublin Bay from 1874. The telegraph cable indicated is probably that laid to Holyhead in 1871 rather than the earlier, failed, cables.

The Irish telecommunications story really begins with this moment on 1 June 1852. For while it had been technically possible to lay a telegraph cable overland for several years – two Irish railways had installed such systems to communicate with their depots – up to now no-one had bothered to lay a single mile of cable for *public* use. After all there was little money to be made connecting an impoverished and decaying Dublin to a countryside that was only just recovering from the effects of the Great Famine. The economics – and politics – were different, however, if messages could be sent not just between different parts of Ireland and Dublin, but between Ireland and London, the Imperial Capital and richest city in the world. And so the first telegraph line provided for public use in Ireland was not between Belfast and Dublin or Dublin and Cork, but between Howth and Holyhead.

The new submarine cable was quickly hooked up to a connecting line from Howth to central Dublin, and another much longer line from Holyhead to London. Within a few days a 15-word telegram could be flashed from London to Dublin in one minute. For the first time information could travel faster than a galloping horse, a train or a steamship. The newspapers were ebullient at the arrival of this communications revolution. The Dublin-based *Evening Packet and Correspondent* (motto 'The Queen – The People – And the Law') hoped that the new rapid communication method with London would lead to 'increasing allegiance to the British Crown'.[3] Unfortunately this jubilation proved short-lived. Within a few days the papers reported that:

> communication between this city and Holyhead is suspended for a few days owing to some accident having occurred to the wire off Holyhead, supposed to have been caused by the anchors of some of the fishing boats having grappled with the cable.[4]

In fact, the cable had failed after only three days of use. But this setback would not stop the race for information. The following year Ireland was connected again by the umbilical cord of a telegraph line, this time successfully. Information has been streaming in and out of the country through cables under the surrounding seas ever since.

Fast forward 160 years: in 2011, an additional cable was laid between Holyhead and Dublin.[5] With a capacity of 1,440 terabytes per second (Tbps), the CeltixConnect-1 fibre-optic cable had a capacity 863,000,000,000,000 times greater than that first cable in 1852.

To understand how it's now possible for every second person walking down Grafton Street in Dublin or Patrick Street in Cork to have their head stuck in their mobile phone, we have to look back to this point and appreciate the impact of that cable landing at Howth. The ability of those first telegraph cables, laid across the land and under the sea, to transmit information at almost the speed of light, unimpeded by the shackles of the physical world, heralded a revolution, creating the first Information Age and setting in motion an unstoppable sequence of events. Soon cables would carry the human voice and, later, data of any kind. The meagre trickle of information would grow stronger and stronger, eventually becoming a flood. And it would not be long before it became possible to transmit voices and data not only along wires but through the ether itself.

The upbeat press coverage from that summer's evening in Howth refers to the coming of the 'electric telegraph'. The word 'electric' was included because the concept of the telegraph had existed long before the ill-fated cable laid by the *Britannia*. The word 'telegraph' is derived from the Greek for 'distance writing'. Historical examples of this abound. Agamemnon erected a 500 km-long chain of beacons for his attack on Troy in 1084BC.[6] In Celtic Ireland, bonfires lit on Lughnasa may have been a way of signalling to surrounding communities that it was time to gather the harvest.[7] Of course such signalling systems could transmit only the simplest of messages over short distances.

The advent of the modern telescope in the eighteenth century overcame many of these limitations, and for a period around the year 1800 Ireland was at the forefront of optical telegraphy technology, thanks to Richard Lovell Edgeworth of Co Longford. Father of the novelist Maria Edgeworth, he was a progressive politician, enlightened land-owner and prolific inventor. Edgeworth developed a telegraph system based on four semaphore arms that could be set to any one of eight positions (Figure 1.2) and used predetermined codes to represent what he considered to be the most useful words and phrases. Thus the code 7275 was short for 'Can the yeomanry be ready to march at an hour's notice?' while 7277 was employed to ask the useful question 'What is the disposition of the peasantry?'[8]

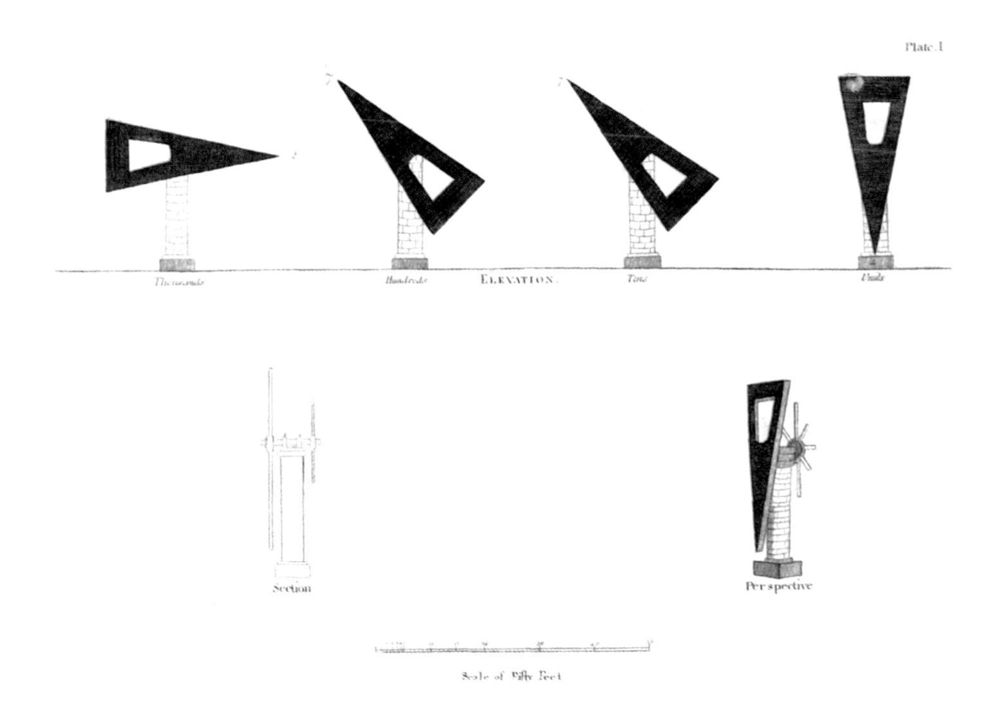

Figure 1.2: Edgeworth's optical telegraph system used four pointers, each of which could be set to any of eight positions.[9]

Edgeworth's telegraph system anticipated several modern communications principles. His use of eight positions was similar to the grouping of eight bits into a byte, which has been used in digital systems since the 1960s, while his use of agreed codes was a form of analogue data compression to maximise the amount of information that could be sent over a low-bandwidth circuit. In August 1795 with the assistance of his son, Lovell, Edgeworth installed his system between Portpatrick in Scotland and Donaghadee in Co Down[10], and sent four messages across the sea – a remarkable feat over a distance of 35 km by optical means. However, despite showcasing the potential of his system to the Irish Parliament by a demonstration between Dublin and the house of the Speaker of the Commons at Collon, Co Louth, Edgeworth received no support from government in Dublin.[11]

Official attitudes changed following the 1798 Rebellion when a French force of 1,000 men landed in Co Mayo and managed to march 120 km across the country before being routed by the British army at Ballinamuck in Co Longford. Amid fears of another French invasion, Edgeworth won a contract to construct a telegraph link across the centre of the country between Dublin and Galway. He was assisted in the two-year project by his brother-in-law and Meath native, Captain Francis Beaufort, more famous as the creator of the Beauford Scale for indicating wind force.[12] After two years of design and construction, the optical telegraph network of 15 stations was opened for service on 2 July 1804 by the Lord Lieutenant with a ceremony at the Royal Hospital in Kilmainham:

> The Lord Lieutenant had the goodness to arrange a message of considerable length for the Telegraph, which was sent to Galway in 35 minutes, and to which an answer was returned by Captain Beaufort, of the Navy, who is a confidential Officer at Galway, in less than five minutes.[13]

Edgeworth designed his optical telegraph system to be secure against both a French invasion and, with memories of the 1798 Rebellion still fresh, restive locals. Each telegraph tower in the network had a well-built guardhouse. The station at Hill of Cappagh, near Kilcock, which was fortified using the ruins of a windmill[8], was described by a contemporary journalist as 'defensible against everything but cannon and would most probably resist six pounders for a considerable time'.[13] No trace of the telegraph tower at Cappagh remains, but a mast for a microwave telephone link was erected at the same location in 1969. The security of the location became a concern again in the 1970s,[14] though, as will be discussed, six-pound cannon balls were no longer the danger. The mast remains in use for mobile telephony and is clearly visible by motorists on the M4 as they slow to pay at the Kilcock toll plaza. While the opening of the motorway brought the car journey from Galway to Dublin down to about two hours, such a journey remains considerably slower than the five minutes taken by a telegraphic message in 1804.

Edgeworth's system was not the only optical telegraph network built at this time. In addition to the chain of round defensive Martello towers, the Admiralty also established a separate network of 81 semaphore-based signal towers from the very northern tip of Ireland at Malin Head in Co Donegal, across almost every headland on the west and south coasts and up the east coast as far as Dublin. However, both the Admiralty signal towers and Edgeworth's system had short lives, being abandoned within a few years as the threat of a French invasion receded.[15] As we will discover, many of the sites chosen for these systems found further use in the telecommunications network in subsequent years (Figure 1.3).

Figure 1.3: The signal tower at Bray Head on Valentia Island originally formed part of the Admiralty network built around 1802-4. Like several other such towers, it was repurposed as a Look Out Post during 'the Emergency' (see Chapter Seven).

Despite the advances of Edgeworth and others, such as the French inventor Claude Chappe, there remained many problems with optical telegraph systems: they had limited capacity, required a relay station every 12-18 km, thus imposing high operating costs, and were useless in poor weather. They were really only of use by government for military purposes or valuable commercial information and were not available to the general public, though Napoleon allowed the Chappe system in France the critical task of carrying the winning numbers in the weekly National Lottery draw.[16] The vast majority of communication was unsuitable for optical telegraphy and could only travel as fast as the best horse or the fastest boat. This was as true for Daniel O'Connell arranging his 'monster meetings' in 1843 as it had been for Oliver Cromwell laying siege to Drogheda in 1649 or Brian Boru when mustering his army to fight the Vikings in 1014.

But all that was about to change.

The mid-nineteenth century was a hotbed of technical innovation, economic progress and social change. In Belfast, flax mills were mechanised, and engineering firms developed markets across the Empire for their machines and tools.[17] The economy was transformed as a growing band of middle-class investors dabbled in the stock market, buying shares in start-up companies involved in new businesses like railways. In parallel came a social transformation, with the national school system providing free education to all children in Ireland from 1831 leading to a corresponding increase in levels of literacy. These developments created a huge demand for communications. Linen merchants in Belfast wanted to know the best market prices for their products, stockbrokers craved up-to-date prices of shares while the surge in literacy created a potential vast new market for up-to-date information to fill the pages of the newspapers.

In tandem with this growing demand for information came the technology to meet it – the electric telegraph. As is so often the case with such inventions, the development of electric telegraphy involved many people, each person building on the discoveries of someone else. So while the first patent for electric telegraphy was taken out by William Fothergill Cooke and Charles Wheatstone in England, their work was built on earlier discoveries by other scientists about electricity, batteries and magnetism.

Cooke and Wheatstone's telegraph was patented in 1837[18] just as the railways were commencing their spectacular expansion from isolated curiosities into a network covering countries and continents. Cooke knew that the railways needed a way to send messages quickly; he also knew that telegraph lines required a route, and so began the symbiotic relationship between the telegraph and the railways in the UK and elsewhere. But he realised that the railways did not have to be the only customer and so in 1842 a commercial telegraph service was launched, allowing anyone who was so inclined to send a message from London to Slough in an instant for a shilling (7c).[19] Cooke and Wheatstone's initial system was based on needles that were deflected by electric current to indicate letters (Figure 1.4).

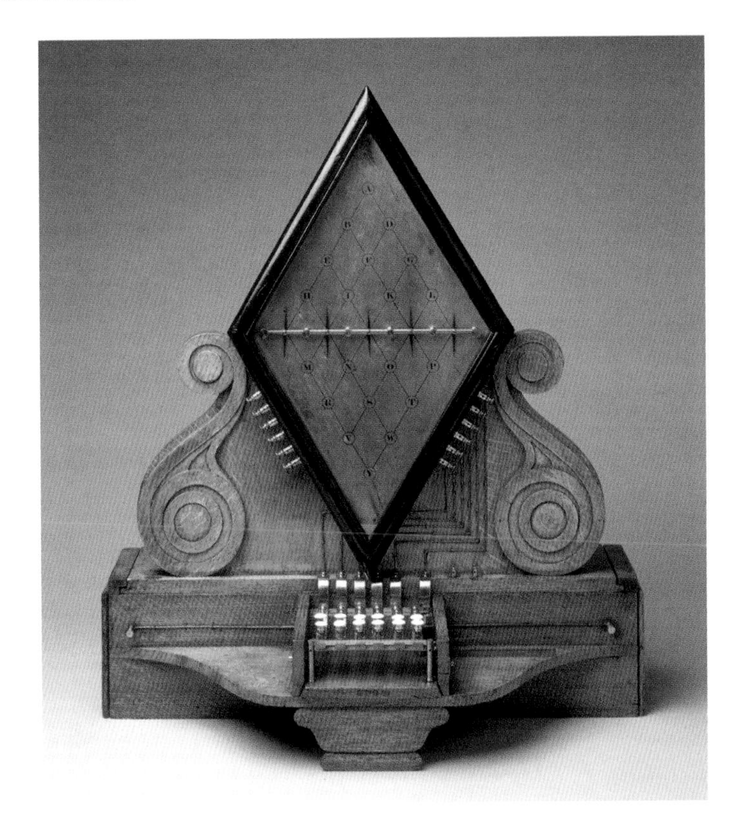

Figure 1.4: Cooke and Wheatstone's five-needle telegraph, patented in 1837, used a device with a grid of 20 letters with 5 needles arranged across the middle. The deflection of any two needles to the left or right would point to specific letters. *(© Science Museum/Science and Society Picture Library)*

An early customer of Cooke and Wheatstone's telegraph was Queen Victoria, who used it to send news of the birth of her son Alfred from Windsor Castle to London. As a result, news of the royal birth was printed in the London *Times* within 40 minutes of the event, the newspaper acknowledging its 'indebtedness to the extraordinary power of the electro-magnetic telegraph'.[20] Newspapers realised that readers would pay for timely information and quickly became a key market for the telegraph industry, using the new technology to supply their growing readership with the hottest political intrigue and up-to-the-minute share prices, under the heading 'By Electric Telegraph'. By 1873, the 17-year-old *Irish Times* had grown from a slim four-page newspaper printed thrice weekly to a 12-page daily[20] filled with stories dispatched by telegraph from around the globe.

The growth in media consumption and acceleration in the speed of news made possible by the electric telegraph affected social attitudes. Up to this point reports of distant wars took weeks or months to be published, their impact lessened by the passage of time. During the Crimean War, which took place in what is now part of Ukraine from 1853 to 1856, Tallaght-born William H. Russell, a reporter for the London *Times*, sent eyewitness accounts of events on the battlefield by telegraph. His reports of the appalling medical and nursing care endured by the British forces involved in the war created a political storm and led to the appointment of Florence Nightingale as superintendent of nursing.[21] In the 12 years since Cooke had opened the first electric telegraph service, the volume and speed of information it permitted was affecting not only newspaper reporting but also the political system.

While Cooke and Wheatstone's invention took off in their native United Kingdom, they did not have the world to themselves. Just 110 days after they had patented their telegraph in England, an initial application for a telegraphy system was made to the patent office in Washington DC by a Samuel Morse. Like several other famous figures in the global story of telecommunications, Morse was not a particularly likeable character. In 1835 he wrote a book called *Foreign Conspiracy against the Liberties of the United States*, a treatise warning Americans against the political influence of Roman Catholicism:

> Foreign emigrants are flocking to our shores in increased numbers, two thirds at least are Roman Catholics, and of the most ignorant classes, and thus pauperism and crime are alarmingly increased. The Irish Catholics in an especial manner clan together, keep themselves distinct from the American family, exercise the political privileges granted to them by our hospitality, not as Americans, but as Irishmen, keep alive their foreign feelings, their foreign associations, habits, and manners.[22]

Morse contested the elections for mayor of New York City in 1836 and 1841 running on an anti-immigrant and anti-Catholic ticket. Having lost both elections by a large margin,[23] he turned his hand, more successfully, to portrait painting. But it was his ventures in the sphere of telegraphy that ensured his name was handed down to posterity.

In the UK, Cooke and Wheatstone got their first break from the railways. In the US, Morse got his first break from Congress when, in 1843, it agreed to provide a grant of $30,000 to build a telegraph link between Washington DC and Baltimore. Telegraph mania soon spread across the USA and by 1850 over 20 separate telegraph companies operated 12,000 miles of lines that connected the main centres on the eastern seaboard.[23] By 1861 these disparate companies had largely been consolidated into one. Known as Western Union, it formed relationships with most of the railroad companies in the US, connecting the east and west coasts of the country[18] using the technology patented by Morse, before expanding into global communications.

Morse's claim over telegraphy did not go unchallenged in his native US. Amongst his many patent battles was one with a rival telegraph proprietor called Henry O'Reilly that went all the way to the US Supreme Court.[24] In a twist that must have irked Morse's sectarian views, his opponent O'Reilly was an Irish Catholic from Carrickmacross in Co Monaghan. The newspapers, by now reliant on the telegraph network for their supply of data, had a vested interest in maintaining competition in the market and devoted many column inches to the case. The court decision in 1854 affirming his rights over telegraphy made Morse a wealthy man and ensured that his name became synonymous with data transmission.

Morse successfully patented his invention in France and his system was quickly adopted across most of Europe. It was simpler to negotiate such deals in mainland Europe as, unlike in the UK and the US where there were multiple competing operators, in most European countries development of the electric telegraph network was the responsibility of a single state agency. Furthermore the UK was the birthplace of needle-based systems such as those devised by Cooke and Wheatstone. However, the need for just a single wire was a huge advantage for Morse's system and it eventually won out with even Cooke admitting to its superiority in 1856.[16]

Although ultimately displaced on most telegraph routes by more sophisticated systems, Morse's system proved extremely durable, being retained for many communications purposes well into the twentieth century. An unwitting harbinger of the on-off binary code used in modern digital telecommunications, Morse code's staccato string of long and short pulses is still used in modern culture to symbolise data flows – in the 2014 film *Interstellar*, the protagonist Cooper uses Morse code to transmit data through space and time.

In Ireland, as in the rest of the UK, Morse code was spurned by early adopters of the telegraph who installed needle-based systems such as those pioneered by Cooke and Wheatstone. The early adopters came from the railway industry where a nascent national rail network was being developed by competing private companies. This railway mania gave rise to some technological oddities, one of which was the Atmospheric Railway that opened in 1844 between Dalkey and Kingstown (now Dún Laoghaire) station, the latter being at that stage the terminus of the railway line from Dublin city. Due to the gradients involved, instead of each train being pulled by a steam locomotive, the new railway adopted an 'atmospheric' vacuum system. A pipe ran between the rails with an opening at its top covered with a leather seal, each train being connected to a piston inside the pipe. A vacuum was created in the pipe by a pumping engine at Dalkey, causing the piston to move and pulling the train with it.[25]

This curious system gave rise to some potential safety issues as staff operating the pumping engine did not know the weight of a train leaving Kingstown and so could not adjust the vacuum to compensate. A particularly heavy train might end up stopping short of the station in Dalkey whereas an unusually light train could be pulled so fast that it would overshoot the buffers. To solve this, within a few months of opening, the railway company had installed a Cooke telegraph[27] so that staff at Kingstown could send a signal to the Dalkey engine house at the appropriate time to start operating the vacuum pump. This was the first installation of an electric telegraph in Ireland, though it could not be used by the public. The same was also true of a telegraph system installed a few years later by the Great Southern & Western Railway between Kingsbridge (now Heuston) station and its depot at Inchicore.[28] These private telegraph systems owned by the railways would later become an essential tool for the safe operation of trains across the world, though as we will soon see, it would take a calamity in Ireland to make their use mandatory.

Along with the big names such as Morse, Cooke and Wheatstone, many other inventors were beavering away on new ideas for telegraphy. Amongst these was Corkman John Nott (Figure 1.5) who patented a system in 1846 that propelled a pointer around a large dial to indicate letters and numbers. It was adopted by a railway company in England but following a legal challenge from a telegraph company[29], Nott emigrated to Australia and took out no further patents in telegraphy.

Figure 1.5: John Nott, a native of Cork city, became embroiled in a court battle with Cooke over patents for items of telegraph apparatus. Nott won in court but emigrated to Australia shortly after, where this photo was taken. *(Courtesy of Angela Sturdy)*

The two small railway systems remained the only telegraphs in Ireland for several years. While Britain was positively criss-crossed by a tangle of wires, one reaching all the way to Holyhead in the furthest corner of Wales, Ireland was considered too small and too poor to justify its own standalone network. In 1850 no one had successfully crossed a body of water anything like the Irish Sea with a telegraph cable. But, as we saw, that was soon to change, a development brought about partly thanks to the efforts of two Irishmen.

The first of these was Limerickman William O'Shaughnessy, who had a wide interest in various aspects of science. In 1838, having moved to India, O'Shaughnessy successfully laid 3 km of insulated wire in the bed of the Hooghly river[6] thus proving that telegraph signals could be carried under water. Appointed Superintendent of Telegraphs for India in 1852, the year of the abortive Holyhead–Howth cable, he oversaw the erection of 7,000 km of telegraph lines across India within four years[6], a feat which earned him a knighthood.[30] O'Shaughnessy's successful experiments under the Hooghly did not, however, solve the problem of laying a telegraph cable under the sea. Seawater is a good conductor of electricity, so any cable carrying a current under the ocean has to be perfectly insulated. Glass, porcelain and wood were suitable as insulators on land but were obviously impractical for use underwater.

This is where another Irishman comes into the story, in the form of Dublin chemist Henry Bewley. A relative of the founders of the well-known café, Bewley was a partner in a soda-water making company called Bewley & Draper of 23 Mary Street. The firm needed stoppers for their soda bottles and used the newly discovered gutta-percha for this.[31] Like rubber, gutta-percha was made from tree sap, in this case the percha tree, but, unlike its chemically-close cousin, gutta-percha did not deteriorate on contact with water.[24] Bewley devised a machine for the extrusion of gutta-percha tubing[18] and helped to found a London-based business called the Gutta Percha Company to make underwater cables. (In addition to soda-water bottler stoppers and cable insulation, gutta-percha was also used for running shoes, hence the continued use of the word 'gutties' for such shoes in Scotland and Ulster.) Through its colonial position, the UK had a huge advantage: the percha tree was native to Malaya, a British colony.[24] As a consequence, the Gutta Percha Company manufactured almost all the submarine cables used during this early period of the telegraph network, including the first cables to span the English Channel, unsuccessfully in 1850 and, successfully in 1851.[6]

The wider expanse of the Irish Sea was more of a technical challenge, but increased competition in the telegraph industry led to a number of newly founded rivals eyeing up the potential of the Irish market. First out of the traps was the same company that had laid the successful Dover–Calais cable. In 1851 it proposed to lay a Dublin–Holyhead cable – provided the government would pay a rent of £1,000 (€1,270) per annum. The *Evening Freeman* considered this to be an excellent deal: 'We cannot for a moment contemplate the possibility of any intelligent government rejecting an offer so spirited, and paralyzing, by a scandalous parsimony, a project of such prodigious international advantage.'[32] Evidently the government did not share this enthusiasm and no more was heard of this particular venture.

The government's apathy towards the subsidy was probably justified as within a few months a different company came forward with much the same project – but without the need for government money. This enterprise, the Irish Sub-Marine Telegraph Company, was the company that laid that first, cursed, cable from Holyhead to Howth. It struck deals with the 'Electric', the largest of the telegraph companies in Britain, to use its London–Holyhead line and with the Earl of Howth for a cable landing site on his estate at Howth.[33] The project had all the features of an overhasty job and from the start the portents were not good. Production of the single-circuit cable was rushed through within four weeks; the cable, at one ton per mile (635 kg/km), was one-seventh the weight of the successful Dover to Calais cable[34] and the company's directors bickered with their engineer. Even before the *Britannia* and *Prospero* had set off from Holyhead for Howth on 31 May 1852, a fault was found in the copper insulation of the cable while it was still in the hold. It was this cable that was landed ashore amid scenes of jubilation at Howth. Having endured such a difficult birth, it was not surprising that this poor cable died after a few days of life.

The failure of this first cable under the Irish Sea, while unlucky, was not exceptional. It was calculated that of the 11,364 nautical miles (21,046 km) of undersea telegraph cable laid between 1851 and 1862, only 25 per cent were successful.[24]

It was a case of third time lucky in 1853 when another company, called the English & Irish Magnetic Telegraph Co., or just the 'Magnetic', laid a cable to link Ireland with its network in Britain. It selected a route over the much narrower North Channel between Donaghadee in Co Down and Portpatrick in Scotland. Its first attempt in October 1852 failed due to bad weather and the company decided to wait out the winter before trying again in May 1853. This time they assigned their young Chief Engineer, Charles Tilston Bright, recently poached from the rival Electric, to supervise the work. Involved with telegraphy since the age of 15 and holder of 24 patents[24], Bright was destined to dominate the early decades of telegraphy in Ireland and beyond. The submarine cable connecting Ireland and Scotland cost £13,000 (€16,507) and contained six circuits[34] – clearly the Magnetic anticipated a lot of business from Ireland. Underground roadside cables were laid to connect the Portpatrick landing point with the rest of Scotland and the north of England and on 23 May 1853 the line carried its first message between Ireland and Britain. There is no record of jubilation at Donaghadee or of cannons being fired remotely. Perhaps the Magnetic, knowing that earlier attempts had ended in failure, did not want to tempt fate.

The Magnetic quickly signed deals with most of the Irish railway companies to erect telegraph wires alongside their lines and a flurry of cable laying ensued over the next few months in order to connect Donaghadee with Belfast and from there to Dublin and beyond. The early telegraph companies were particularly keen to reach Galway where lucrative transatlantic telegram traffic was expected if a long-mooted plan to develop a transatlantic port in the city came to fruition. As a result the country's first inland telegraph line was erected alongside the brand-new Midland Great Western Railway between Dublin and Galway.[28] The two-wire telegraph line, which cost £11 *0s. 7d.* per mile (€8.75 per km) to build, opened for business in July 1852[28] with the Galway telegraph station at the railway station.[35] The failure of the first cross-channel cable rendered it an isolated offshoot for over a year until the rest of the network caught up. Stitching together this network caused some disruption to the lives of Dubliners:

> The centre of Sackville-street, Carlisle-bridge, the quays, and other of our leading thoroughfares, presented a busy scene yesterday, several hundreds of workmen being busily employed in laying down the wires of the magnetic telegraph intended to connect the termini of the Great Southern and Western [Heuston] and the Drogheda Railways [Connolly]. The wires were laid nearly in the centre of Sackville-street, in a trench about a foot deep.[36]

The commercial life of Dublin has continued to be disturbed by cable laying ever since.

The Magnetic was also busy back in Britain, moving southwards on London. By September, Belfast and Dublin were linked to Liverpool and Manchester.[37] Finally, on 16 January 1854, Dublin and London were connected[38] and a message could be sent within minutes between the Home Office in London and the Castle in Dublin.

The Magnetic's exclusive agreements with the railway companies in Ireland gave it a huge advantage over potential rivals but it was not content with that alone, deploying soft power as well as poles and wires in order to retain dominance in the Irish market. One example of this was when Queen Victoria toured Ireland in 1861. The Magnetic's manager in Dublin, J. H. Sanger, ensured that press reports about the royal visit were transmitted without charge in return for repeated mentions of the company in the articles published. He also forwarded messages and press comment from London to the Royal party in Ireland, with the Queen sending her thanks for this 'public service'.[35]

Another example of the Magnetic's soft power was its approach to employment and training, displaying remarkably progressive attitudes towards gender roles for the period. Perhaps because many of the company directors employed thousands of women in their Lancashire cotton factories, the Magnetic was one of the first telegraph companies to employ female telegraph operators.[18] Spotting the potential for vocational female education in Ireland, the Magnetic sponsored the pioneering Queen's Institute for the Training and Employment of Educated Women in Dublin from its establishment in 1861.[35] The first of its kind in Europe, the institute offered classes in a variety of skills including telegraphy, engraving, architectural drawing, bookkeeping and woodwork. The telegraph company supplied equipment, provided an engineer to conduct the classes, and paid a small grant to a number of students who were subsequently employed as telegraph clerks.[39] In time, all the telegraph companies became substantial employers of women, a practice that continued into the telephone era.

Meanwhile competitors to the Magnetic suffered not only from defective undersea cables but also criminal activities. The rival Electric, which owned an underground telegraph cable between Dublin and Belfast laid alongside the roadway, was the target of a heist in November 1861 when a long section between the Co Louth towns of Drogheda and Dunleer was lifted out of the ground.[35] Soon after, detectives from the Dublin Metropolitan Police (DMP) arrested two men in a yard off Dublin's Capel Street who had melted down the cable to separate the gutta-percha insulation from the copper wire inside.[40] The police then uncovered 540 lbs (245 kg) of gutta-percha in the elegant Kenilworth Square home of Mr John W. Elvery[41], sporting goods retailer, who had bought it from the two thieves. The two men were convicted but the telegraph company asked the court to apply a lenient sentence and they got away with a fine of £20. Mr Elvery received a severe rebuke from the judge[42] but was not charged with any crime: the business he founded prospered and remains one of Ireland's best known retail chains. Theft of copper and, occasionally, fibre-optic cables continues to be a problem for telecommunications companies though sportswear retailers are no longer a major market for such stolen goods.

As the telegraph networks spread across Ireland, public awareness of the new technology grew and costs fell. A service to transfer money by telegraph was introduced in 1854 between a number of cities, including from Dublin to London.[35] Meanwhile the voracious appetite of the newspaper industry for information continued unabated. By 1869 the Dublin offices of the three main dailies, the *Irish Times*, *Freeman's Journal* and *Saunders's Newsletter*, had installed leased lines allowing them to receive news reports directly at their premises for publication with the minimum of delay. The new communications method began to spread further, beyond the initial early adopters of the newspaper and railway industries. The military, ever concerned about restive natives, decided in 1859 to connect every garrison in Ireland and strategic points on the coast to the telegraph network,[35] starting with the Curragh. The first Information Age was in full swing, thanks to this growing network of telegraphs.

Part of the Magnetic's original business plan in 1852 was to carry transatlantic messages over its network to Galway and thence by steamship across the Atlantic. With a transatlantic telegraph cable no more than a pipe-dream at this point, such a service could enable news of cotton prices in the US to be transmitted to cotton millers in Lancashire[37] within five days, a saving of many days over the regular mail. However, despite its prime position on the west coast, Galway port had remained undeveloped and only shallow vessels could dock there. Numerous proposals for a transatlantic port in the western city had been mooted for many years. Understandably the Midland Great Western Railway, which operated the Dublin–Galway line, also backed such a plan, but support came from all walks of life with the most vociferous champion being local priest Fr Peter Daly, described by the London *Weekly Register* as 'the foremost business man in Ireland'.[43]

Pressure became more focused in 1851 when a government report recommended against the development of Galway, partly because of the lack of a telegraph connection.[44] There was thus a big incentive to connect Galway to the telegraph network, an incentive the Magnetic was happy to meet. In 1858 their investment finally looked as though it would be justified when a new shipping line called the Galway Line commenced a transatlantic liner service from the western port. For a period, the company earned considerable revenue through messages carried across the Atlantic via Galway,[27] but its ships proved unsuitable and suffered frequent delays. Disaster befell the new service in August 1859 when the *SS Argo*, under the command of 24-year-old Wicklow native Robert Halpin, struck an iceberg and sank. While no lives were lost, it was another nail in the coffin of the beleaguered Galway Line, which went out of business soon after. Halpin fared better: despite his master's ticket being suspended, he continued to forge a successful career in shipping. In any case, as we will see later in our story, thanks to Halpin the days of forwarding telegraph messages across the Atlantic by steamship would soon be over.

About the same time, some transatlantic liners began to call at Queenstown (now Cobh) or Derry en route to and from British ports such as Liverpool. A liner arriving at Queenstown from New York could send data about cotton prices in Mississippi to a cotton miller in Manchester or news of the American Civil War to eager journalists in London and Paris 22 hours before the liner finally arrived in Liverpool.[45] The Magnetic's route between Queenstown and London ran via Dublin, Belfast and Portpatrick, so rivals saw an opportunity for business by providing a more direct line. Thus in 1862 a new four-circuit cable was laid under the Irish Sea from Abermawr in south Wales to Greenore Point in Co Wexford along with overhead wires from Wexford to Cork, mostly laid along the roadway due to the deal between the Magnetic and the railway companies. The new cable from South Wales reduced the length of the route between London and Queenstown from 750 miles (1200 km) to 465 miles (750 km)[27] providing faster and more reliable communications. But there were bigger potential markets to serve than just those ships that called at Queenstown, Galway or Derry.

The telegraph companies realised that, if they established stations on coastal headlands, even liners that were not calling at a port in Ireland could have the messages they had carried across the Atlantic transmitted by telegraph from an Irish port. The method adopted was for the liners to make several copies of their messages, seal them in waterproof canisters and drop them into the sea as they steamed past Irish coastal telegraph stations. Tenders with nets attempted to catch the canisters and race back to shore to transmit news and personal messages from the New World to an eager audience over the telegraph network. The Magnetic was the first to exploit this market when, on 2 April 1863, it opened a telegraph station at the small harbour town of Greencastle in Co Donegal to catch canisters jettisoned from liners using the northern transatlantic route. Within eight months, the rival Electric had installed a new line to Roche's Point in Cork harbour[28] and quickly extended this to Crookhaven to pick up messages from liners using the southern approach. The Magnetic, not to be outdone, set up its own Co Cork station at Cape Clear.

Building stations on headlands did not solve the fundamental problem that the rate the information moved at was ultimately constrained by the speed of a ship over the long journey between North America and Ireland. For example in 1865, on the morning of the death of President Lincoln, a message with the news was telegraphed from Washington DC to Portland, Maine and put on board the SS *Nova Scotian* just before it set off for Derry and Liverpool.[46] It was 12 days later before the message was dropped into the outer reaches of Lough Foyle[47] and reached Greencastle from where it was flashed over the Magnetic's network to London and to the rest of Europe. As we will see in the next chapter, by the following year, this system of canisters and nets soon seemed positively antiquated.

Like the system of canisters and nets, rivalry between telegraph companies such as the Magnetic and Electric was also drawing to a close by the mid 1860s. The two companies were effectively a private-sector duopoly, with competition between them constrained by legal and other constraints. For example, the Magnetic's exclusive rights with most of the Irish railway companies meant that when the rival Electric wanted to connect Wexford with Dublin, its cables had to be strung along a circuitous route following the Barrow Navigation and Grand Canals[28] rather than along the railways.

A debate ensued about ownership of the telegraph network, spurred on by a price increase applied in concert by all the telegraph companies in 1865. A subsequent government report recommended purchasing the telegraph companies, placing the service within the Post Office and charging a uniform rate across the whole country. Accordingly, on 28 January 1870, the Magnetic and its rivals disappeared. It was the first time in modern history that the government had provided a service that had hitherto been the preserve of private enterprise.[48] Arguments about the benefits or otherwise of state ownership of the telecommunications network have continued, on and off, ever since.

The Post Office set about rationalising and improving the network it had inherited with some impressive innovations. Amongst these was a system of pneumatic tubes in Dublin, allowing a telegram message written out at the GPO, Four Courts or Custom House[30] to be inserted in a cylinder and blasted by compressed air to the Central Telegraph Office at College Green along tubes under the city's streets. The plans also included a major expansion of the network of telegraph cables and an increase in the capacity of the cross-channel lines.

Some of this network expansion occurred as a result of events outside the control of the Post Office, however. While the railway companies made considerable use of the now nationalised telegraph service, they did not use it as extensively as some would have liked. Several reports in the 1870s and 80s had urged the railway companies to implement 'block signalling' by dividing railway lines into discrete sections or 'blocks'. Under this system, when a train passes through a block of track, the signal box at the end of the block tells the signal cabin at the start of the block that the section is clear by sending a sequence of bell rings over telegraph wires. Until the signal cabin at the start of the block has received this message, it does not clear its signal and no other trains are allowed to enter the block. However, the government, which had a policy of light-touch regulation over the railways, did not make use of this system mandatory. The private railway companies were anxious to avoid the extra costs required to install the telegraph equipment required for block signalling. They thus largely continued to use time-based control methods, so that a train was allowed to proceed a certain number of minutes after the previous one. An event in 1889 changed this light-touch approach to regulations governing signalling for ever.

On 12 June of that year, a Great Northern Railway excursion train to Warrenpoint crowded with children and their parents from Armagh Sunday School collided with a regular passenger train.[49] Eighty people were killed and 260 injured, a third of them children, making it the worst rail disaster in the United Kingdom in the nineteenth century, and Ireland's worst ever railway disaster. The tragedy had an immediate effect on public and political opinion. Within two months of the crash, Parliament had enacted the Regulation of Railways Act 1889 across the United Kingdom.[50] The new regulations made mandatory the use of the block system of signalling and the interlocking of all points and signals. Implementing the block system required a network of signal boxes across the railways, each connected to the next by telegraph wires, making the railways even more dependent on telecommunications infrastructure.

While the government was willing, when prodded by public outrage, to extend its powers over industries like the railways, it seemed less capable of applying control over its own departments. Nationalisation of the telegraph service had cost far more than forecast, with the bill for buying out the private companies ending up at £7.8m (€9.9m) instead of the estimated £2.4m (€3m).[11] The Treasury began to keep a close eye on the telegraph department's budget and the rate of network expansion slowed significantly.[51]

The government introduced a policy of requiring revenue guarantees, whereby a local authority or similar body would have to underwrite any losses incurred by extensions to locations considered unprofitable. Sometimes creative accounting was employed to help matters, as was the case in 1883 when the Treasury was petitioned to approve the extension of a telegraph circuit to Belmullet in Co Mayo on humanitarian grounds. In a sign of just how poor the area was, the argument used was not that the service would benefit local business or allow medical services to be summoned but rather that a telegraph station would 'greatly facilitate the arrangements … by Mr Tuke's Committee for the emigration of poor persons'.[52] With estimated revenues covering only a quarter of the running costs the Treasury was unmoved so that Mr Tuke's Committee had to continue its work, dealing with 2,420 applicants to emigrate from Belmullet alone,[53] without the help of the telegraph. James Hack Tuke's continued activities to aid the West of Ireland led to the formation of the Congested Districts Board, which was, indirectly, to have a major effect on the rollout of telecommunications in rural areas for decades to come.

The Treasury's attitude to the Belmullet telegraph changed, however, when it was identified that such a line would eliminate the need to maintain the mounted police at Bangor Erris and Belmullet who were used to summon backup from Ballina if required. With the forecast savings of £300 a year practically covering the running costs of the extended telegraph service,[52] the line was duly erected to Belmullet and the police horses withdrawn. This telegraph line remained the only telecommunications service in this part of Co Mayo for 54 years until a telephone line was finally installed in 1940.

The policy whereby the Post Office required underwriting of losses was gradually phased out[54] and the telegraph network became a means to unite the disparate parts of the United Kingdom. At the time of nationalisation there were 115 telegraph offices in Ireland; within 18 months this had almost quadrupled to 439 offices.[55] By 1898 this had grown to 581 telegraph offices[56], covering the length and breadth of the country (Figure 1.6) from Ardara in Co Donegal, to Taghmon in Co Wexford and Inishmore, largest of the Aran Islands.[57] To assist shipping, lines were extended to headland lighthouses, while the insurer Lloyd's of London had telegraph equipment installed at their lookout stations including Tory Island and Malin Head. The insurer's post on Rathlin island was left, for the moment, to communicate by carrier pigeons.

It was truly a United Kingdom of Telegraphs, with the telegraph lines forming the sinews binding the nation together. In a speech in 1887 to celebrate 50 years of telegraphs, Sir William Thomson, the Belfast-born scientific genius responsible for many innovations in telegraphy, commended this expanding national network:

> Dublin can now communicate its requests, its complaints, and its gratitudes to London at the rate of 500 words per minute. It seems to me an ample demonstration of the utter scientific absurdity of any sentimental need for a separate Parliament in Ireland.[58]

While many in his native Ireland would have disagreed with Thomson's conclusion, there was no question that the Post Office was a powerful symbol of the British state in Ireland. Not only did its telegraph cables criss-cross the entire kingdom, its buildings were adorned with royal insignia and its 20,000 staff in Ireland[59] were UK civil servants.

With such a large organisation, it was sometimes difficult to ensure policies were applied consistently. For example, while there was no prohibition on telegrams written in Irish, staff at post offices were sometimes uncertain of how to treat them. In 1898 a telegram written in Irish was charged by a clerk at Balbriggan at a higher rate on the grounds that it was in cipher.[60]

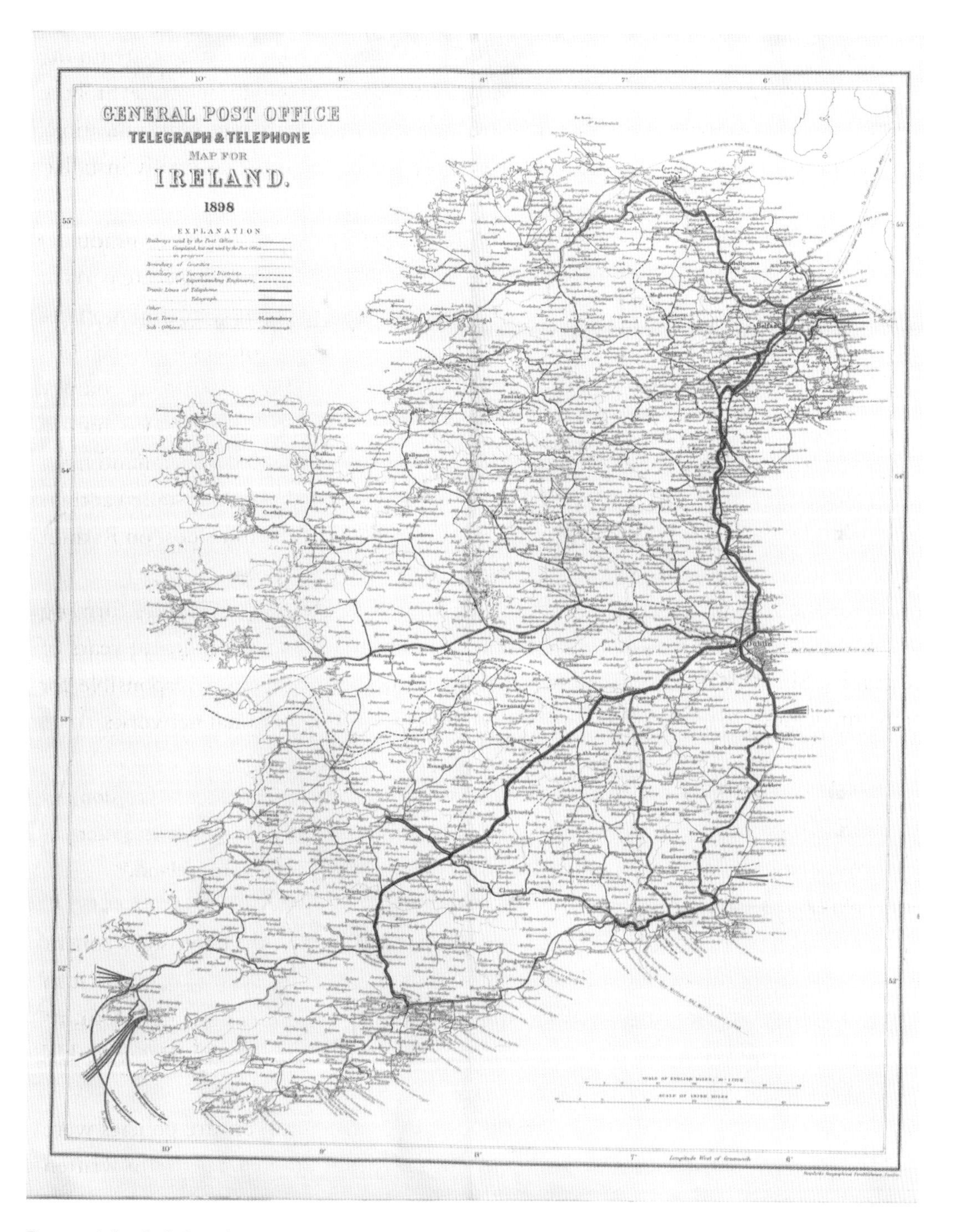

Figure 1.6: By 1898 the telegraph network, shown in red, covered the entire country from Donegal to Wexford.

(© By permission of The National Archives of the UK)

The pressure to constrain costs provided an impetus to deploy technology that reduced the number of specialist staff required. For this reason, the ABC telegraph, patented by Wheatstone in 1858, remained popular for decades.[61] This device had a dial with 30 keys around the edge and a pointer. To work it, the person operating the device pressed the key for the letter they wanted and wound a generator. At the receiving end, the pointer moved around a dial one letter at a time until it reached the letter being sent (Figure 1.7). While slow to use, it was widely deployed on less busy routes and private circuits such as those between the Dublin Metropolitan Police headquarters in Dublin Castle and police stations around the city.[62]

Figure 1.7: An ABC telegraph of the type used at police stations in Dublin. *(© Science Museum/Science and Society Picture Library)*

The impact of cost-saving measures such as the use of the ABC telegraph was, however, negated by reductions in charges pushed through by politicians anxious to win public support. The net effect was that by the 1890s the service was consistently loss making. But the biggest threat to the telegraph proved to be a technological one.

Within 30 years of nationalisation, the telegraph service was in decline, eclipsed by a new telecommunications method – the telephone.

In parallel with this national telegraph network, which connected London and Dublin with towns and villages throughout the land, another telegraph network was developing, an international one with Ireland at its centre. So the focus of our story moves from Dublin to what would seem to be a most unexpected location for a global communications hub – the Ring of Kerry.

CHAPTER TWO

KERRY, COMMUNICATIONS HUB OF THE WORLD

One January evening in 1854, the New York paper-making tycoon Cyrus Field received an unexpected visitor at his luxurious home overlooking Gramercy Park.[1] His visitor was telegraph engineer Frederic Gisborne, who proceeded to reveal his plan to link St John's, on the most easterly part of Newfoundland, to the telegraph network of North America. The project would require 650 km of telegraph cable across the unforgiving barren eastern fringes of the North American continent and two submarine cables to connect the island of Newfoundland with the Canadian mainland. Gisborne explained that the object of this ambitious plan was to hasten the flow of information between the new and old worlds. A steamship to Europe would call at St John's and pick up any urgent messages flashed there by telegraph from New York, Boston or Montreal, shortening the time needed for news to cross the ocean by days. Gisborne's idea was a mirror image of the Magnetic's aim to develop Galway as a transatlantic port on the other side of the ocean.

Legend has it that after Gisborne took his leave, a polite but apparently unimpressed Cyrus Field twirled the globe in his office and experienced an epiphany. Field realised that if you could lay a telegraph cable, partly underwater, connecting the USA to Newfoundland, then surely you could lay one between *Ireland* and Newfoundland, and if you could do that you could shorten the time needed for news to cross the Atlantic not by a couple of days – but by a couple of weeks.[2] A project to erect a telegraph cable to Newfoundland was merely foolhardy: a plan to lay a cable across the Atlantic was simply mad. The distance between Ireland and Newfoundland was 3122 km (Figure 2.1), 88 times longer than the longest submarine cable then in use.

No one knew if it was possible to transmit an electrical impulse such a long distance under the ocean. The 34-year-old Cyrus Field, however, possessed a great advantage that freed him from such worries: he knew nothing about telegraphy nor the obstacles that would be involved in the scheme – he just thought it would be a great idea.

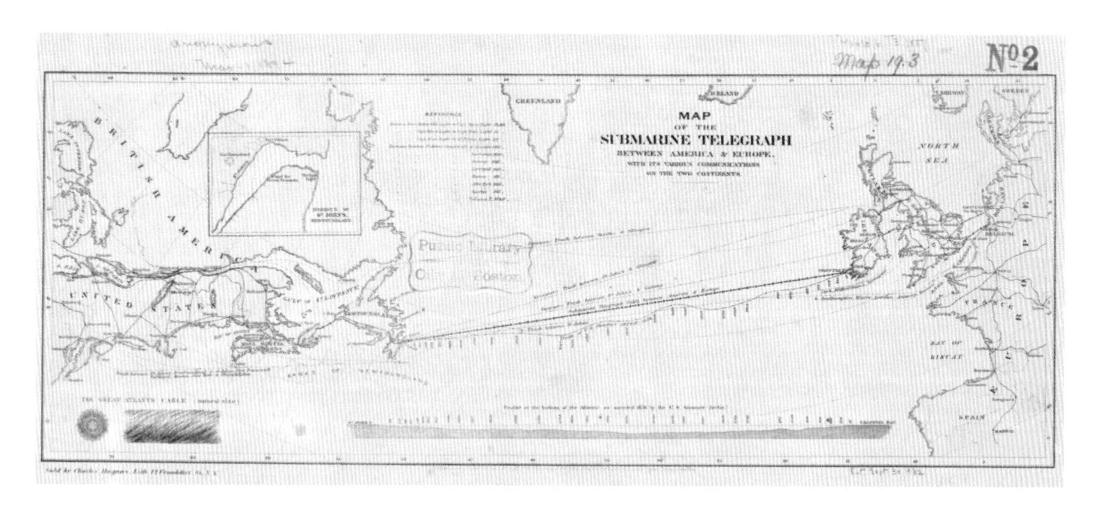

Figure 2.1: Map from 1857 showing the route of the first transatlantic cable. *(Reproduction courtesy of the Norman B. Leventhal Map & Education Center at the Boston Public Library)*

The next day, Field dashed off letters seeking advice about his mad transatlantic idea. One letter went off to Samuel Morse,[2] who replied to offer his enthusiastic support. Field seized upon Morse's affirmation as proof of the scheme's assured success, even naming his second son Edward Morse Field in his honour[2]. However, with no technical knowledge himself, and possibly blinded by his own enthusiasm, Field was oblivious to the fact that Morse knew little of the technical challenges involved in communicating over long submarine cables.

Within months, Field had established a company to lay a transatlantic cable, persuading the authorities in Newfoundland to grant his company exclusive rights for 50 years to lay telegraph cables on or touching its territory, in addition to grants of land and money.[2] That was the easy part, for next he had to raise $1.5m to lay the cable. It was a huge sum – at the time, the entire Federal budget of the US amounted to $58m – but it was nowhere near enough.

In fact most of the capital came in the form of pounds rather than dollars as the investors were overwhelmingly from the eastern side of the Atlantic, with the Magnetic, the main telegraph company in Ireland and Scotland, taking a substantial share. In view of the mammoth size of the project and Cyrus Field's own lack of experience of telegraphy, one might have expected him to try to surround himself with the best experience available. However, while his choice of Charles Tilston Bright, who had overseen the first successful telegraph cable between Britain and Ireland, as chief engineer was a wise one, his other choices were less prudent. His chief electrician was a surgeon by the name of Wildman Whitehouse. Although Whitehouse knew more about telegraphy than his boss, that was not a particularly high bar.[3] Meanwhile the talented physicist William Thomson of the University of Glasgow remained just a company director with no technical position within the company. Likewise, Frederic Gisborne, whose meeting in Cyrus Field's New York home on that cold January evening had inspired the whole enterprise, had fallen out with Field before a single cable had been laid.

Nevertheless the rather motley crew Field had assembled did not deter the UK and US governments. The British government, which had spurned the idea of a subsidy to construct a telegraph line to facilitate communications between Britain and Ireland, was happy to commit £14,000 a year to facilitate communications between the UK and North America. Naval support came in the form of a ship each from the British and American navies.[4] In return for their support, both governments received guarantees that their messages would be given priority – assuming the enterprise worked, of course. When Cyrus Field was asked by Lord Clarendon, British foreign secretary, what he would do if the cable was lost at sea, Field replied, 'Charge it to profit and loss, and go to work to lay another.'[5] The question – and answer – proved prophetic.

In addition to lobbying governments, Cyrus Field was also subject to lobbying himself. This was in the form of lengthy correspondence with Sir Peter Fitzgerald, nineteenth Knight of Kerry and resident of Valentia Island. Unlike many of his nineteenth-century contemporaries, the Knight was not an absentee landlord but lived at his family's ancestral home at Glanleam close to his tenants, whose welfare the family tried to improve. The slate mines established by his father in 1816 provided employment for local people and, perhaps as a result, the island had been spared the worst ravages of the Famine. The name Valentia derives from the Irish *Béal Inse* meaning mouth of the island, a reference to the sheltered harbour entrance north of the island that Sir Peter Fitzgerald proposed to Cyrus Field as the European landing point of any transatlantic cable,[6] viewing this a means to further develop the island's economy. Of course, his promotion of Valentia was not completely selfless for, as a landowner, he would benefit from wayleaves from cable landings and leasing of land for buildings.

The knight's perseverance paid off and he became not just Field's local partner for the enterprise[5] but also a personal friend. More importantly for Valentia, its harbour was indeed selected to be the eastern terminus of the transatlantic cable. Often described as the 'next parish to America', this corner of Kerry would never be the same again. For the next 100 years Valentia and the rest of the Iveragh Peninsula in Co Kerry would be the communications hub of the world.

With money and ships committed and locations selected, Field's fantastical project commenced on 3 August 1857 with a huge media event. The Great Southern and Western Railway, keen to promote its new line to Killarney and the tourist potential of Kerry to a wide audience, laid on transport to whisk dignitaries and reporters from Dublin to Valentia. On the latter part of the journey, the guests undoubtedly noted the line of telegraph poles being installed by the Magnetic to connect the terminus of the transatlantic cable with the existing telegraph network at Killarney[7] and from there to London and the rest of Europe. The next day, the Knight of Kerry hosted an 'elegant déjeuner', the guests including the Lord Lieutenant, the Superintendent of Telegraphs for India Sir William O'Shaughnessy and clergymen from the two main denominations. The large party of journalists present were happy to report on the festivities, noting that in the room:

> the words of the Irish welcome, 'Cead Mille Faite' [*sic*] were prominently displayed, and at either sides of the wall, immediately over the chairman, were placed the national flags of the United States and the United Kingdom.[8]

The event was designed to portray Ireland as a loyal and integral component of the United Kingdom that was doing its part to foster closer communications between England and America. The Irish language mottos on the walls were a token gesture: while the majority of the island's inhabitants at the time spoke Irish,[9] all of the speeches were delivered entirely in English.

On 5 August 1857, the cable-laying fleet arrived off Valentia. In a gesture of commercially inspired diplomacy, it was the American sailors from the *USS Niagara* who carried the cable ashore at White Strand at Ballycarbery (Figure 2.2). In a brief speech delivered to the crowd on White Strand, Cyrus Field, who had arrived at Valentia in time to see the fleet leave, hoped that 'what God has joined together, let no man put asunder'.[8]

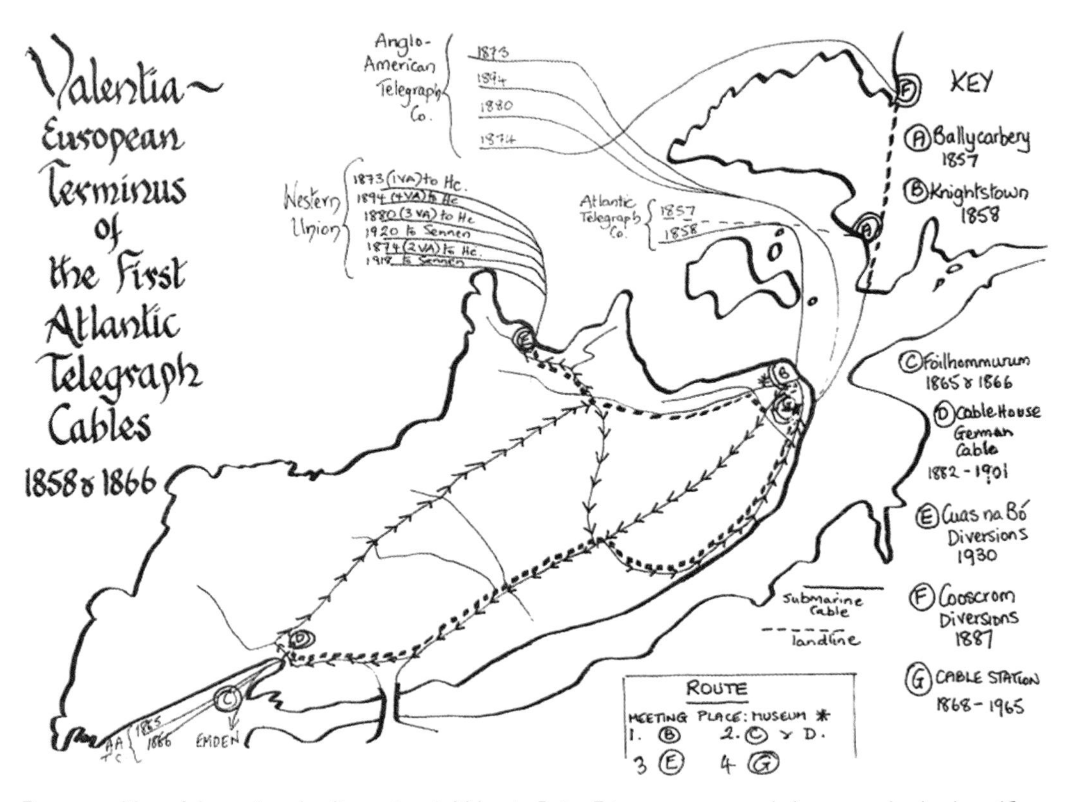

Figure 2.2: Map of the various landing points in Valentia. Point E is more commonly known as Lacknabau. *(Courtesy of the Valentia Island Heritage Centre)*

God evidently had not been listening on White Strand as, three days into the voyage, the cable snapped and 500 km of it was lost to the ocean floor. Field wasted no time on regrets, ordering the remaining cable to be brought back to England for storage and immediately planning for the next voyage. For the second attempt in June 1858 a new plan was adopted whereby the two ships proceeded together to a point midway across the Atlantic. There the two sections of cables were spliced together before the ships headed off in opposite directions, one east and one west. On this occasion one of the two ships encountered a storm so vicious that it left 45 members of the crew in the ship's sick bay with ailments ranging from broken arms to temporary insanity.[2] More importantly for the company's shareholders, the storm also damaged the cable. It appeared to be a case of third time lucky when, on 5 August 1858, the natives of Valentia were roused by gunfire[10] as the British navy's *Agamemnon* sailed into the harbour, carrying the eastern end of the 3,400 km long cable.

On 16 August 1858, with the cable ready for use, the first public message was transmitted under the Atlantic. The sender was Queen Victoria, whose message to President

Buchanan sought 'to congratulate the President upon the successful completion of this great international work, in which the Queen has taken the deepest interest'.[11] President Buchanan's upbeat response described the telegraph as 'an instrument designed by Divine Providence to diffuse religion, civilization, liberty, and law throughout the world'.[12]

In New York, news of the first transatlantic telegraph message led to wild celebrations, including a fireworks display so huge that it ignited the dome of City Hall, causing $50,000 of damage.[13] An official celebration in New York for the cable hero Cyrus Field followed on 1 September. Towards the end of the celebration, Cyrus Field advanced to the edge of the stage and addressed the huge crowd, saying: 'Gentlemen, I have just received a telegraphic message from a little village, now a suburb of New York, which I will read to you.'[11] The message was one of congratulations from the directors of the telegraph company, on their way from Dublin to Valentia on the morning of the New York reception. The 'little village' was Mallow in Co Cork, from where the directors had presumably dashed off the message while waiting for their connecting train to Kerry. Thanks to the transatlantic telegraph, the fine market town of Mallow, like all of Ireland, was now 'a suburb of New York'.

In Ireland too there were days of celebration, the number of official events partly reflecting the religious-political divide besetting the country. The first official event was a banquet for the company's Chief Engineer, Charles Bright, at Dublin's Manson House that took place almost simultaneously with the reception for his boss in New York. Amid some controversy, the Dublin banquet was snubbed by the Lord Lieutenant and other representatives of the government due to the attendance of Cardinal Wiseman, Catholic Primate of England, who was touring Ireland. In his speech honouring Bright and the transatlantic cable, the Cardinal anticipated a world united by instant communications, though his vision sounds more like the world of Facebook comments and family WhatsApp groups than nineteenth-century telegraphy:

> I can imagine a poor mother in the west of Ireland ... sitting on the furthest crag that juts into the Atlantic, no longer contemplating that waste of water as a desolate wilderness which separates her from those she loves, but as a means of instant communication with them – as a way of making known to them at once her joy or her distress, and of receiving back from them in a few hours messages of consolation and of promise.[14]

The Lord Lieutenant arranged a private ceremony three days later[15] at the Viceregal Lodge (now Áras an Uachtaráin) at which the 26-year-old Bright was knighted.

Celebrations continued with a banquet on 7 September 1858 at the recently opened Railway (now Great Southern) Hotel in Killarney. Needless to say the Knight of Kerry was amongst those present; with no Catholic clergy present, the Lord Lieutenant was also happy to attend. However, newspaper reports commented on the relatively small attendance at the dinner, attributing this to the high price of two guineas (€2.66), and, more ominously, problems with the cable.

The perils of the premature celebrations that accompanied the first Holyhead – Howth cable in 1852 had been forgotten. The inadequately specified and hastily built transatlantic cable behaved erratically with messages becoming increasingly fragmentary and incomprehensible. The log of all messages sent from Newfoundland to Valentia during the sixth day of service conveys the problems being experienced:

> 'Repeat, please.'
> 'Please send slower for the present.'
> 'How?'
> 'How do you receive?'
> 'Send slower.'
> 'Please send slower.'
> 'How do you receive?'
> 'Please say if you can read this?'
> 'Can you read this?'
> 'Yes.'
> 'How are signals?'
> 'Do you receive?'
> 'Please send something.'
> 'Please send Vs and Bs.'
> 'How are signals?'[16]

One of the few messages to be transmitted successfully[17] was from the British Government to its forces in Canada, cancelling a previous order to move two regiments from there to quell the Indian Rebellion.[18]

On 1 September, as Cyrus Field was about to be feted in New York for his contribution to transatlantic communications, the station at Trinity Bay received a message addressed to him saying 'please inform American government we are now in position to do best to forward'. The message did not make sense. It was incomplete: the words 'their government messages to England' had been sent by Valentia but they never arrived.[17] The cable, unreliable from the start, had stopped working. As New York prepared the official celebration for the new Atlantic cable, the directors were forced to issue a press release suspending service. It was the first official admission of the problems that had been ongoing since the cable was landed. Only a few more words were ever transmitted. The cable was dead.

Retribution was swift: the company's share price plunged and the media speculated that the whole project was faked. The chief electrician, Whitehouse, was dismissed on 17 August on rather flimsy grounds[19] and went down in posterity as the villain solely responsible for the failure. It was widely reported that in a desperate measure to push signals through Whitehouse applied a voltage of 2,000 volts thereby incinerating the cable's insulation.[20] The reality was more complex. The cable designed by Whitehouse was supposed to be 16 mm in diameter, no thicker than a man's index finger, with a copper core of 2 mm^2 but sections of it did not even meet this inadequate specification. In the period between the 1857 and 1858 voyages, the cable had been left in the open air, where its gutta-percha covering had dried out and cracked.[21] Though Whitehouse was undoubtedly responsible for many of the problems, his vilification allowed him to become a convenient scapegoat. It was much safer to put the blame on him than to implicate figures more central to the continuing cable enterprise, such as Cyrus Field and Charles Bright, all of whom arguably shared at least as much responsibility for the failure[22] by setting unrealistic dates for the construction of the cable.[3]

The field of electricity was a new one and many principles were little understood. This included the specifics of transmitting data by cable over long distances, where electrical phenomena we now know as resistance and capacitance were experienced for the first time. The first of these, resistance, causes a signal to become weaker as it passes along a wire. If the resistance of a telegraph cable is too great it renders the signal impossible to discern at the receiving end. This affects all wires, whether strung between poles along the side of a railway track or encased in gutta-percha insulation and dropped to the seabed. The resistance of a cable is directly related to its length and its composition. It was discovered in the early days of experiments with electricity that copper has a low resistance, with a thick wire of pure copper offering the least resistance of all.

However, as cables began to be laid underground, in trenches along roadsides or, increasingly, on the seabed, a new and different phenomenon known as capacitance was observed. Electrically speaking, when a cable is laid under the ground or under the sea, capacitance occurs because the electric charge in the wire induces an opposite charge in the surrounding earth or water. As the two charges attract each other, the exciting charge – the message being sent – is retarded. This retardation effect is proportional to the length of the cable and is much greater for a cable under the sea than for one placed under the ground. Capacitance would be a minor irritation if the delay was constant: if the dots and dashes merely emerged at the other end some seconds or even hours later they would still have beaten the fastest ship by many multiples. But the retardation is not uniform and instead has the effect of blurring together the signals (Figure 2.3) so that before a 'dash' has emerged from the far end, a 'dot' may already be treading on its heels.[23] Under such circumstances the only way to keep the message readable is to send the signals more slowly.

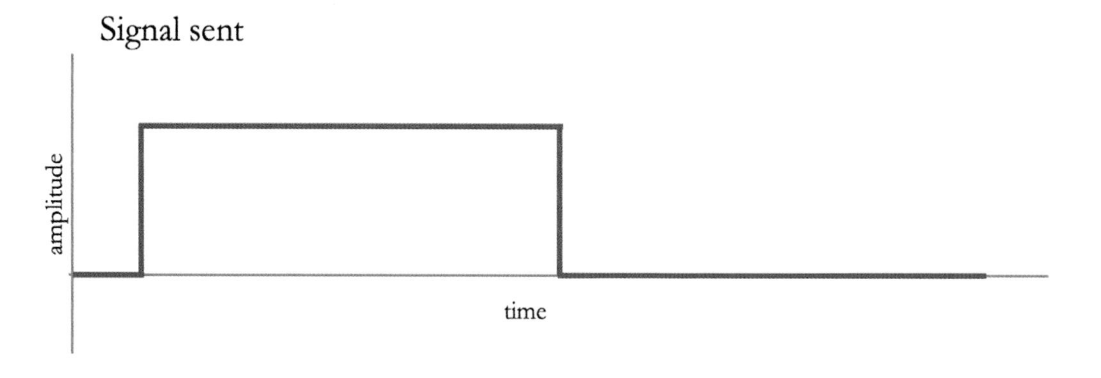

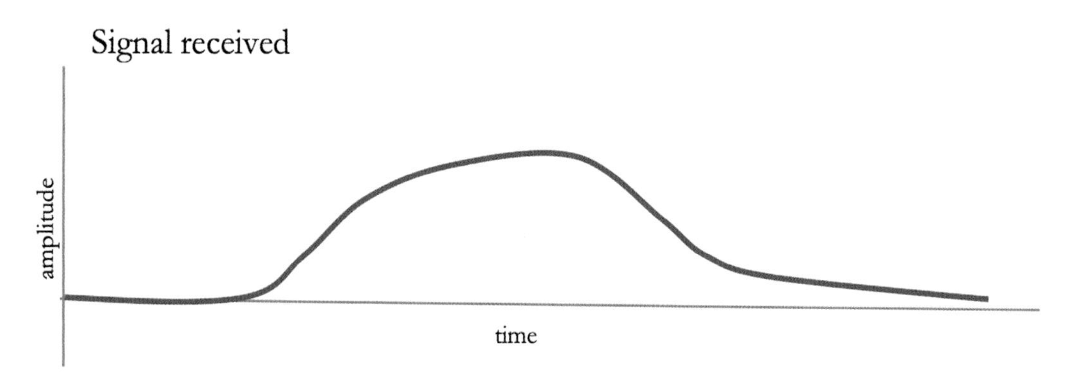

Figure 2.3: The effect of capacitance is to blur the transmitted signals making them difficult to decipher.

This problem was just emerging as Field and his partners were making their plans to span the Atlantic but there was no consensus as to the solution. Whitehouse considered that capacitance would be minimised by using a cable of the thinnest possible copper. In fact such a strategy would not have any effect on capacitance but would increase resistance, thus reducing the strength of the signal at the receiving end. Samuel Morse agreed with Whitehouse and dissenting voices, such as that of company director William Thomson, were ignored. Field had promised that the transatlantic cable would be open for business by the end of 1857 and so, like the doomed Holyhead–Howth cable five years before, both the design and manufacture of the transatlantic cable were rushed. The surprise was not that the cable had failed so quickly but that it had worked for so long.

As we saw in Chapter One, most submarine cable projects in the mid-nineteenth century, even under much shorter stretches of sea than the Atlantic, ended in failure. The British government, which had committed both money and naval resources to the Atlantic cable, had seen the potential benefits of telegraphy, albeit briefly: the telegram sent to Canada to cancel the movement of two regiments to India had saved the government £50,000.[24] The failure of another high-profile project, also partly funded by the British government, to link Europe and India via cables under the Red Sea spurred the government to establish a joint committee of inquiry in 1859 to investigate all aspects of long-distance submarine telegraphy and establish guidelines for future cables.[21]

The inquiry's star witness was William Thomson. Born at College Square East in Belfast on 26 June 1824, he moved to Scotland at age six after his father was appointed Professor of Mathematics at Glasgow University.[25] A veritable child prodigy, Thomson started attending Glasgow University himself at the age of ten before studying at Cambridge. After his appointment to the Chair of Natural Philosophy at Glasgow, he established a physics laboratory where students learnt by performing practical experiments instead of merely learning theory, a technique that became the norm across the world.[2] He extended this hands-on approach to public lectures too: in 1857 when he was invited to deliver a lecture to open the new museum of the Royal Dublin Society (RDS) on Merrion Square (now the Natural History Museum), Thomson adorned the room with a portion of telegraph cable and other electric apparatus.[26] Though initially unpaid by the cable company, Thomson personally participated in all of the major cable expeditions,[27] while Whitehouse was laid low by an illness that apparently prevented him from sailing, forcing him instead to accept the hospitality of the Knight of Kerry.[28] During his 53 years[25] as professor, Thomson made major contributions to the fields of electricity, magnetism and optics and laid the foundations of the new science of thermodynamics.[23] He was no ivory-towered academic and happily used his insight to solve urgent practical problems, such as those besetting the infant telegraph industry, netting him some handsome returns.

Thomson introduced two interlinked innovations to long-distance submarine telegraphy. The first was to modify Morse's code. Rather than representing letters by means of short 'dots' and long 'dashes', Thomson used an idea from German scientist Carl August von Steinheil so that a dot was represented by a positive impulse of current and a dash by a negative impulse of identical duration[29]. It became known as 'cable code' (Figure 2.4).

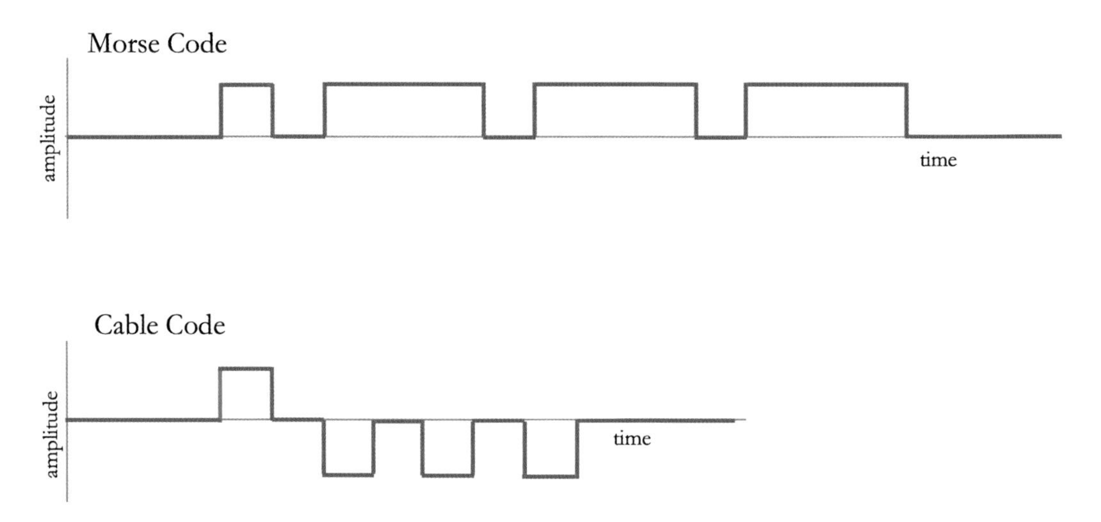

Figure 2.4: The letter 'J' in conventional Morse code (above) and in cable code (below) where the dashes are represented by signals of negative polarity.

The second innovation was his own invention, born of his skills in precision instrumentation. Thomson's mirror galvanometer (Figure 2.5) consisted of a tiny mirror attached to a magnet that was suspended in the centre of a coil. The incoming signal moved the magnet clockwise or anti-clockwise depending on its polarity, so that a positive 'dot' would turn it clockwise while a negative 'dash' would turn it the other. A fine spot of light reflecting on the mirror and focussed onto a screen would then be deflected left or right, indicating a dot or a dash.[29]

Thomson was knighted on 10 November 1866 for his contribution towards the success of the Atlantic cable and in 1892 he was appointed to the House of Lords, taking the title of Lord Kelvin and so giving the name to the temperature scale he first proposed. The first scientist in British history to be named a Lord, his elevation was not only due to his outstanding scientific mind and contribution to telegraphy, but also a reward from the then Conservative government for his opposition to Irish Home Rule.[30]

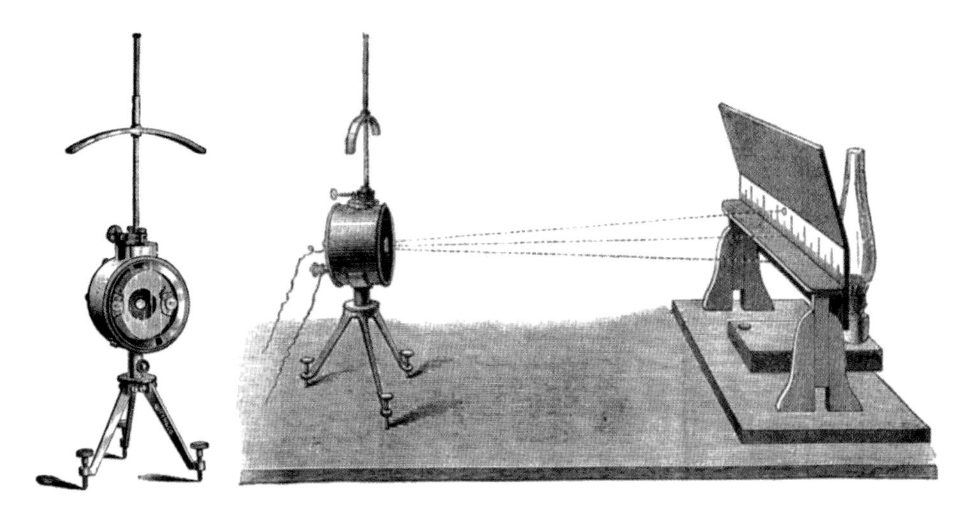

Figure 2.5: Thomson's mirror galvanometer.

Back in 1858, it was Thomson to whom Cyrus Field wisely turned for technical advice following the spectacular failure of that year. Field pressed on with a plan for another cable, declaring that the doomed cable 'did more by his death than his life'.[10] The seasick prone Field is said to have crossed the Atlantic 64 times in his efforts to gain support on both sides of the ocean[18] for a new expedition. Progress was slowed by the need to solve the technical issues that had arisen with the earlier cable and by the American Civil War[31] but Field pressed on.

In July 1865 a new expedition commenced, its cable nearly four times as bulky and almost twice as heavy as its predecessor and benefitting from improvements both in the purity of copper and in protective sheathing (Figures 2.6 and 2.7).[32] This time, rather than making the cable in two pieces and splicing them together in the middle of the turbulent Atlantic, a single cable was made and loaded onto the *Great Eastern*, the largest ship in the world. There were also changes to the landing points with the Newfoundland terminus moved a few kilometres to the delightfully named Heart's Content. On the Irish side, the landing point remained on Valentia Island but now at Foilhomurrin Bay (Figure 2.8) close to the old Admiralty signal tower at Bray Head. A substantial wooden building was constructed for £350[33] (€444) at the top of the cliff to receive the cable, house the instruments and provide living accommodation.[34] The press corps on board the *Great Eastern* to witness the enterprise included Tallaght-born William H. Russell,[10] in what was presumably a more enjoyable assignment for him than the Crimean War. As the *Great Eastern* steamed westward from Kerry, progress reports were dispatched over the cable, which stretched out behind the ship all the way back to Valentia[35] and from there on to an eager public across Europe. However, despite all this careful planning, the cable snapped during a storm, dropping to the bed of the ocean.

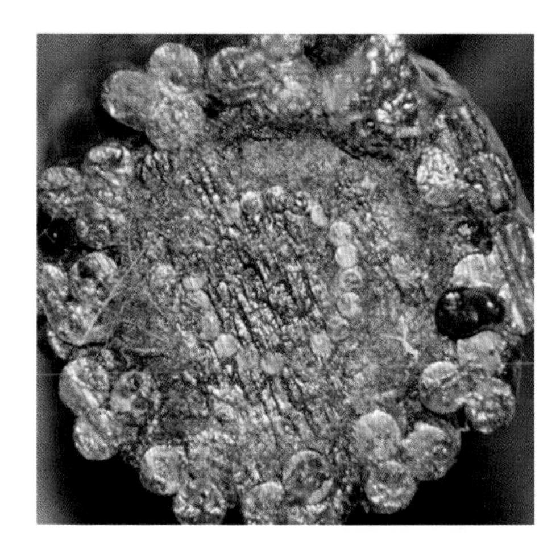

The 1865 cable was considerably more advanced than its predecessors. Figure 2.6 (left): This contemporary drawing shows the seven copper conductor wires surrounded by layers of protection; Figure 2.7: This fragment of the same cable, recovered over 150 years later, demonstrates its robust construction. *(Courtesy of Derek Cassidy)*

The indefatigable Field was back in business again within 12 months of the failure of the fourth cable. The new financial investors for his company, now called the Anglo-American Telegraph Company – or 'Anglo' for short – included Dubliner Henry Bewley[36] who must have been delighted at all the gutta-percha being utilised by these cable attempts. And so, in July 1866, the *Great Eastern* set off towards Newfoundland laying yet another cable, the fifth attempt since 1857.[37]

On 27 July 1866, twelve and a half years after Cyrus Field had twirled the globe in his New York office, the *Great Eastern* arrived in Trinity Bay bringing with it a working cable that stretched 3400 km all the way back to Co Kerry. Cyrus Field sent a telegram from Newfoundland to his friend the Knight of Kerry with the words 'Ireland and America are united by Telegraph'[36] (Figure 2.9). There was much less hoopla than in 1858, though again the first official message carried under the ocean was from Queen Victoria who congratulated President Johnson 'on the successful completion of an undertaking which she hopes may serve as an additional bond of Union between the United States and England'.

Figure 2.8: 'Landing the Shore End of the Atlantic Cable' Foilhomurrin Bay, Valentia in 1866. (*The Metropolitan Museum of Art, New York: Gift of Cyrus W. Field, 1892*)

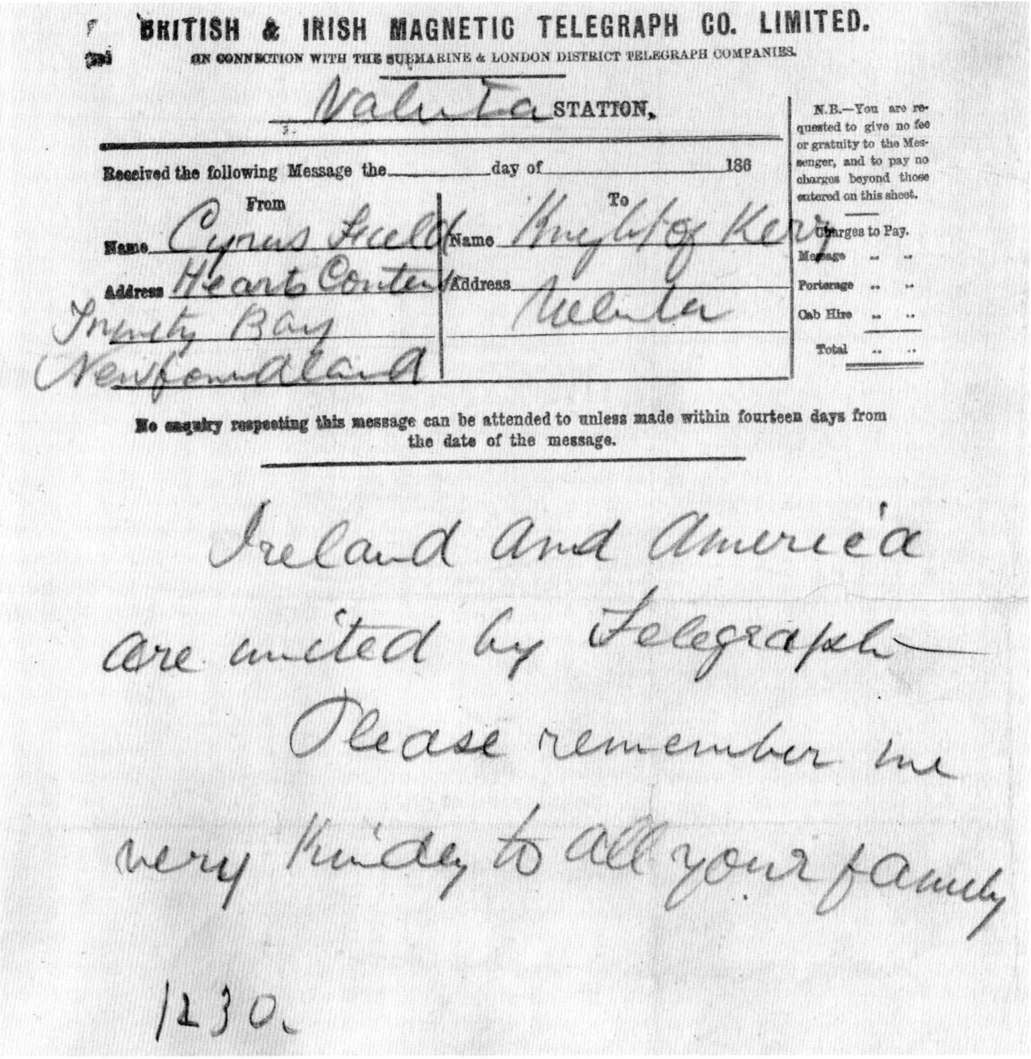

BRITISH & IRISH MAGNETIC TELEGRAPH CO. LIMITED.
IN CONNECTION WITH THE SUBMARINE & LONDON DISTRICT TELEGRAPH COMPANIES.

Valentia STATION,

N.B.—You are requested to give no fee or gratuity to the Messenger, and to pay no charges beyond those entered on this sheet.

Received the following Message the ______ day of ______ 186

From
Name Cyrus Field
Address Heart Content Trinity Bay Newfoundland

To
Name Knight of Kerry
Address Valentia

Charges to Pay.
Message .. ..
Porterage .. ..
Cab Hire .. ..
Total .. ..

No enquiry respecting this message can be attended to unless made within fourteen days from the date of the message.

Ireland and America are united by Telegraph Please remember me very kindly to all your family

1230.

Figure 2.9: Telegram from Cyrus Field to the Knight of Kerry in 1866. The friendship they had developed over the preceding ten years is evident. *(Courtesy of the Institution of Engineering and Technology Archives)*

The initial charge for a message from the UK to the US was £20 (*c.* €25) for 100 letters: Victoria's brief message would have cost £39 (*c.* €50). Had her Majesty decided to convey her message to President Johnson in person, she could have sailed from Liverpool to New York for £22 (*c.* €28) in Chief Cabin class[38], and only £5 5*s.* (€6.60) in the – admittedly unlikely – event that she had elected to travel steerage.[39] Despite these high costs, within three months the company had transmitted 3,000 messages across the Atlantic and was taking in $2,500 per day.[37] The broken cable laid in 1865 was successfully located and spliced together so that soon two cables were in operation. Unlike the 12 days it took for news of the assassination of President Lincoln to reach Europe in 1865, news of the death of President Garfield in 1881 was carried in the Irish papers within hours, under the banner 'by telegraph'. The Information Age had gone global.

The high cost of those early telegrams was due not just to the huge capital investment involved in the repeated attempts to lay the cables but also the labour-intensive operating system initially adopted. At Valentia, one telegraph operator watched the deflections of the tiny spot of light generated by the mirror galvanometer in response to the codes being tapped out by the operator in Newfoundland, and dictated the letters to a writer behind him. The message was then carried across to a separate room where it was retransmitted to London by another operator. This isolation of the transatlantic cable from the wires to London was deliberate. Overhead lines, such as those that stretched from Valentia to the Wexford coast, were prone to lightning strikes. If such lines were connected directly to the transatlantic cable, a bolt of lightning could cause even more damage to the precious cable than Whitehouse had managed in 1858.

Those lines connecting Valentia with London were of almost equal importance to those crossing the Atlantic, for London was now the capital not just of the United Kingdom but of a global empire covering a third of the world's territory. While other European powers also scrambled to carve up the world, none were as successful as the British in building a telegraph network to bind their empire together. Many Irishmen were involved in this empire building, their contributions largely forgotten in an Ireland that sees itself as the colonised rather than the coloniser.

That first truly successful cable was laid by the *Great Eastern*, under the command of first officer Wicklowman Captain Robert Halpin.[40] Born in 1836, Halpin had gone to sea aged just 11. As we saw in Chapter One, while his term as captain of the *Argo* for the Galway Line ended in disaster, Halpin was regarded as a talented mariner. Thus, some years later, he was able to obtain a position as first officer aboard the mammoth *Great Eastern*. With Halpin at its helm, the *Great Eastern* went on to lay an estimated 41,800 km[41] of subsea cable – more than enough to circle the globe. His efforts at creating a global communications network earned him both honours from many governments and institutes, and a small fortune with which he was able to build Tinakilly House, a large and elegant country house just outside Wicklow town.[42] Truly a cable man, Halpin married Jessie Munn of Heart's Content[43], the village at the western end of the first Atlantic cables.

Other Irish people were involved in connecting the UK with its empire in Asia. Chapter One introduced Limerickman Sir William Brooke O'Shaughnessy, Superintendent of Telegraphs for India. Partly thanks to his efforts, by 1870 a 11,000 km telegraph line stretching across central Europe, Georgia, Persia (now Iran) and the Persian Gulf[20] connected Britain with its Indian possessions. But O'Shaughnessy was not the only Irishman involved in this enterprise. In a neat twist of fate, the Indian Government gave responsibility for the portion of this line in Persia to Tipperary-born John Joseph Fahie. Thus two men from neighbouring counties in Ireland were in charge of the telegraph networks in neighbouring countries in Asia.

Kerry was at the hub of this global network (Figure 2.10), its telegraph stations visited by royalty anxious to view this technological marvel first-hand. Prince Alfred, the Duke of Edinburgh, news of whose birth at Windsor Castle had been sent by the world's first public telegraph line, visited the Valentia station in 1880. The Duke was introduced to station staff by Maurice Fitzgerald, who had inherited the title of Knight of Kerry,[44] and a similarly hospitable disposition, from his late father.

For most of the world's inhabitants, however, this early Information Age was an irrelevance. Despite Cardinal Wiseman's vision in 1858 of the 'poor mother in the west of Ireland … receiving messages of consolation and of promise', it remained too expensive for most private individuals to send a telegram across the Atlantic.

While all the private telegraph companies in the UK, including the Magnetic and the Electric, were nationalised in 1870, international telecommunications remained largely the domain of private companies. The field thus remained open for competitors to lay their own transatlantic cables – and two of the new companies also chose Kerry as their hub. In 1874 a rival to the Anglo emerged, much to the delight of the media, who considered that competition would lead to reduced rates. The founders of the new company, called the Direct United States Telegraph Company, included bankers, stockbrokers and also a politician, Henry Labouchère,[45] better known as the author of the infamous 'Labouchère Amendment' to the UK's homosexuality laws, which subsequently ensnared Oscar Wilde and Alan Turing amongst many others. With the exclusive licence awarded to Cyrus Field's company by the Newfoundland government still in force, its western terminus was at Tor Bay in more distant Nova Scotia. Electrical testing of the new cable was directed by William Thomson,[27] who had evidently finished his association with the Anglo. The eastern landing point was in Co Kerry at Ballinskelligs, just 11 km from Valentia as the crow flies.

This period of competition proved, however, short lived as within three years the Direct United States Cable Company had come under the control of the rival Anglo,

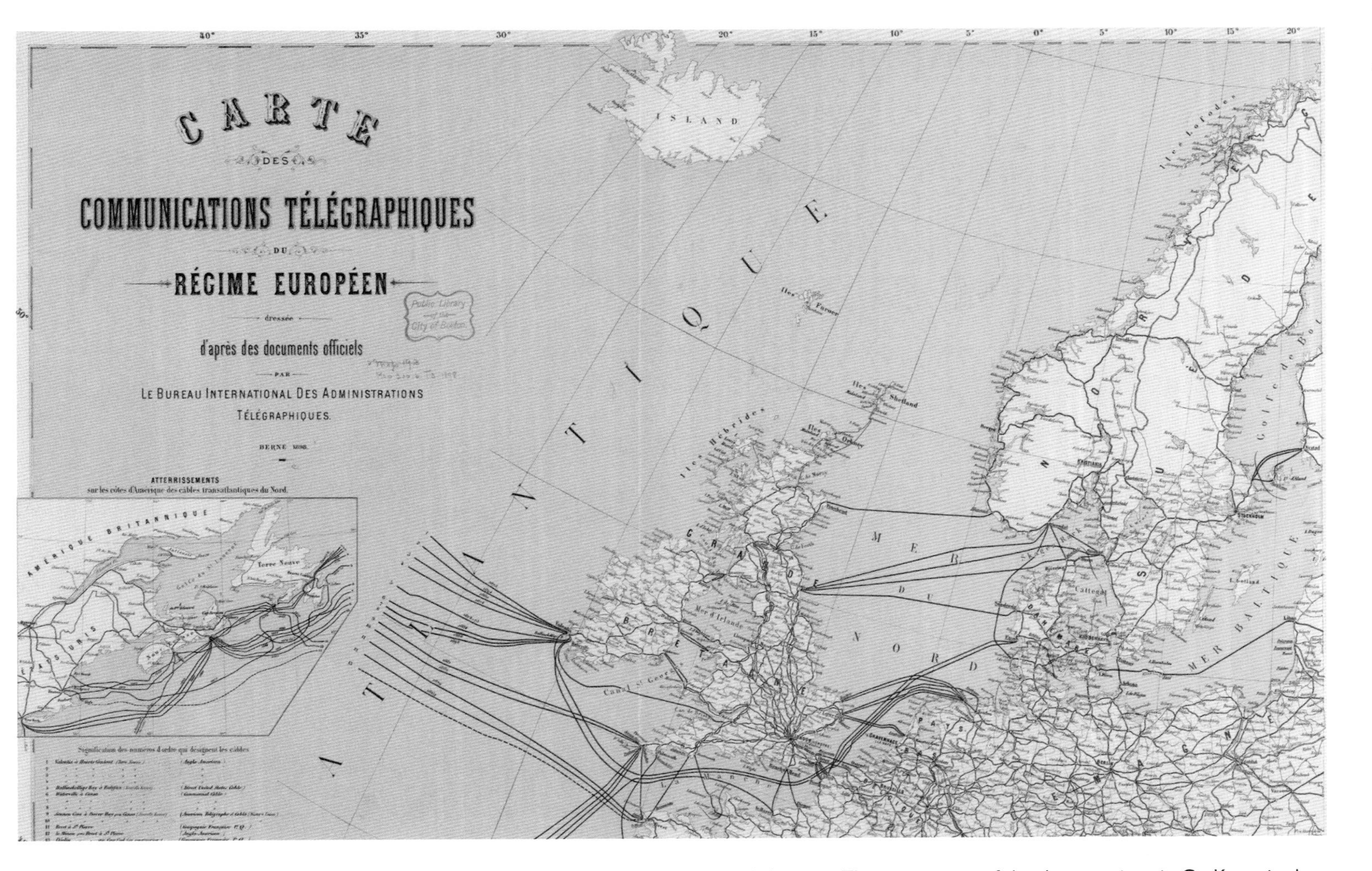

Figure 2.10: 1887 map showing the submarine telegraph lines between Europe and North America. The importance of the three stations in Co Kerry is clear. *(Reproduction courtesy of the Norman B. Leventhal Map & Education Center at the Boston Public Library)*

leaving Ballinskelligs as an isolated quirk, its route network never expanded by its new owner. By 1882 the Anglo, the Direct and the two other companies operating telegraph cables across the Atlantic had agreed a revenue sharing deal called the Joint Purse. Part of the reason for the arrangement was operational: the companies agreed to carry each other's traffic in the event of damage to a cable, thus providing an early form of network redundancy. The disadvantage for the consumer was that the companies effectively operated as a cartel, all charging the same rates.[46]

Into the fray came the Commercial Cable Co., founded by James Gordon-Bennett and John W. Mackay. The Mackay family had left Dublin in 1840 to seek a better life in the USA when John W. was nine years old. Their move paid off and he became one of the richest men in America through his involvement in silver mining in Nevada. Mackay became interested in the new field of telegraphs and helped establish the Postal Telegraph Company, a rival of Western Union in the USA.[47] Mackay's wife did not care for life in the mining towns of Nevada and spent most of her time in Paris, communicating with her husband by cable.[48] Mackay's business partner, James Gordon-Bennett, had inherited the *New York Herald* from his father but had moved to Paris and managed his business by telegram. Thus, for neatly symmetrical reasons, both men had reason to dislike the existing cable companies and the rates they charged. Mackay and Gordon-Bennett regarded themselves as buccaneers, ready to shake up the cosy world of transatlantic telegraphy but they followed the established order in one respect, choosing Co Kerry as their European hub.

Their Commercial Cable Co. opened for business on Christmas Eve in 1884 with its station at Waterville in Co Kerry connected by submarine cable to Dover Bay, Nova Scotia. In turn Dover Bay was linked by their own cable to New York. Within a few months Waterville was connected by undersea cables to Weston-super-Mare in England and Le Havre in France, thus bypassing the sometimes congested telegraph circuits that linked Ireland with Britain. Their Waterville station lay 14 km from the Anglo station at Valentia as the crow flies and several staff were poached to work in the new operation with offers of better pay.[29] Unlike its older rivals, the new company undertook its own cable repair work, its ship being modestly named the *Mackay-Bennett* after the firm's founders.[48]

A price war ensued after opening with the cost per word for eastbound telegrams falling from 50¢ to 12.5¢. This had the desired effect: within ten months of opening, the new station was receiving 900 messages a day from Canada[49], which were dutifully transcribed and relayed to their destinations in England, France or further afield. Within two years the Commercial had garnered half the transatlantic telegram traffic but, as with the Direct a few years earlier, this era of competition was again ephemeral. By 1887 the upstart Commercial Cable Co. had agreed prices with the incumbents – the poacher had become the gamekeeper.[47] As we will discover later in our story, real competition did finally emerge in 1907 – not from another cable company but from the new technology of radio.

There was plenty of potential business for transatlantic communication as emigration had created large Irish communities in the New World. For the vast majority of people in Ireland, however, sending a transatlantic telegram even at these reduced rates was inconceivable. In 1870 a 10-word telegram from Ireland to Washington DC cost the equivalent of almost a month's wages for the average farm labourer.[50] As a result, most emigrants kept in touch by post, sometimes sending what became called the 'American Letter' containing dollars for family at home. Instead, the main business for the telegraphs came from carrying news, financial data, government and military messages[47]: by the end of nineteenth century a stockbroker in New York who telegraphed an order to London expected to receive a reply within five minutes.[20]

Even if too expensive for individual use, emigrants did benefit indirectly from the expanding telegraph network that relayed news from the Old World to emigrant readerships in the New World. For example the *Southern Cross* newspaper, established in Buenos Aires in 1875, took advantage of a new submarine cable between Portugal and South America to bring news from the 'old sod' to the then-substantial Irish community in Argentina.

Transmitting newspaper copy across the oceans by telegraph was still relatively expensive and newspapers were careful to minimise the number of words transmitted. As a result, the floral elaborate prose style of newspapers in the mid-nineteenth century gave way to a more clipped 'telegraphic' style that in turn seeped into the general English language, simplifying syntax and punctuation[51] on both sides of the Atlantic. Journalists, particularly in the USA, adopted their own language of abbreviations when sending copy by telegram. Some of these abbreviations, such as the use of the term POTUS to mean President of the United States, have been given a new lease of life in the twenty-first century thanks to text messaging and Twitter.

It was not just the text of the message that could be abbreviated. Businesses could apply for a Telegraphic Address, a unique identifier registered with the telegraph service, to be used in place of the full name and address and thus saving the sender money. Telegraph staff would use the telegraphic address to determine the actual name and location of the recipient, rather in the manner that a URL like http://facebook.com/ is used by the internet to know which IP address to use to reach Facebook's servers.

A telegraphic address was a valuable part of a company's corporate identity and some were quite ingenious: in the 1920s, one could order a taxi in Cork by telegramming 'CAB, Cork' (Figure 2.11) or book a room at the Royal Marine Hotel in Kingstown (now Dún Laoghaire) at 'Comfort, Kingstown'.[52]

Figure 2.11: An advertisement for a taxi company in Cork from the 1928 telephone directory.

Comfort in Kerry was available to employees of the cable companies, who enjoyed more money and security than almost any other worker in the county. The generosity started early. In 1873 James Graves, who had been appointed superintendent of the Valentia telegraph station at the time of the fourth cable-laying attempt in 1865, received a salary of £500 a year.[44] More junior staff were on lower, but still handsome, salaries. In addition, skilled staff were provided with accommodation while unmarried employees were looked after by a housekeeper and servants. The cable houses were connected to the station's electrical supply and enjoyed piped water long before anyone else on the island. The social lives of staff were also well catered for – the station at Valentia boasted three grass tennis courts, a cricket pitch, library,[9] a full-sized billiard-table, no fewer than thirteen pianofortes and one harmonium.[53] Social activities included formal dinners and cricket matches between the three cable stations.[9]

Des Lavelle, who started work in 1950 in the power station of Western Union's Valentia station, considered the island as having two economies: the cable people and other local people. Interviewed in 2020 he recalled the 'charmed' life of cable employees with tennis in the afternoon and a library supplied with *Life* magazine and the *Illustrated London News*, concluding 'You wouldn't have found it anywhere else in the county of Kerry.'[54]

The cable companies contributed to local communities, sometimes in surprising ways. At Valentia, the cables landing at Knightstown were often fouled by fishing gear. In an effort to discourage fishermen from cutting cables in order to release tangled nets and lines, the cable company offered generous compensation if the fishermen were prepared to cut their gear and spare the cables. It is alleged that the company's deal was so generous that it was exploited by fishermen who wanted to replace their old nets and lines.[29] Tired of the expense involved in paying such compensation, around 1930 the company decided to move the cables to Lacknabau (Figure 2.2), well away from fishing nets and anchors. The reaction of the fishermen is not recorded.

The Valentia station quickly outgrew the original wooden hut at Foilhomurrin and new permanent buildings were built in 1868 at Knightstown (Figure 2.12) at the eastern tip of the island with its school and church. The arrival of a staff of well-paid workers on Valentia brought benefits to everyone in the local area as facilities at Knightstown post office were upgraded to include money transmission and a savings bank.[44] The staff body was decidedly cosmopolitan for a small town in rural Ireland: the 1901 census for Knightstown includes members of the Church of England, Plymouth Brethren, Methodist Wesleyan and Welsh Calvinist faiths[55], the latter possibly descendants of slate miners. While women were employed extensively as operators in the domestic telegraph and telephone services, the transatlantic cable operators were exclusively men. Sons often followed their fathers and sometimes grandfathers into the stations and in turn were sent around the world by their global communications employers.

HEAD OFFICE, CABLE STATION. WATERVILLE. CO. KERRY. 8295. W.L.

Figure 2.12: Despite the photo caption, this image is of the cable station at Knightstown on Valentia Island about 1880–1900, before the addition of two wings. The building still exists, along with the staff houses on either side. *(Image courtesy of the National Library of Ireland)*

As the global cable network expanded, Valentia gained its first submarine link with Europe in 1882 when a cable was laid to Emden in north-west Germany. Telegrams between Germany and North America no longer had to face delays, and possible covert scrutiny, in transit via London. The supervisor at Valentia, James Graves, received an additional annual payment of £100 from the *Vereinigte Deutsche Telegraphen-Gesellschaft* for the extra responsibility involved. Equipment at the Kerry telegraph stations was also constantly upgraded to increase the amount of traffic that the cables could carry and reduce the amount of human intervention required. One of the earliest such changes was the siphon recorder, another invention of Professor Thomson. Introduced in 1873, this printed the received signals onto a tape, reducing the possibility of human error, improving productivity[20] and increasing capacity tenfold from 2 to 20 words per minute.[56]

Duplex working, pioneered by Tipperaryman John Joseph Fahie in Persia, was introduced to the transatlantic cables in 1879,[44] effectively doubling their capacity again.[57] The cables themselves had also become vastly more durable than their finger-thick predecessors: the '1VA' cable that was laid in 1873 between Valentia and Heart's Content remained in service until 1966, carrying messages under the ocean for 93 years. Around the 1920s, Western Union introduced time-division multiplexing (Figure 2.13) which allowed several operators to transmit messages over a single cable. The extra capacity this provided allowed them to handle vastly more traffic without laying expensive new lines.

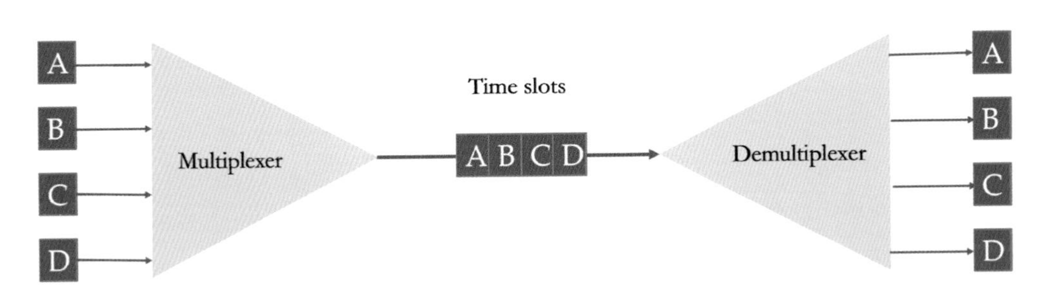

Figure 2.13: With time-division multiplexing (TDM), time on the common channel, such as a single submarine cable, is shared among the many data sources, such as telegraph operators. TDM was later to become a cornerstone of digital communications.

President Buchanan's optimistic view in 1858 of the telegraph as 'an instrument designed by Divine Providence to diffuse religion, civilization, liberty, and law throughout the world' unfortunately did not prove accurate. Wars continued and, as telegraph services expanded, their potential use by enemies to transmit intelligence began to be realised. In December 1913, with war on the horizon, the British War Office drew up a set of 'Regulations for censorship of submarine cable communications throughout the British Empire'.[57] The private cable companies had little option but to cooperate with any conditions imposed by the British authorities since government permission was required to land cables throughout the Empire. The regulations included a clause to ensure that no telegrams could be originated or received at cable repeater stations, both Post Office and private, though with a little planning such a restriction could be overcome, as was proved a few years later.

The onset of the Great War saw the three cable stations in Co Kerry surrounded by barbed wire and kept under military guard. Initially, censors were installed at each of the three stations but after October 1914 censorship of telegrams from the Western Union stations was centralised in London while a censor was maintained at Waterville to deal with messages between France and North America. While the British cut all German cables within hours of the war being declared,[47] the cables from the Kerry stations remained virtually intact through the war. The volume of traffic increased dramatically. There was extra government traffic and, of course, news about the war itself was transmitted across the Atlantic to eager newspaper readers. By the end of the Great War the Valentia station alone had about 200 staff.

The period around the First World War was the high-watermark of Kerry's role as communication hub of the world. Not only were its three cable stations working flat out, but the county also played a part in the development of the new technology of radio communications. As we will see later in the story, however, those wireless experiments in Kerry were not continued and Ireland, including the 'Kingdom', soon lost its pivotal role in that early global radio network.

More seriously, after 1919, a period of slow decline set in for the transatlantic cable stations. Western Union laid a new undersea cable between Valentia and Sennen Cove near Penzance in Cornwall, which meant that communications between Britain and North America could still continue even in the event of any problems at Valentia.[48] A plan for such a cable had been made in 1918, apparently due to fears of possible instability in Ireland following the Rising, and, despite wartime shortages, the proposal was given priority by the UK authorities.[58] The following year Western Union offloaded its operation at Ballinskelligs onto the Post Office which was anxious to have a state-owned link to Canada.[59] However, again sensing a new political order, the Post Office decided to divert the eastern end of the cable to Mousehole in Cornwall[9] and the station at Ballinskelligs was taken out of service in November 1922. Such foresight proved wise as the Civil War disrupted operations at both Valentia and Waterville.

Following Irish independence the cable stations settled into a long, slow decline with no new cables laid after 1923 (see Appendix 1). This process of decline was brought about by a combination of political and technical factors. Memories of Civil War disruption left a lingering distrust of Ireland as a communications hub while independence led to uncertainty about which government was responsible for granting licences for cable landings.[60] From the 1920s automated repeaters eliminated the need for staff at the Kerry stations to relay messages received over the transatlantic cables to their European destinations and staff numbers at Valentia fell from 200 to 20.[60] Technical developments meant that by the mid-1930s the transatlantic cables were competing with airmail, radio telegraphy and, for the super-rich, radio telephony. Kerry was slowly being bypassed.

This decline was temporarily interrupted by the onset of another war. During the Emergency, the stations continued as before, largely relaying information between Britain and the US, but now protected by the army of a neutral state who maintained bases at the two stations during the war. It was a busy time for the telegraph operators as, for security reasons, warring governments preferred to use cables rather than radio for communications.[21] With fuel in short supply, wind turbines were erected at Waterville to charge the batteries used to transmit messages. After the war, the introduction of submerged repeaters further increased the number of messages that could be squeezed through the aging cables.

It was just a temporary fillip, however. Submerged repeaters also made it possible to maintain audible speech over a long submarine telephone cable and thus in 1957 the British Post Office and AT&T opened TAT-1, the first transatlantic telephone cable. Like its telegraphic predecessors, its North American terminus was Trinity Bay in Newfoundland. However, memories of the disruption to services at the Co Kerry stations during the Civil War led to the choice of Oban in Scotland as the European landing point.[29]

Faced with competition from this expanding telephone network, the telegraph companies adopted the approach of 'if you can't beat 'em, join 'em'. Since each telephone circuit was capable of carrying 22 telegraph circuits, the telegraph companies could free themselves from the burden of maintaining their own cables simply by leasing a single telephone line. In 1961 when the Commercial Cable Company arranged to lease capacity on the new TAT-1 cable, all the cables from Waterville, some which had been in use from 1884, fell silent and the station was closed. At its peak, 300 staff had been employed; by 1961 this had fallen to 33. Ten staff members were made redundant with the remainder redeployed to Britain.[61] Unlike at Ballinskelligs, the buildings were retained and sold off with the cable staff residences fetching £800–1,300 each in 1962.[49] The houses and former telegraph station remain an attractive feature of Waterville, faring better than the Commercial Cable Company itself which quietly disappeared in the 1980s.

Western Union also decided to lease capacity on the new telephone cables, closing its station at Valentia on 3 February 1966.[62] The loss of the station and the jobs it provided was felt not only on the island but across Kerry when former cable man and Kerry football captain Mick O'Connell emigrated to England – though luckily for the team he was able to return to Kerry and Croke Park soon after.

The ability of a single telephone circuit to replace 22 telegraph circuits was the penultimate salvo in the battle between telephone and telegraph. Within a few years, with more transatlantic telephone cables and the first communications satellites in operation, the cost of intercontinental telephone calls plummeted and the remaining raison d'etre for the transatlantic telegraph service evaporated.

After serving the world for a hundred years, the revolutionary telegraph gave way to the new upstart technology, the telephone. To understand this transition we have to go back to 1881, when the telephone was an expensive and temperamental fad, newly arrived in Ireland.

CHAPTER THREE

MISS AGNES DUGGAN, TELEPHONE OPERATOR AND REVOLUTIONARY

On the morning of 18 July 1881, 16-year-old Agnes Duggan (Figure 3.1) climbed the stairs of Commercial Buildings on Dublin's Dame Street to start her new job as a 'telephone operator'.[1] Her destination was the United Telephone Company on the top floor, a newcomer amid the venerable insurance companies and accountancy practices that also occupied Commercial Buildings.[2]

Miss Duggan was not the first telephone operator employed by the London-based United Telephone Company. When the exchange opened 15 months earlier a young boy was employed in the role but with only five subscribers to this new 'network' he quickly became bored and adjourned to the courtyard to play marbles, leading to his dismissal.[3] A woman – the company's first – was then employed as an operator and in the intervening 12 months business had grown so much that a second operator was required, and Agnes Duggan got the job.

On arriving at the top floor, Miss Duggan was led into the 'exchange' room, where a jumble of wires looped down from the ceiling before disappearing behind an upright wooden board punctuated with rows and rows of small single-hole sockets, some of which had cords dangling from them. In front the board was a table with even more contraptions (Figure 3.2).

Figure 3.1: Miss Agnes Duggan, one of the first telephone operators in Ireland, pictured in 1909. *(Courtesy of BT Heritage & Archives)*

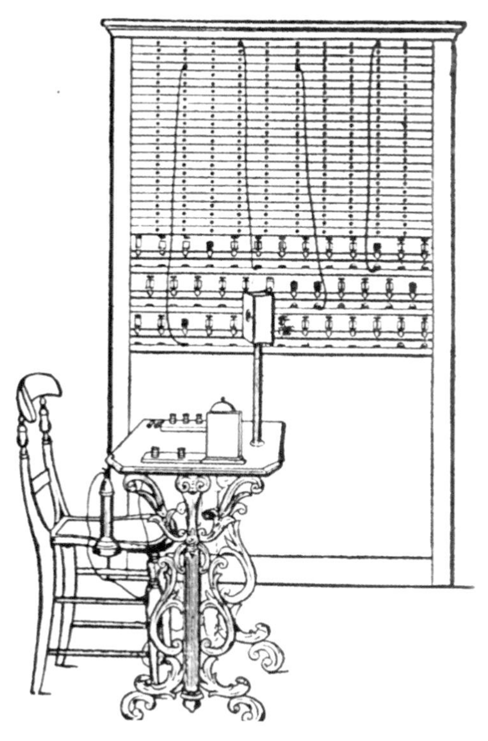

Figure 3.2: An illustration of the switchboard at Coleman St in London; the first Dublin exchange was of a similar design.[4]

The apparatus must have appeared fearsomely complicated and Miss Duggan was perhaps daunted by the prospect of learning how to master such a contraption. But she might also have been exhilarated at the idea of talking to people all over Dublin, their voices carried through the telephone wires that stretched out over the streets from Commercial Buildings. And amid all this novelty, she probably wondered if this new-fangled telephone was merely an expensive and temperamental fad and was grateful that she had her certificate from the Royal Irish Academy of Music to fall back on.[5] Either way, on that morning in 1881, Miss Duggan certainly could never have guessed the power she was unleashing.

The dauntingly complicated-looking switchboard Miss Duggan was employed to operate was most probably built by Charles Williams, Jr, of Boston,[4, 6] on whose premises Alexander Graham Bell had his laboratory. It was an impressive pedigree, linking the Dublin telephone exchange directly to the very men who had patented the telephone only a few years before. The United Telephone Company was anxious to promote this lineage, advertising itself as 'the Owners of all the patent rights in this country granted to Professor Alexander Graham Bell and Mr Thomas Alva Edison'.[7]

This anxiety over patents and pedigree was understandable as the early years of the telephone saw many patent battles. Indeed the question of who was the 'inventor of the telephone' does not have a simple answer. While conventionally attributed to the Scottish-born Alexander Graham Bell, there are many other potential claims to the title. Like the telegraph before, and radio after, there were many experimenters beavering away, largely oblivious to each other's efforts.

Possibly the person with the greatest claim to the title 'inventor of the telephone' was German schoolmaster Philipp Reis. In the early 1860s he produced a device he called a *telephon* which consisted of a microphone linked by a wire to a speaker. Music or speech could – just about – be transmitted between the microphone and speaker. Reis failed to garner much scientific or commercial interest in his invention, though it did go into small-scale production in Frankfurt in 1863.[8] One device made its way to Stephen Mitchell Yeates, whose family ran an instrument-making and opticians' business on Dublin's Grafton Street.[9] Yeates set about improving the *telephon*, replacing the microphone with one of his own invention (Figure 3.3).

Figure 3.3: In 1888, Yeates made this facsimile of his 1865 'improved Reis telephone'. *(© Science Museum/Science and Society Picture Library)*

On 14 March 1865, Yeates packed up his improved device, crossed the street to Trinity College and set it up in front of the audience attending a debate organised by the Dublin University Philosophical Society.[10] It was the first demonstration of a telephone in Ireland, and one of the first in the world.

Crucially, however, neither Reis nor Yeates patented their inventions, with the first patent application for a telephone not made until 11 years later. That patent application was, of course, from Alexander Graham Bell. On the very same day, American electrical engineer Elisha Gray also made a preliminary application for a telephone device.[11] In the legal cases that followed, the claims of Gray and Bell came into direct conflict, but in the end Bell was awarded the patent and it is his name the world remembers.

In 1877, a year after he had filed a patent for his 'telephonic device', the newly wed Bell embarked on a year-long tour of Europe with his wife Mabel to promote his new invention.[8] It might have been an unusual honeymoon choice but as a promotional tactic it was a great success with Bell registering a company in England to sell his invention in the UK.[3] The Bell's honeymoon itinerary did not include Ireland but that did not impede the Scotsman's invention reaching the country. Indeed, even prior to Bell receiving his patent, it seems that he had sent a pair of telephones to his friend Mr Holdbrook, headmaster of Aravon School in Bray, in 1876.[12] The following year a demonstration of Bell's telephone was arranged by the RDS in the packed council chamber of their Leinster House building. The other phone was installed in the lecture theatre at the far end of the building, though the newspaper reports were mixed, noting that:

> as a consequence, however, of the crowded state of the room, and the continual entrances and exits of the audience, the experiments could scarcely be said to be very successful, although enough was shown to prove that the telephone might ultimately become a useful instrument.[13]

In 1922, the government of the newly formed Free State acquired Leinster House, with the lecture theatre becoming the Dáil chamber. Incidentally, the telephone, in a different guise, returned to this part of Leinster House in the twenty-first century, provoking an exasperated Ceann Comhairle to declare that 'it's particularly sad that legislators are so addicted to their mobile phones that they cannot leave them outside when they are transacting the important business of the Dáil'[14].

In December 1877, a few weeks after the RDS demonstration, Ireland was at the fore again in a more ambitious experiment to test the potential of telephone communications under the sea. Seven telegraph circuits between Howth and Holyhead were taken out of regular service, connected in parallel and a telephone attached to each end. Despite the primitive equipment and the distance involved 'the voices were distinctly heard and intelligible'.[15] It was the longest distance the human voice had been carried across the sea.

Meanwhile the telephone patent wars continued on both sides of the Atlantic. Within seven months of Bell being granted his US patent, the American inventor Thomas Alva Edison was granted a patent for 'acoustic telegraphy'. By 1879 Edison's telephone instruments had also been shipped across the Atlantic and the Edison Telephone Co. of London was formed.[16] As it happens, one of its employees was a 22-year-old George Bernard Shaw, who described his brief employment there as 'my last attempt to earn an honest living'.[3]

Those early patents were just for a telephone device, with initial installations simply consisting of two phones permanently connected to each other, such as between different parts of a factory. It did not take long for the concept of a telephone exchange to be developed, allowing a telephone to be connected to any other telephone – by a telephone operator, like Miss Duggan.

The world's first exchange opened in New Haven, Connecticut, on 28 January 1878.[16] Within 18 months the first telephone exchanges were installed in Europe, with one opening in Dublin on 29 March 1880[17] followed by two in Belfast. While the Bell and Edison companies in the UK merged on 13 May 1880 to form the United Telephone Company[18] (Agnes Duggan's employer), rivalry between companies competing for telephone business continued in a number of cities. This was well illustrated in Belfast, where the two exchanges just alluded to opened for business within a month of each other, causing some chaos. The Scottish Telephonic Exchange Company was first off the mark, opening an exchange at Castle Chambers in May 1880.[19] It claimed that its devices did not infringe on the patents of either Bell or Edison. Within weeks, the newly formed UTC expanded from Dublin to open its Belfast telephone exchange at 3 High Street in June 1880, warning testily in its advertisements 'that legal proceedings will be taken against all persons infringing their Patent'.[20] The Scottish company stole a march on its rival by offering three months free rental to the first 50 customers[21], a marketing tactic that was rediscovered in recent years by competing telephone companies.

With no interconnection between the two rival systems, a subscriber who wanted to be fully connected was required to rent a line from both companies with the result that, by August 1880, Belfast Fire Brigade had two telephones.[22] The duplication in

Belfast was solved when the Scottish company merged with its rival in July 1881, presumably reducing confusion at Belfast's fire station. Perhaps to compensate for the initial chaos, the fire station was allocated the number Belfast 1.[23]

Down in Dublin, by the end of 1880 the subscriber base had grown to 32 (Figure 3.4) and when Miss Duggan started the following July she joined a team of seven: Mr Morgan the Managing Director, the other 'lady operator' mentioned at the start of this chapter, three clerks, a handyman who served as electrician and linesman, and a boy who carried all the handyman's tools in a straw basket on his back.[4]

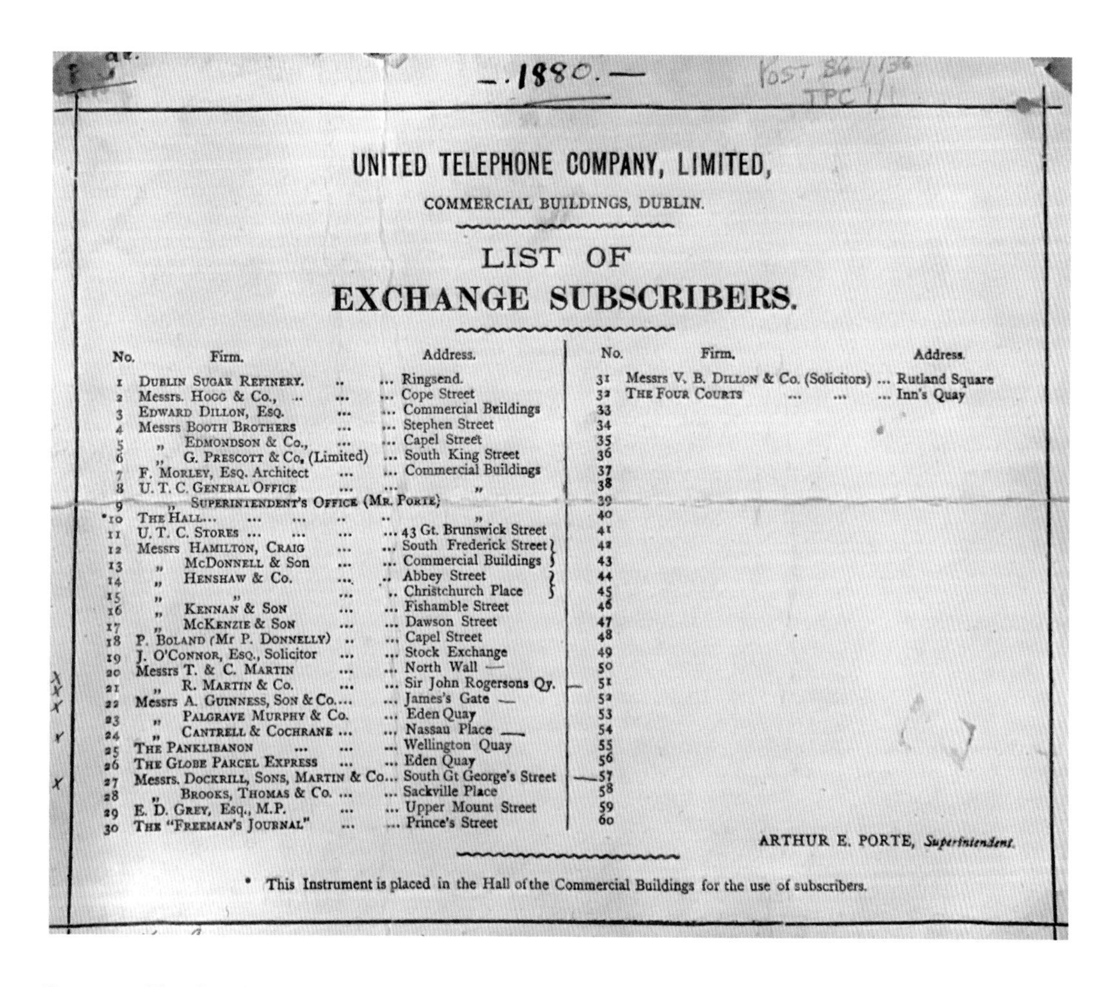

—1880.—

UNITED TELEPHONE COMPANY, LIMITED,

COMMERCIAL BUILDINGS, DUBLIN.

LIST OF

EXCHANGE SUBSCRIBERS.

No.	Firm.	Address.
1	Dublin Sugar Refinery.	Ringsend.
2	Messrs. Hogg & Co.,	Cope Street
3	Edward Dillon, Esq.	Commercial Buildings
4	Messrs Booth Brothers	Stephen Street
5	" Edmondson & Co.,	Capel Street
6	" G. Prescott & Co. (Limited)	South King Street
7	F. Morley, Esq. Architect	Commercial Buildings
8	U. T. C. General Office	"
9	" Superintendent's Office (Mr. Porte)	
*10	The Hall...	"
11	U. T. C. Stores	43 Gt. Brunswick Street
12	Messrs Hamilton, Craig	South Frederick Street }
13	" McDonnell & Son	Commercial Buildings }
14	" Henshaw & Co.	Abbey Street }
15	" "	Christchurch Place }
16	" Kennan & Son	Fishamble Street
17	" McKenzie & Son	Dawson Street
18	P. Boland (Mr P. Donnelly)	Capel Street
19	J. O'Connor, Esq., Solicitor	Stock Exchange
20	Messrs T. & C. Martin	North Wall
21	" R. Martin & Co.	Sir John Rogersons Qy.
22	Messrs A. Guinness, Son & Co.	James's Gate
23	" Palgrave Murphy & Co.	Eden Quay
24	" Cantrell & Cochrane	Nassau Place
25	The Panklibanon	Wellington Quay
26	The Globe Parcel Express	Eden Quay
27	Messrs. Dockrill, Sons, Martin & Co.	South Gt George's Street
28	" Brooks, Thomas & Co.	Sackville Place
29	E. D. Grey, Esq., M.P.	Upper Mount Street
30	The "Freeman's Journal"	Prince's Street

No.	Firm.	Address.
31	Messrs V. B. Dillon & Co. (Solicitors)	Rutland Square
32	The Four Courts	Inn's Quay
33		
34		
35		
36		
37		
38		
39		
40		
41		
42		
43		
44		
45		
46		
47		
48		
49		
50		
51		
52		
53		
54		
55		
56		
57		
58		
59		
60		

ARTHUR E. PORTE, *Superintendent.*

* This Instrument is placed in the Hall of the Commercial Buildings for the use of subscribers.

Figure 3.4: The first Dublin telephone directory, 1880. Amongst the businesses that survive to this day are brewers Messrs A. Guinness, Son & Co. and wine merchant Edward Dillon Esq. Clearly selling alcohol in Ireland is a sustainable business model. *(Courtesy of BT Heritage & Archives)*

The subscriber growth picked up even more after Agnes Duggan joined so that by 1883 the customer base had grown to 300 entities, making an average of 11 calls per day, the highest of anywhere in the UK.[1] Unfortunately for the company it derived no financial benefit from its loquacious Irish customers as the annual line rental of £20 (*c.* €25) per year included unlimited local calls. While the early telephone users were overwhelmingly from the business community, there was also a small but increasing base of residential customers. The telephone companies were not averse to preying on potential concerns about security and welfare in their attempts to boost sales to residential users as demonstrated by the 1912 advert (Figure 3.5) with the slogan 'If you had a Telephone at Home Your Wife could do most of the Shopping with it. It would save Time and Worry.'

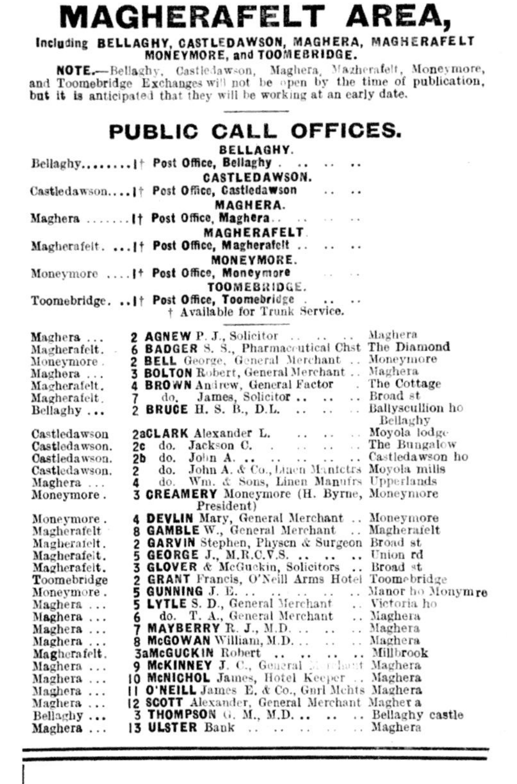

MAGHERAFELT AREA,

Including BELLAGHY, CASTLEDAWSON, MAGHERA, MAGHERAFELT MONEYMORE, and TOOMEBRIDGE.

NOTE.—Bellaghy, Castledawson, Maghera, Magherafelt, Moneymore, and Toomebridge Exchanges will not be open by the time of publication, but it is anticipated that they will be working at an early date.

PUBLIC CALL OFFICES.

		BELLAGHY.
Bellaghy........	1†	**Post Office, Bellaghy**
		CASTLEDAWSON.
Castledawson....	1†	**Post Office, Castledawson**
		MAGHERA.
Maghera	1†	**Post Office, Maghera**
		MAGHERAFELT.
Magherafelt. ...	1†	**Post Office, Magherafelt**
		MONEYMORE.
Moneymore	1†	**Post Office, Moneymore**
		TOOMEBRIDGE.
Toomebridge. ..	1†	**Post Office, Toomebridge**

† Available for Trunk Service.

Maghera ...	2	**AGNEW** P. J., Solicitor	Maghera
Magherafelt.	6	**BADGER** S. S., Pharmaceutical Chst	The Diamond
Moneymore.	2	**BELL** George, General Merchant	Moneymore
Maghera ...	3	**BOLTON** Robert, General Merchant	Maghera
Magherafelt.	4	**BROWN** Andrew, General Factor	The Cottage
Magherafelt.	7	do. James, Solicitor	Broad st
Bellaghy ...	2	**BRUCE** H. S. B., D.L.	Ballyscullion ho Bellaghy
Castledawson	2a	**CLARK** Alexander L.	Moyola lodge
Castledawson.	2c	do. Jackson C.	The Bungalow
Castledawson.	2b	do. John A.	Castledawson ho
Castledawson.	2	do. John A. & Co., Linen Mantctrs	Moyola mills
Maghera ...	4	do. Wm. & Sons, Linen Manufrs	Upperlands
Moneymore.	3	**CREAMERY** Moneymore (H. Byrne, President)	Moneymore
Moneymore.	4	**DEVLIN** Mary, General Merchant	Moneymore
Magherafelt	8	**GAMBLE** W., General Merchant	Magherafelt
Magherafelt.	2	**GARVIN** Stephen, Physcn & Surgeon	Broad st
Magherafelt.	5	**GEORGE** J., M.R.C.V.S.	Union rd
Magherafelt.	3	**GLOVER** & McGuckin, Solicitors	Broad st
Toomebridge	2	**GRANT** Francis, O'Neill Arms Hotel	Toomebridge
Moneymore.	5	**GUNNING** J. E.	Manor ho Monymre
Maghera ...	5	**LYTLE** S. D., General Merchant	Victoria ho
Maghera ...	6	do. T. A., General Merchant	Maghera
Maghera ...	7	**MAYBERRY** R. J., M.D.	Maghera
Maghera ...	8	**McGOWAN** William, M.D.	Maghera
Magherafelt.	3a	**McGUCKIN** Robert	Millbrook
Maghera ...	9	**McKINNEY** J. C., General [illegible]	Maghera
Maghera ...	10	**McNICHOL** James, Hotel Keeper	Maghera
Maghera ...	11	**O'NEILL** James E. & Co., Gnrl Mchts	Maghera
Maghera ...	12	**SCOTT** Alexander, General Merchant	Maghera
Bellaghy ...	3	**THOMPSON** G. M., M.D.	Bellaghy castle
Maghera ...	13	**ULSTER** Bank	Maghera

If you had a Telephone at Home

Your Wife could do most of the Shopping with it.

It would save Time and Worry.

(A 6) (Ulster) 3

Figure 3.5: Extract from the 1912 directory. It is not clear whether the advertisement is aimed specifically at potential domestic customers in the Magherafelt area or more generally. *(Courtesy of BT Heritage & Archives)*

To cater for this growing demand, the company opened a new, purpose-built, telephone exchange for central Dublin on Crown Alley in Temple Bar.[3] The exchange, housed in a building (Figure 3.6) designed by Thomas Manley Deane, whose father had been responsible for the Anglo-American Telegraph Company's station at Valentia Co Kerry[24], was opened by its longest serving operator, Agnes Duggan, on 4 June 1900.[4]

Figure 3.6: Crown Alley exchange, which opened in 1900. Now surrounded by pubs and restaurants, the building is still in use as a telephone exchange, making it possibly the longest-serving such building in Europe.

The new switchboard was very different to the primitive predecessor that had greeted Miss Duggan on that July morning in 1881. In the intervening years, the design and operation of switchboards had been streamlined and standardised (Figure 3.7). When a call came in, the indicator disk above the calling subscriber's socket would drop down. This was the signal for the operator to insert a cord with a brass plug into the relevant socket that enabled her to speak to the calling subscriber, who would tell her the number they wanted. She would then take another plug, insert it into the socket of the required number and depress a key in order to ring the bell of the requested telephone. This mode of operation changed little over the decades that followed and Miss

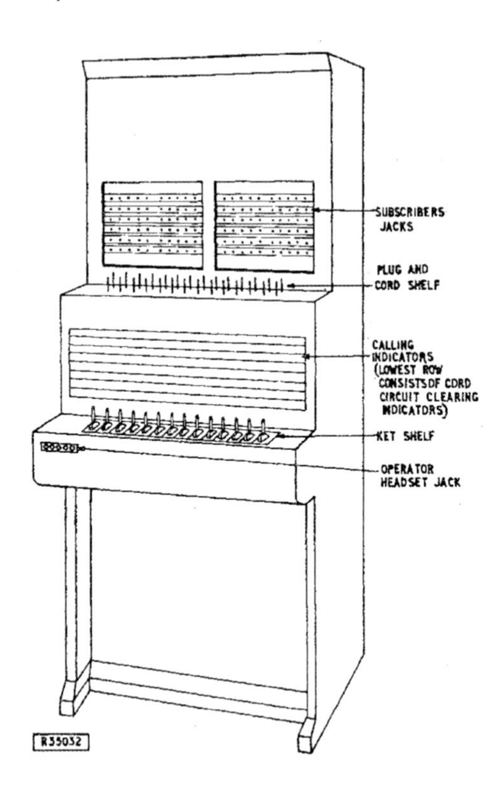

Duggan would have had little difficulty operating the Magneto 'cord board' in the Co Clare village of Mountshannon, which remained in use until 1987.

Figure 3.7: A typical Magneto switchboard of the type used at Crown Alley in Dublin in 1900, Mountshannon in Co Clare in 1987 and many places in between. *(Courtesy of BT Heritage & Archives)*

One Dublin business that did not rush to install a telephone was Yeates's instrument-making and opticians' business on Grafton Street. It is first listed in the 1904 phone book[25], three years after the death of telephone pioneer Stephen Mitchell Yeates. Perhaps he did not enjoy the idea of paying royalties to an old rival.

The introduction of the telephone had major social impacts, with the new technology popularised by popular culture and literature. The hit Gilbert and Sullivan comic opera *HMS Pinafore*, which played Belfast's Theatre Royal in the month before that city's first telephone exchange opened[26], refers to the newly invented telephone in song. Describing the prospective imprisonment of a comrade, the chorus sing 'No telephone/ Communicates with his cell!'[27]

The telephone appears in more serious literature too with several references to it in *Ulysses*, providing an interesting insight into the use of the telephone in the life of Joyce's imaginary Dubliners in and around 1904. In the *Aeolus* episode, Leopold Bloom uses the phone in the office of the *Freeman's Journal* to call one of his advertising clients. Bloom has the number (almost) at the tip of his tongue, asking the operator for 'Twenty eight ... No, twenty ... Double four ... Yes'. Bloom is not an employee of the *Freeman's Journal* but feels entitled to ask to use their phone. It is likely that the *Freeman's Journal* was still on a tariff that included free local calls and, with street-side telephone kiosks still 20 years away, a request to use a phone in this manner would not have been considered unusual.

Earlier, in the *Proteus* episode, Stephen Dedalus thinks of the telephone wire as an umbilical cord stretching back to Adam and Eve and imagines asking a telephone operator to put him through to the Garden of Eden: 'Hello! Kinch here. Put me on to Edenville. Aleph, alpha: nought, nought, one'.[28] This imaginary telephone number for the Garden of Eden starts with the first letter of the Hebrew alphabet and the first letter of the Greek alphabet, representing the beginning of time. It is also an unwitting harbinger of the two-letter exchange name codes that were used in many cities from the 1920s to the 1960s.

What we now term the emergency services were also affected by the telephone as the utility of the phone as a tool to summon help at any time was quickly realised. As we saw, Belfast fire brigade was an early adopter, while by 1883, at the request of Dublin Corporation, 24-hour service was introduced at the Dublin exchange[1] to facilitate emergency calls. Continuous service was soon extended to all the larger centres but such a facility could not be justified at smaller exchanges, a source of discontent that took over 100 years to eliminate. By 1884 Dublin's two fire stations,[29] most of the city hospitals[30] and the headquarters of the Dublin Metropolitan Police (DMP) at Dublin Castle were connected up.

Curiously, however, the DMP rejected a proposal from the United Telephone Company to connect local police stations with headquarters at reduced rates in favour of retaining its ABC telegraph system.[21] As a result, as late as 1894, if a fire was reported to a police station, the procedure was for the station to send the details by the tediously slow ABC telegraph to DMP headquarters, which in turn contacted the Fire Brigade by phone.[31] This convoluted method of communications was later modernised, with a Central Police Telephone Office set up in Dublin Castle. This office was to prove a vital hub for government communications during the Easter Rising.

The technology used within the telephone industry also changed as a result of discoveries in other areas. Initially, telephone circuits consisted of a single wire with an earth return, replicating telegraph practice. Such circuits were very prone to problems of cross-talk so customers were often able to hear their neighbours' conversations and it was soon discovered that much better results could be obtained if a pair of wires was used. A revolution in metal-making made it possible to produce thin wire from copper, while new forms of insulation made feasible the production of underground cables with capacity for hundreds of lines. Around the same time, thanks to new legislation and agreements with local authorities, it became permissible for the private telephone companies to place lines underground. From the 1890s, overhead circuits typically consisting of a single steel wire were replaced with pairs of copper wires, linked to the exchange through underground ducting. By 1908, 20 km of cable ducting had been laid under the streets of Dublin, some of which still remains visible on the streets of the capital (Figure 3.8). The copper pair remains a common method of carrying voice and data traffic into homes and businesses to this day. The broadband revolution of the twenty-first century can trace its roots back to the Victorian telephone revolution 123 years earlier.

Figure 3.8: Manhole cover in Dublin dating from the period of the National Telephone Company 1893–1912.

Despite these advances in cable technology, there remained constraints, both technological and commercial, on early telephone development. The first telephone exchanges established in Dublin, Belfast and elsewhere were initially isolated from each other and could connect only local calls. As the network expanded, 'junction' circuits were provided to adjoining exchanges operated by the same company, so that by 1883 it was possible to make a call between the main exchange in Dublin and the newly opened exchange in suburban Rathmines[1] – but no further.

The ability to make long-distance 'trunk' calls took a long time to become universal, for largely technical reasons. With no way to amplify signals available at the time, trunk circuits had to use heavy gauge copper wire to minimise resistance so that the call remained audible. This added greatly to the cost of building a trunk network and, as a result, it was not until 5 April 1892 that Ireland's first trunk telephone line went into operation between Dublin and Belfast.[32] To maintain adequate transmission quality, the wire used was the thickness of a pencil and weighed a whopping 800 pounds per mile (227 kg/km)[3]. The line was inaugurated by a call between the mayors of the two cities and calls were free of charge for the first week.[32]

Even before it had carried its first call, however, the new trunk line was affected by a change in government policy. Concerned that the ability to make long-distance telephone calls over the privately owned telephone network would adversely affect telegram revenues, the Post Master General announced in March of 1892 that he would purchase the nascent trunk line network from the private telephone companies, who would henceforth be confined to operating the local networks.[3] This was a further manifestation of more active government involvement in the telephone network, a policy that had begun in 1882 with the opening by the Post Office of telephone exchanges in cities that had been hitherto ignored by the private sector[33] such as Derry, Cork, Limerick and Waterford.[1]

With the Post Office now in charge of the trunk network, a submarine telephone cable was soon laid between Portpatrick in Scotland and Donaghadee in Co Down. The 23.5 nautical mile long (43.5 km) cable cost £20,000 (*c.* €25,000) and allowed telephone users in Ireland to make calls to Britain for the first time. The new lines were tested by the *Irish News* who were 'exceedingly gratified to find how marvellously the voice is carried by the new wires'.[34] The fact that the first trunk connections in Ireland established under Post Office control were to Britain, rather than any of the many cities and towns in Ireland without such a facility, was no accident. This new connection was part of a 'Backbone Scheme' to create a national trunk line network to link all the larger cities of the UK. Connections from Dublin via Belfast across the Irish Sea were part of the scheme but no other Irish city was included.

Unlike the telegraph system, the Post Office did not make a concerted attempt to create a national network across the whole of the United Kingdom. Treasury rules required that the provision of new trunk lines would not impose a financial loss. This gave rise to a catch-22: the Post Office would not expand its trunk network unless there was an exchange and an exchange would not be viable unless there was a trunk network connection.[21] This policy had a particularly adverse effect on development in Ireland and the slow rollout of trunk lines was the subject of regular criticism by Irish MPs. For example, in 1902 in response to questions from Cork MPs about new trunk lines, the following exchange occurred in the Commons:

> Captain Donelan (MP for Cork East): Do I understand that the Hon. Gentleman can hold out no hope of the extension from Midleton to Youghal?
> Mr Austen Chamberlain (Postmaster General): Not unless a guarantee is given which will secure the Post Office Revenue against loss.
> Mr J. F. X. O'Brien (MP for Cork City): Are these guarantees required in England?
> Mr Austen Chamberlain (Postmaster General): Yes, under similar circumstances they are required in all parts of the United Kingdom.[35]

As a consequence of this financial policy, exchanges tended to follow the slowly expanding trunk network, which in turn tended to follow the railway lines. Where there were no trunk lines, there were generally no exchanges and thus many important towns had no telephone service. This gave rise to an east–west divide: Freshford, Co Kilkenny (population 446 in 1911) and Cloyne, Co Cork (population 756) had telephone service by 1910, but much larger towns in the west including Ballina, Co Mayo (population 4662), or Tuam, Co Galway (population 2980)[36] did not. There was no understanding, at least by the governing Conservative party in Westminster, that providing infrastructure could lead to economic and social development in the most deprived parts of the land. Rules were set in London and applied nationally without flexibility. This underdevelopment of the telephone service in Ireland provided additional grist to the mill of the nationalist movement in its quest for greater Irish independence.

These financial strictures were not the only source of difficulty. Expansion of the trunk network was also thwarted by technological limitations. For instance in 1898 the Post Office wanted a more direct telephone connection between Dublin and London and laid a new submarine cable between Newcastle, Co Wicklow, and Nevin (now Nefyn) in North Wales. The distance involved was considerably further than any existing telephone cable and a brand-new design using an air-space cable with gutta-percha insulation was employed. Unfortunately, after the cable was laid it was found that serious overhearing occurred between the two telephone circuits it carried[1] and the cable was relegated to providing additional cross-channel telegraph lines. It took 16 years for another attempt to be made but eventually in 1914 a new two-circuit telephone route between Dublin and Manchester was successfully inaugurated. Costing £78,000 (*c*. €99,000)[37], the Irish terminus of the cable was at the Martello tower in Howth, continuing the relationship between telecommunications and the north Dublin harbour town.

Aside from the exchanges opened by the Post Office in a handful of cities in 1882, the government was happy to leave the local telephone network to private enterprise. The private telephone companies became consolidated so that by May 1893 all exchanges throughout Ireland had come under the ownership of the National Telephone Company (NTC).[32] This pattern was replicated across the UK creating what was effectively a nationwide private sector monopoly. There was considerable dissatisfaction with the quality of service offered and the number of telephones per capita was adversely compared to the US and other places in Europe.[38] As a result, the concept of the government taking over the telephone service, as it had done with the telegraph service in 1870, began to be debated.

There was little discussion at the time about the possibility of the telephone network becoming a semi-state company, owned by the state but independent of government on a financial and operational basis. This was unsurprising, as the concept of a state-owned enterprise was little known at the time in any sphere. There was an assumption that nationalisation of the telephone network would lead to it becoming part of the Post Office, which already operated the trunk telephone lines, several local telephone exchanges, the national telegraph service, and a savings bank in addition to its traditional mail business.

The licences of the private phone companies, now largely consolidated into the NTC, were due to expire at the end of 1911, providing an opportunity to resolve the ownership question. With telegraph revenues being eroded by competition from the telephone, the Post Office saw its chance to pounce. In August 1905, after prolonged negotiations and intense haggling, the Post Master General and the NTC signed an agreement by which the state was to take over the latter's operations throughout the UK on the expiration of its licence for the sum of £12.5m (*c.* €14m).[39] The formal handover would take place on 1 January 1912 with NTC staff becoming civil servants employed by the Post Office.

The telephone network was expanded rapidly after nationalisation with many new exchanges opened, often in post office premises where the postmistress tended to the switchboard in between serving customers at the counter. The trunk network was also extended, finally reaching Galway in 1913. By 1918 the telephone system in the 26-county area comprised of about 212 exchanges. But there remained a distinct geographical bias in the telephone network. Of the 12,500 lines in use, 6,400 (51.2 per cent) were in or around Dublin. There were no exchanges at all in the counties of Mayo, Leitrim or Roscommon[3] (Figure 3.10). A new country was in the making but its foundations, telephonically speaking, were shaky. One element was, however, firm. The telephone network was a government-controlled monopoly. It would remain so for 81 years.

The influx of telephone operators and other ex-NTC employees into the Post Office reinforced its position as a significant employer of women in a time where other opportunities were limited. In 1913 the Post Office employed 306 female telephonists and 539 female 'sorting clerks and telegraphists' in Ireland alone. While such jobs provided secure employment, the salary of about £26–45 (*c.* €33–57) per year[40] was

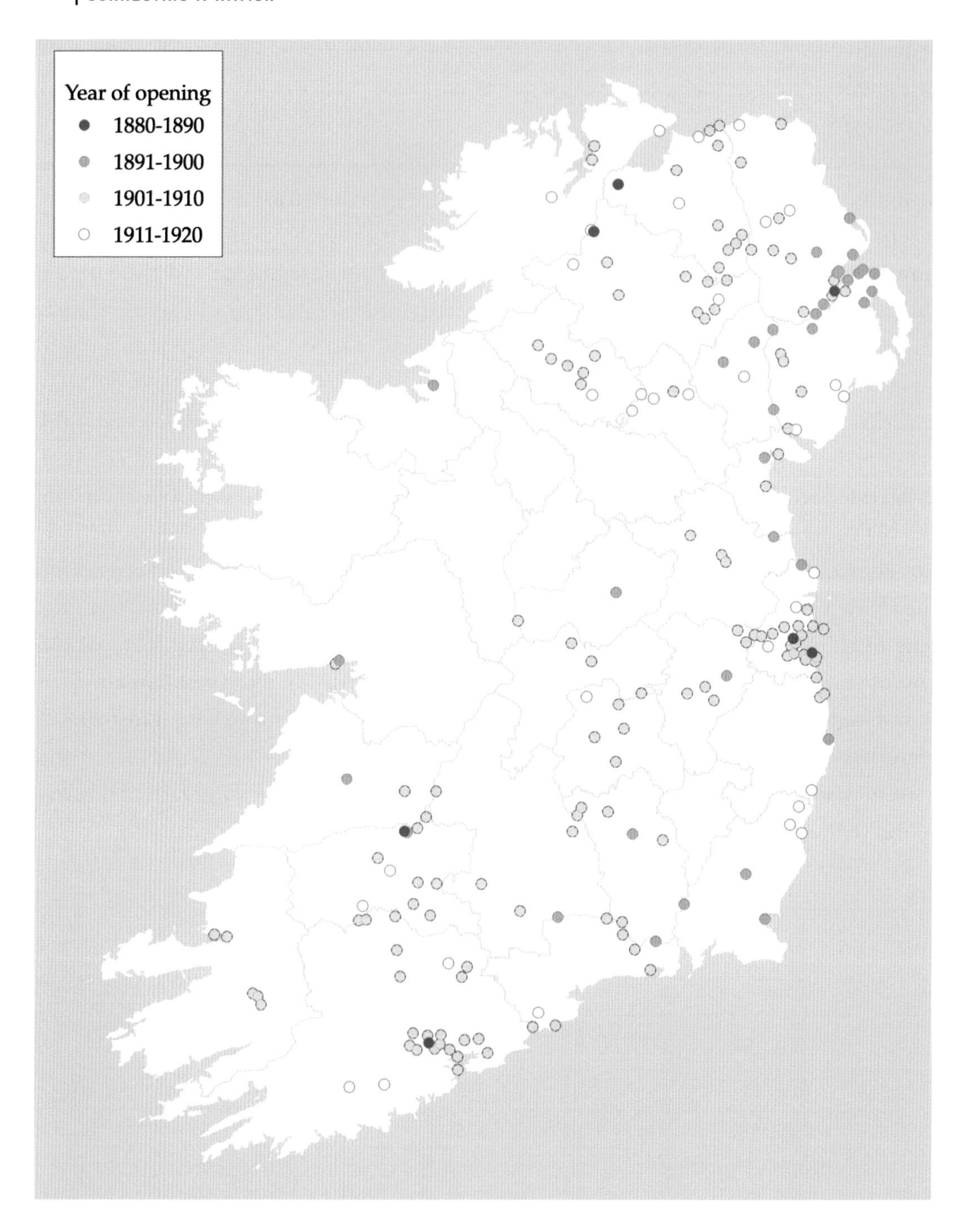

Figure 3.9: Telephone exchange opening dates, 1880–1920. Source: BT Archives, Lists of telephone exchanges, TPF 3/1.

equivalent to just seven times the line rental charged by their employer. The concept of the telephone operator as a 'woman's job' persisted: as late as 1977, the Irish phone book instructed callers who phoned the operator to 'ask her for the number required'.[41]

As it was considered unseemly for ladies to have to travel to and from work late at night, in exchanges with 24-hour service the night shift was operated by men on a separate and, needless to say, higher-paid grade.[42] Even the male operator grades, however, earned only a fraction of the £400 (*c.* €508) per year salary of the Dublin District Manager[43] in 1912, Mr Percy F. Currall. Indeed Mr Currall earned almost enough in a single year to buy outright the eight-room house in Ranelagh with garden advertised in the *Daily Express* for sale at £450 (*c.* €571)[44].

Miss Agnes Duggan avoided the fate of her marble-playing predecessor and remained a telephone operator for over 30 years. We know that by 1907 she had progressed from putting through a handful of calls to the role of Chief Operator[45] in the NTC, supervising a team of 29 operators handling 33,000 calls a day.[46] History does not record if she moved to the Post Office at the time of nationalisation. However, the Crown Alley exchange building she opened in 1900 remains in use, still handling tens of thousands of calls each day, automatically and digitally, without the aid of any operator. Any doubts Miss Duggan had on her first day were unfounded. The new-fangled telephone had proved to be not a fad but an invention that transformed countless aspects of life.

On that morning in July 1881 Agnes Duggan had been at the vanguard of a revolution that was to sweep not just Ireland but the world.

CHAPTER FOUR

THE ITALIAN AND THE HUNGRY HAWKS

Lloyd's of London had a problem with hawks. Their station at Torr Head in Co Antrim (Figure 4.1), part of a network of lookout posts around the coastline, was supposed to inform head office in London by telegraph of shipping movements but the ships were often invisible due to the mist and rain that regularly descend on this isolated headland. The keepers at the East lighthouse on Rathlin island, a few miles to the north-west, had better visibility of the passing ships but, with no telegraph or telephone connection from Rathlin, it was difficult to forward this valuable information

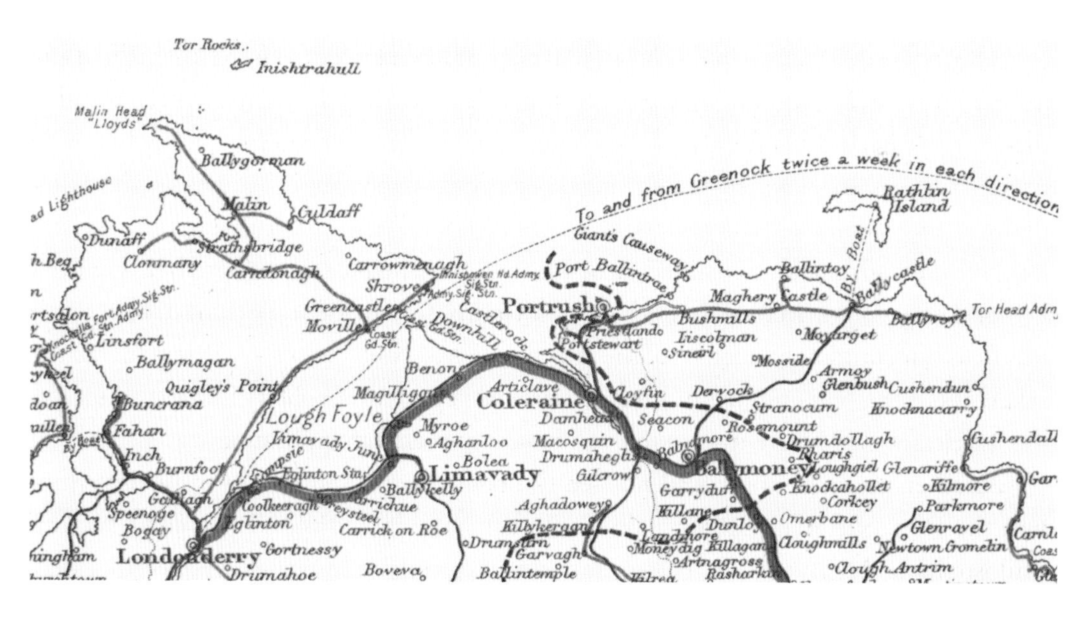

Figure 4.1: The North Antrim coast where the world's first radio transmissions were made. The telegraph network, shown in red has reached Ballycastle and Torr Head, but not Rathlin. *(By permission of The National Archives of the UK)*

to Lloyd's in London. Semaphore signals from the lighthouse to the mainland of Co Antrim were also interrupted by poor weather.[1] Lloyd's resorted to training a flock of carrier pigeons to fly from Rathlin to Ballycastle Coastguard Station, but that solution came a cropper when the carrier pigeons were eaten by the hungry hawks that inhabited the island.[2] Salvation seemed to be on the horizon at last in 1897, when a young Italian man pitched up in London with his mother – and a radically new idea.

The young Italian man was Guglielmo Marconi. Born in Bologna in 1874, he was the son of Enniscorthy-born Annie Jameson whose grandfather John Jameson had founded the famous whiskey distillery in Dublin. The Jameson family had Scottish roots, naming their home in the Donnybrook area of Dublin after the Scottish town of Montrose. (In a neat historical coincidence, Montrose was later to become the home of the Irish radio and television service, though sadly there is no evidence that Guglielmo visited the house.)[3] Marconi's Irish mother had a huge influence on his early life, raising him to be fluent in English as well as Italian and nurturing his interest in technology. When he failed to impress the Italian authorities with his ideas about something called 'wireless communication', his mother encouraged him to try his hand in the UK, travelling with him to London in 1896. There, aged 22, he demonstrated his transmitter to several prospective customers in the UK including the Post Office and the defence forces. A veritable nineteenth-century tech entrepreneur, Marconi had filed his first patent,[4] for wireless telegraphy, by the age of 22 and formed his first company when he was 23.

The first technological breakthrough to be promoted by Marconi was the radio transmitter. It is difficult to grasp what a revolution this was. The telegraph had been in public use for 55 years and the telephone for about 17 but both remained completely bound by wires. Long distances were difficult and expensive to span by cable, while ships at sea and many islands remained completely incommunicado. The concept that you could send a signal through the air remained at best a theory. Brilliant scientists such as Heinrich Hertz, James Clerk Maxwell and Edouard Branly had made important discoveries in this field but could not conceive of the uses their ideas could be put to. When Hertz was asked about the potential use of his discovery, he reportedly replied, 'I do not think that the wireless waves I have discovered will have any practical application'[5].

Another pioneer who could not see the commercial application of his invention was the Rev. Nicholas Callan from Co Louth. In 1836, while Professor of Natural Philosophy at Maynooth, he discovered how to produce high-voltage electric pulses from a low-voltage DC supply, such as that produced by a battery. Father Callan could, however, see no application for his invention, apart from using it to literally shock the seminarians under his tutelage.[6] His induction coil, later perfected by German instrument-maker Heinrich Ruhmkorff, formed a vital component of Marconi's spark transmitter.

By contrast, some other inventors could see the value of radio even more than Marconi but failed to woo investors and politicians. One such visionary was Nicola Tesla who in 1926 predicted a future when:

> wireless is perfectly applied and the whole world will be converted in to a huge brain … [and] we shall be able to communicate with one another instantly, irrespective of distance.[7]

Despite his uncannily accurate prediction of the internet-connected smartphone world of the twenty-first century, Tesla died in obscurity in a New York hotel room, his legacy largely overlooked for decades.

Unlike Tesla or Hertz, Marconi was the consummate marketeer who understood not only the value of a network but also the value of networking. While often considered to be the father of radio, and certainly responsible for popularising a variety of new technologies, Marconi's genius was not as an inventor but in adapting and exploiting the earlier discoveries of others. As his biographer, Marc Raboy, put it, 'Marconi's greatest invention was himself.'[8]

The first technological breakthrough to be promoted by Marconi was the radio transmitter. Marconi's transmitter was based on spark gap technology that worked by generating radio frequency electromagnetic waves using an intermittent electric discharge – literally a big electric spark controlled by a Morse key. At the receiving end, a device called a coherer (Figure 4.2) was used to detect the distant sparks.[9] This consisted of a tube or capsule containing two electrodes spaced a small distance apart with metal filings in the space between them. When a radio signal, such as that generated by the spark-gap transmitter, is detected by the device, the resistance of the filings reduces, allowing an electric current to flow through it and producing an audible click. Spark transmitters and coherers were the technologies behind the first decades of wireless, until both were replaced by devices based on continuous wave (CW) technology from 1907 onwards.

While his demonstrations to the Post Office and the defence forces were to reap rewards later on, Marconi's first real break came when Lloyd's of London arrived at the door of his London office in 1897. In the nineteenth century, travel by sea was still a hazardous undertaking. As the world's leading maritime insurer, Lloyd's of London compiled details of worldwide shipping movements and casualties, and to this end Lloyd's utilised lookout stations positioned all around the coasts including many in

Ireland. Most of the shipping traffic from the US and Canada en route to ports such as Liverpool, Glasgow and Belfast passed along the north coast of Ireland. Lloyd's had lookout stations at places along this coastline including Malin Head in Co Donegal and Torr Head in Co Antrim to report this traffic back to London, but as already noted, Lloyd's had no luck connecting Rathlin island's East lighthouse to this network. Colonel Sir Henry Hozier, secretary of the famous insurer, having heard of Marconi's 'wireless' device, arranged a meeting[8] where the problems at Rathlin island and its hungry hawks were discussed. Hozier agreed to finance an experimental wireless link from the island and thus in 1898 Marconi ended up in the north-eastern corner of the country of his mother's birth.

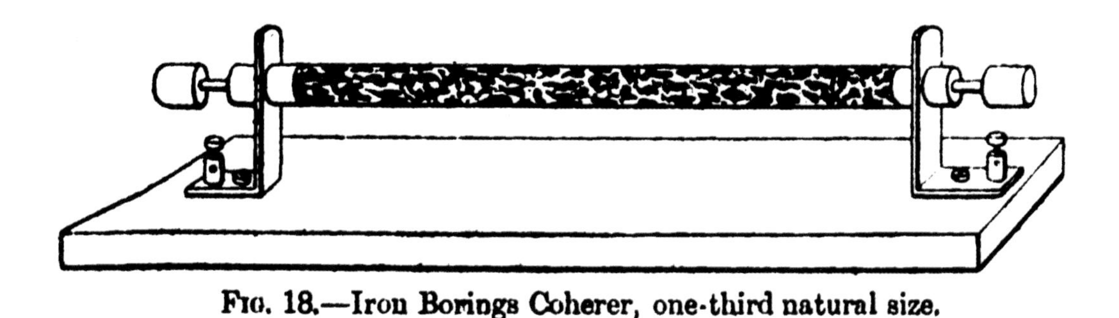

FIG. 18.—Iron Borings Coherer, one-third natural size.

Figure 4.2: An 1894 illustration of a coherer, a device used by Marconi (and many others) to detect radio signals.

Marconi's plan was to erect a transmitter and mast at Rathlin East lighthouse that would send reports to Ballycastle on the mainland from where they would be relayed to Lloyd's by telegram. He made a preliminary survey but the main work was given to ex-naval instructor George Kemp, his right-hand man. Kemp added Edward Glanville, a 25-year-old science graduate from Trinity College Dublin, to the team and hired islander Johnny Cecil as a labourer.[10] Various sites were tried around Ballycastle for the receiver location with particularly good reception noted when the mast was attached to the spire of the town's Catholic church.[10] Apart from the Marconi staff, no-one knew Morse code, so Kemp provided some quick lessons to the principal keeper[11] at the East lighthouse, Galwayman Michael Donovan, and his sons John and Charles.[12] On Wednesday 6 July 1898 Kemp could make out a few 'V's being sent in Morse from Rathlin, thus making the north Antrim coast the birthplace of radio for commercial use.[1] Tragedy befell the experiment on 21 August 1898 with the untimely death of Edward Glanville, who was interested in geology and stumbled when out exploring the island and fell down a cliff.[13] The trials continued undaunted and by the end of August the experimental system was working reliably, reporting shipping movements from Rathlin to Lloyd's via Ballycastle until the experiment was ended in September.

Marconi, never one to turn down an opportunity for publicity, took a break from the experiments in Co Antrin for a few days to demonstrate the power of wireless to a larger audience. Each summer, the Royal St. George Yacht Club in Kingstown (now Dún Laoghaire) organises a regatta. This was a popular event in the social calendar of Victorian Dublin and attracted considerable media attention. There was a commercial advantage to the newspaper that could print the results of the weekend's events first. Unrelated to its London-based namesake, but adopting a similarly pro-Union political stance, the Dublin *Daily Express* considered Kingstown Regatta to be of natural interest to its readership. Thus in 1898 it commissioned Marconi to send real-time reports of the regatta using wireless telegraphy.[14]

Marconi and Kemp travelled from Ballycastle to Kingstown and chartered a tug, incongruously called the *Flying Huntress*, equipping it with a transmitter and aerial. The harbourmaster at Kingstown loaned Marconi a room at the rear of his house for the receiving apparatus, which was operated by Kemp. Wireless reports sent from the *Flying Huntress* were printed on a Morse tape machine, decoded and then relayed by telephone to the *Daily Express* office on Parliament Street. The newspaper posted updates in its window as the reports were received. Marconi, a keen sailor, followed the regatta out into Dublin Bay and up to 40 km offshore to the Kish Lightship and beyond[15], transmitting more than 700 messages during the two-day event. This was the first live transmission of a sporting event in the world.[16] As well as the yachting results, the *Daily Express* gave ample coverage to Marconi's technology, and news of the innovation spread globally,[17] creating an unstoppable publicity machine.

A few weeks after his success at the Kingstown Regatta, Marconi was back in Dublin to assist at a lecture about wireless hosted by the RDS. It was quite the media lovefest with newspaper reports describing how 'Signor Marconi a young gentleman of handsome, pleasing, and attractive appearance … was received with loud applause'[18]. Marconi reciprocated by informing the crowd that he was 'proud of being half an Irishman'.[19] As ever, Marconi knew his audience.

Typical of the difference between the two men, a few months after Marconi's successes in Ireland, Nicola Tesla demonstrated a radio-controlled miniature boat at New York's Madison Square Garden but failed to gain the interest of US Navy or the local media.[20] In the following October, the *New York Herald* reported on the America's Cup yacht races by radio – using Marconi's technology.[17]

This transatlantic traffic in radio technology was not just one-way. The American inventor Lee De Forest set up a radio-telegraph demonstration in 1903 between Holyhead and Howth in an attempt to woo the British Admiralty and Post Office.[21] However, despite the faster speeds offered by De Forest's technology, the Admiralty was unconvinced. Back home in the US, a few years later, De Forest patented a device that became known as the triode: a glass bulb containing a vacuum and electrodes (Figure 4.3). A small voltage applied to one of these electrodes causes a much larger voltage difference between the other two, so that the valve acts as an amplifier. This was a crucial innovation, eventually enabling the transmission of speech over radio waves and making long-distance telephone calls economic. But while Marconi could see the potential in the inventions of others, De Forest could not see the potential in the inventions of his own and, like those of Tesla, his contributions were completely overshadowed by Marconi's marketing machine.

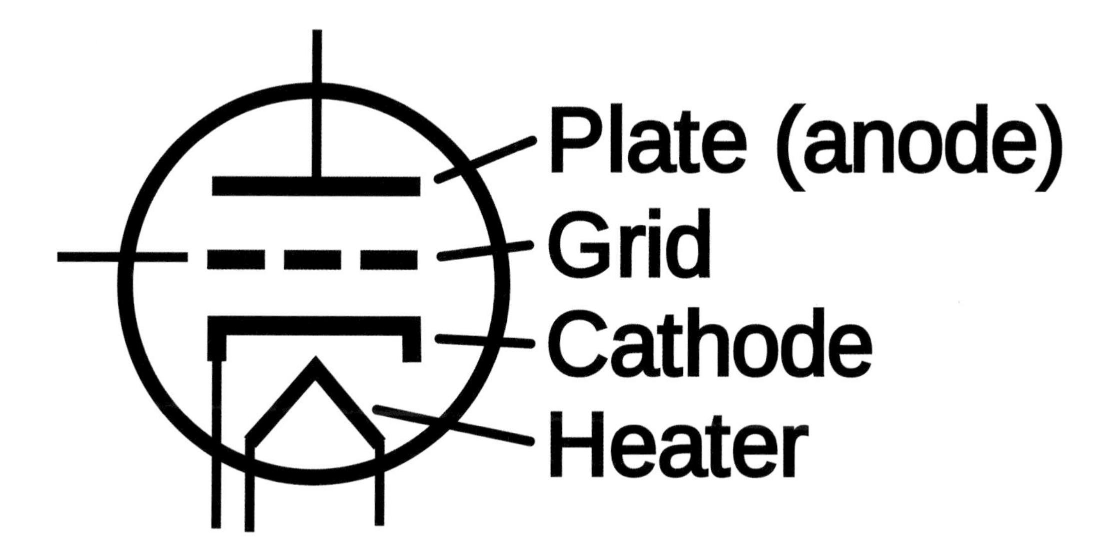

Figure 4.3: The 'Audion', invented by Lee De Forest, and later largely referred to as a 'triode valve'. A small voltage applied to the grid causes a large change in the voltage difference between the cathode and the anode. *(Chetvorno, public domain via CC-0)*

De Forest's attempts to woo the Admiralty were understandable as there was a huge potential market for ship-to-shore communications. The Royal Navy had begun to equip some of its ships with Marconi's wireless telegraphy apparatus in 1899. In the following year commercial shipping lines began to follow suit. Most of the systems were supplied by Marconi's firm with shares in the company quadrupling in value in the immediate aftermath of the sinking of the *Titanic*.[22]

Not surprisingly, following their successful experiments from Rathlin Island, Lloyd's undertook to install wireless equipment in many of its shore stations along the coastline of the UK using Marconi equipment. In Ireland these were at Rosslare, Crookhaven and Malin Head.

The wireless telegraphy station established by Lloyd's at Malin Head in January 1902 lay right beside the signal tower that had formed part of the optical telegraph network established 100 years before. Connected by telegraph line to the post office 5 km away at Ballygorman[9] and thence to London and the wider world, the station became a vital link for commercial and naval communications. After the Post Office took over responsibility of the Lloyd's stations around the coast in 1909 there was a programme to upgrade the facilities and a more powerful 5-kilowatt (kW) spark transmitter was installed at Malin Head in 1913. Three houses for staff were also built and, in an unusual move for the normally parsimonious Post Office, the London retailer Harrod's was invited to tender for the supply of curtains and floor coverings.[23] Thanks to its location on marshy low ground with the sea on two sides, the station at Malin Head could provide excellent radio coverage of the North Atlantic and a maritime radio service has continued there ever since.

A sister station to Malin Head was operated by Marconi at Brow Head near Crookhaven, Co Cork on the south-west tip of Ireland. After the Post Office take-over, it was decided to replace the Crookhaven station with a new wireless station at Valentia, close to the transatlantic cable station. The Post Office acquired the land from the Knight of Kerry, Sir Maurice Fitzgerald. Maurice must have inherited an interest in telecommunications from his father Peter for, in return, he sought free telephone rental at the family's house at Glanleam for 90 years. The Post Office, anxious not to incur such an uncertain expense, negotiated more conventional rental terms[24] and the new station opened in August 1914, just after the outbreak of the Great War.

Marconi's ambitions were larger than just communicating with ships at sea, however. His real aim was to span the Atlantic by radio waves. There was money to be made: the transatlantic cable companies, including the once-pioneering Commercial Cable Co., operated as an effective cartel, charging 1*s* (€0.07) per word for a telegram from the UK to the US.[25] A few months after his marriage in London to Beatrice O'Brien from Dromoland Co Clare (Figure 4.4), Marconi was in Ireland again to view a site at Derrigimlagh near Clifden in Co Galway for a new, permanent station for transatlantic telegraphic communications. The site was picked as it had a clear line of sight west to the Atlantic and a bog to provide turf for the electrical generators.[8]

Figure 4.4: Guglielmo Marconi and his wife Beatrice O'Brien pictured outside the Houses of Parliament in London in 1905. *(© National Portrait Gallery, London)*

By October 1905 construction of this vast undertaking was underway.[17] The station, which occupied a site of 2.5 km^2, included a turf-fired 300 kW electricity generator to power its 20,000 volt spark transmitter (Figures 4.5 and 4.6), the most powerful in the world.[8] It had huge directional antennae, orientated for maximum radiation towards the corresponding station at Glace Bay in Nova Scotia on Canada's eastern extremity. The noise of the equipment and the sparks emitted by its aerial system[26] must have been a wonder in this remote part of Connemara.

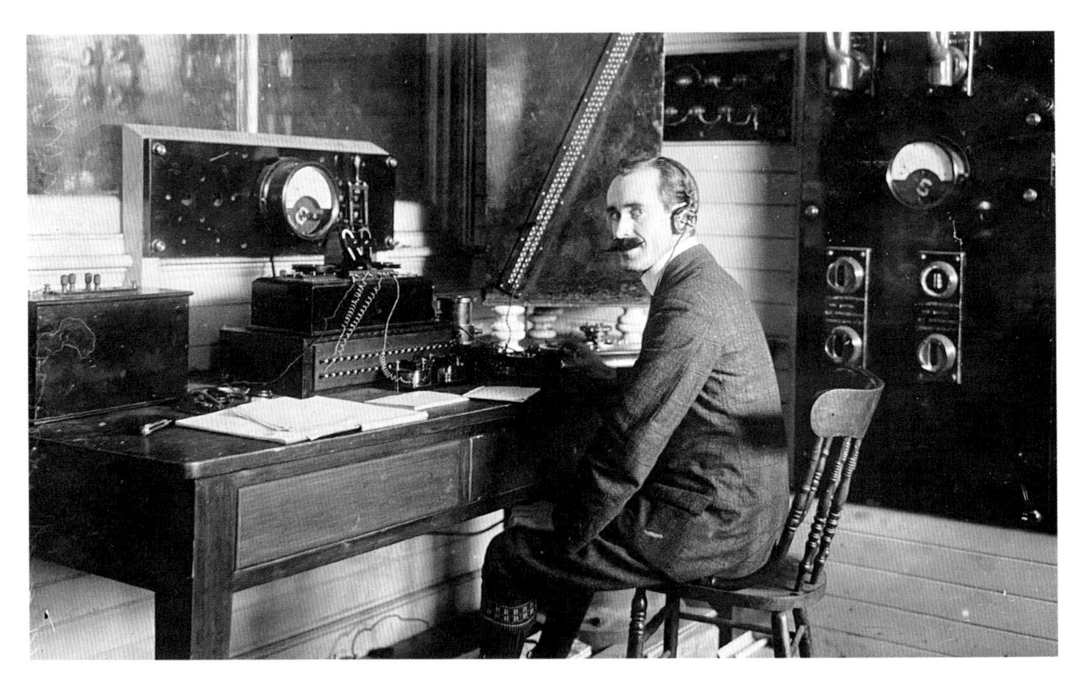

Figures 4.5 and 4.6: Two views of the Marconi station at Derrigimlagh outside Clifden, Co Galway. Figure 4.5 (top) shows the condenser house, the scale of which is clear when compared to the two men standing outside. Figure 4.6 (bottom), marked 'Clifden 1905 Fracy', shows a wireless telegraph operator at work. *(The Bodleian Libraries, University of Oxford, MS. Photogr. d. 46)*

Two years later, on 15 October 1907, following test transmissions from Clifden to ships, the station was launched. The first message sent from Clifden was destined for the *New York Times*, and thus began a close relationship between the newspaper and Marconi, a relationship that helped launch Marconi as a global communications brand.[27] In a 1912 article, the *New York Times* described how 20,000 words each week written by its correspondents in London, Paris and Berlin were transmitted across the Atlantic via Clifden – each one of them a veritable advertisement for Marconi.[28]

With the Clifden station connected to London over a telegraph cable and, at the other side, Glace Bay linked to New York and, later, Montreal, a 'Marconigram' could be sent from London to New York in minutes at a charge of 4*d* (€0.02) per word[29], one-third of the cost levied by the cable companies. The service was enhanced in 1911 with the construction of a separate set of receiving antennae at Letterfrack, about 12 km from the existing station. With a similar additional facility on the Canadian side, this allowed duplex working so that messages could be transmitted from east-to-west and west-to-east at the same time, increasing bandwidth.[17] An agreement reached in 1912 allowed a Marconigram to be sent from any post office in the UK, increasing the availability of the transatlantic wireless service.

Despite these developments, and all the news stories carried for the *New York Times* and other papers, there is conflicting evidence about the success of the link between Clifden and Glace Bay. The proportion of transatlantic telegram correspondence sent by wireless never threatened the dominance of the cable companies.[30] Furthermore, as technology improved it became possible to span longer distances by radio and the need for a station to be located on the remote west coast of Europe reduced in importance. Starting in 1918, the Marconi company began to use its newly built sites in Britain for transatlantic telegram traffic. As we will see later, the Clifden station was one of the many infrastructural facilities targeted by anti-Treaty forces during the Irish Civil War. This damage, combined with technological improvements and hostility to the Marconi company from some sections of the newly independent Irish state, conspired to kill off the station.[8]

The technology used by the Clifden station was also a factor. Marconi's prowess at marketing disguised the fact that the technology he promoted was crude and inefficient: his noisy, power-hungry spark gap transmitters such as the one at Clifden splattered their clicks across the radio spectrum, interfering with each another. There was another technology, however, that could overcome these problems: the continuous wave transmissions mentioned above. Significantly, CW could also be used to carry the human voice and other sounds, rather than just the dots and dashes of Morse code.

For many years this remained merely a theory as there was no means to generate CW transmissions. In 1904 that looked set to change with the invention by Danish engineer Valdemar Poulsen of a carbon arc transmitter that produced a continuous signal at a specific frequency. In 1907, just as the Marconi station in Clifden was being tested, a rival station using Poulsen's CW transmitter was being built at Knockroe on Ballyheigue Bay in Co Kerry. While the station was designed to transmit Morse code, the potential of this new form of wireless technology was well understood with the *Kerry Evening Post* reporting that 'While Herr Poulsen will not at present commit himself to any definite announcement on the subject, he believes it will be possible to telephone right across the Atlantic.'[31] A triangle of aerials over 100 m high was built but within months the masts were blown down in a storm[32] and the station was abandoned. Six years later, another Poulsen station was built in another Co Kerry town. The Ballybunion station, about 20 km north of its predecessor at Knockroe, was built as part of a Canadian government funded system to provide a commercial wireless telegraph service between Newcastle in New Brunswick and Ballybunion. The Canadian government wanted to encourage commerce by charging less than the cable companies and Marconi.[33] However, the radio link proved unreliable and on the outbreak of the Great War the Ballybunion station was taken over by the Royal Navy. By 1915 Poulsen's patents had been bought out – by Marconi.

Like Poulsen and De Forest, Marconi saw the incredible potential of transmitting voice by radio waves and he now had the technology to achieve it. The wireless station at Ballybunion, now under Marconi ownership, was re-equipped with new apparatus for voice transmissions based on thermionic valves similar to those patented by his rival De Forest. The words 'Hello Canada. Hello Canada. This is the Marconi valve transmitter station, Ballybunion, Ireland calling…' picked up by the corresponding Marconi station at Louisburg in Nova Scotia on 19 March 1919 were the first spoken words transmitted across the Atlantic from east to west.[34] Describing these experiments from Kerry, the managing director of the Marconi company prophesised that 'business men in New York would soon be able to converse clearly and easily with their equals in London by wireless telephone'.[35] His prophesy was proven correct. But Kerry, while a global hub for telegraph wires, would not become a hub for wireless communications.

The newly independent Irish state did not provide fertile soil for Marconi. In the early 1920s the concept of broadcast radio emerged, exploiting the ability of this new technology to transmit speech and music to a large audience. Accordingly, at the start of Horse Show Week in August 1923, his company installed Ireland's first official radio broadcast transmitter at the Royal Marine Hotel[8] in what was now Dún Laoghaire. Intended as a marketing ploy, the station transmitted musical performances to receivers set up in the RDS.[36] Two days after the broadcasts commenced, the Irish Postmaster General withdrew his previously granted permission and the station was forced to close down early.[37] The reason for the volte-face was not given but, like the abandonment of Clifden, was perhaps due to antagonism from the new government. To complete the picture, the station at Ballybunion was closed and it too had its equipment sold for scrap in 1925.[38]

In any case Ireland was now small fry in Marconi's successful global empire. His decision to purchase the patents for Poulsen's inventions and his adoption of the thermionic valve proved wise. The noisy spark gap transmitters were quickly replaced and transmission of voice became commonplace. By the 1930s thanks to radio telephone stations such as that at Rugby in England, it was possible to make a telephone call not just to New York but to almost anywhere in the world. One of Marconi's final personal projects was the installation in 1932 of a radio telephone link to span the 20 kilometres between the Vatican and the Pope's summer residence in Castel Gandolfo.[39]

Two years later, in a rather different setting, the UK Post Office inaugurated an 'ultra-shortwave' radio telephone link between Ballygomartin in Belfast and Portpatrick in Scotland using VHF frequencies (Figure 4.7) of 50–75 MHz.[40] This radio link was claimed to be cheaper to implement and maintain than an undersea cable though intermittent problems with the system could sometimes make cross-channel conversations sound like 'a mixture of Scottish and Irish bagpipes'.[41] The radio connection across the North Channel was a forerunner of the microwave links over land and sea that were to become a key component of the telephone network in the post-war period. Indeed the Ballygomartin site was re-equipped many times and remained part of the telecoms network until the early 2000s, being considered such a sensitive piece of infrastructure that the file about it in the Public Record Office of Northern Ireland remains closed until 2039, 104 years after the station opened.[42]

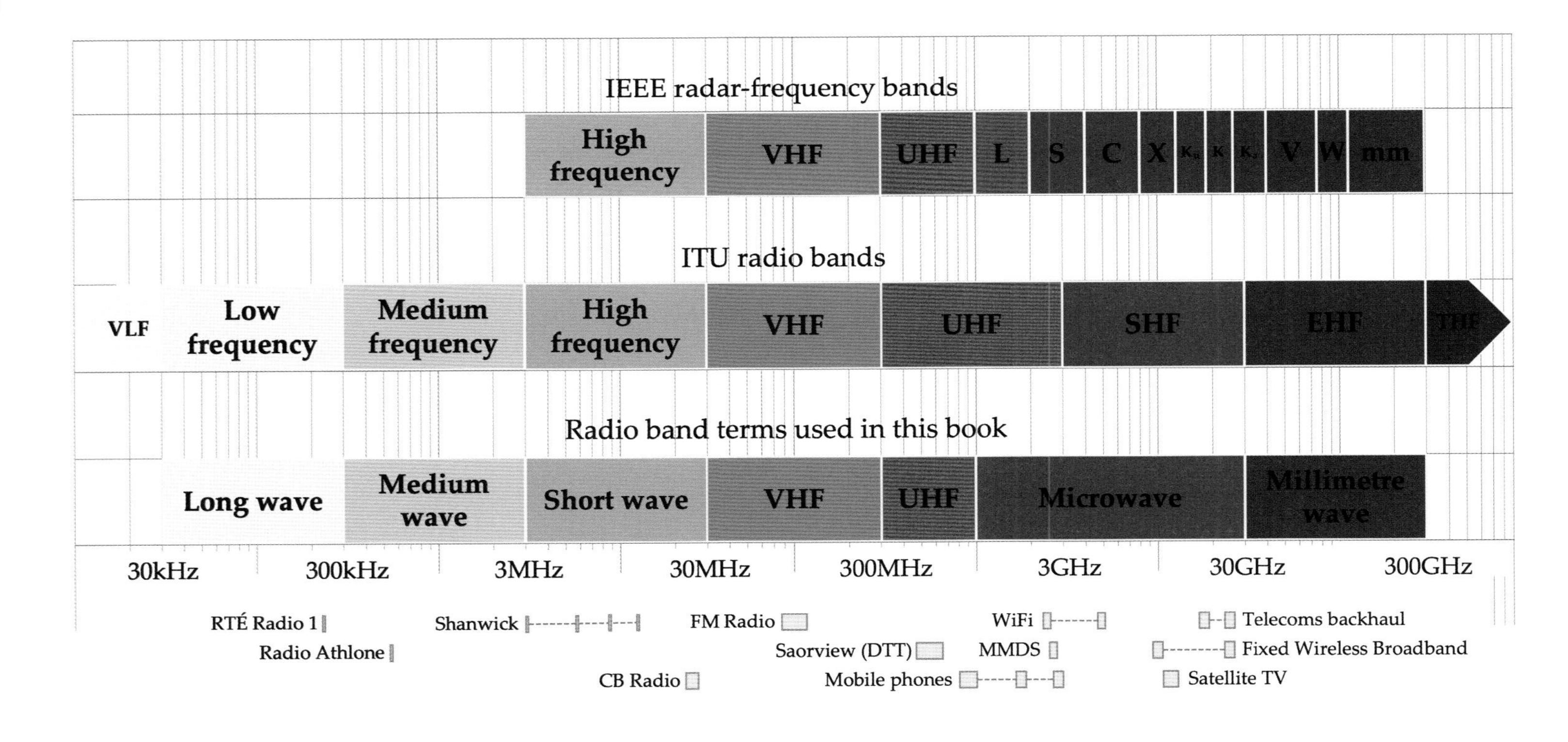

Figure 4.7: The Radio Frequency Spectrum.

By the mid 1920s, an independent Ireland was peripheral to Marconi's now global business, merely a small market for the company's equipment. Marconi's later personal life seemed to mirror his business disengagement from Ireland; he failed to attend the funeral of his Irish-born mother Annie in London in 1920[43] and divorced his Irish wife Beatrice in 1924.[8] His name and influence persisted, however, in the cultural sphere. In Brian Friel's play *Dancing at Lughnasa*, set in August 1936, the temperamental radio owned by the Mundy sisters brings music, dance and delight to their home just outside the fictional Donegal village of Ballybeg. The radio is a key household member, referred to by the name emblazoned across its front – Marconi.[44]

Guglielmo Marconi died on 20 July 1937. As an Italian Senator and confidante of Benito Mussolini, Marconi's state funeral in Rome was orchestrated with military-style pomp and the many wreaths adorning his coffin included one sent by Adolf Hitler.[45] As a mark of respect, radio stations went silent across the world – though not in Ireland.[46] By the time of his death, the country of his mother had been transformed from an integral, if sometimes reluctant, part of the United Kingdom into an independent state. Meanwhile the radio technology Marconi had championed had become commonplace, not just for broadcasting, but also within the telecommunications network. However, even Marconi himself probably did not foresee that within 90 years most of the world's inhabitants would have a telephone in their pocket based on the wireless technology he had pioneered.

CHAPTER FIVE

'GET BACK, SON, GET BACK, THE BRITISH ARE IN THE TELEPHONE EXCHANGE!'

The morning of Easter Monday 1916 was unexpectedly busy at the Central Police Telephone Office in Dublin Castle (Figure 5.1). At 11:20 the Castle took a call to say that about 50 Volunteers had been observed travelling 'by tram car 167 going in direction of the city'. Based on this report, the Castle phoned all stations to be on the lookout for volunteer movements.[1]

At 11:50 a call came in reporting that Volunteers were turning everyone out of St Stephen's Green. Then, just after midday, Store Street station rang to report that a group of Volunteers were leaving Liberty Hall, on their way, as was quickly learnt, to the GPO. The pace of calls grew faster and faster. At 12:10 Broadstone railway station phoned about Volunteers digging up the railway tracks; five minutes later there was a report that Christchurch Cathedral had been occupied. Soon after that Superintendent Murphy of Kevin Street was on the phone to the Castle to say that the Volunteers had 'taken South Dublin Union, Marrowbone Lane Distillery and other buildings here'.[1]

The switchboard staff were busy making outbound calls too. At 12:25 the Superintendent of G Division, in charge of intelligence, called the representative of the British Monarchy in Ireland, Lord Lieutenant Wimborne (Figure 5.2), to inform him of the mounting unrest and advise him to 'take all necessary precautions'.[1] Calls were also made to two army barracks in the city to report the attacks.[2] By 12:40 the Castle had received further phone calls about attacks on the Four Courts[1], Jacob's biscuit factory and Westland Row railway stations.[3] The Easter Rising was underway and, thanks to the still-functioning telephone network, the authorities knew all about it.

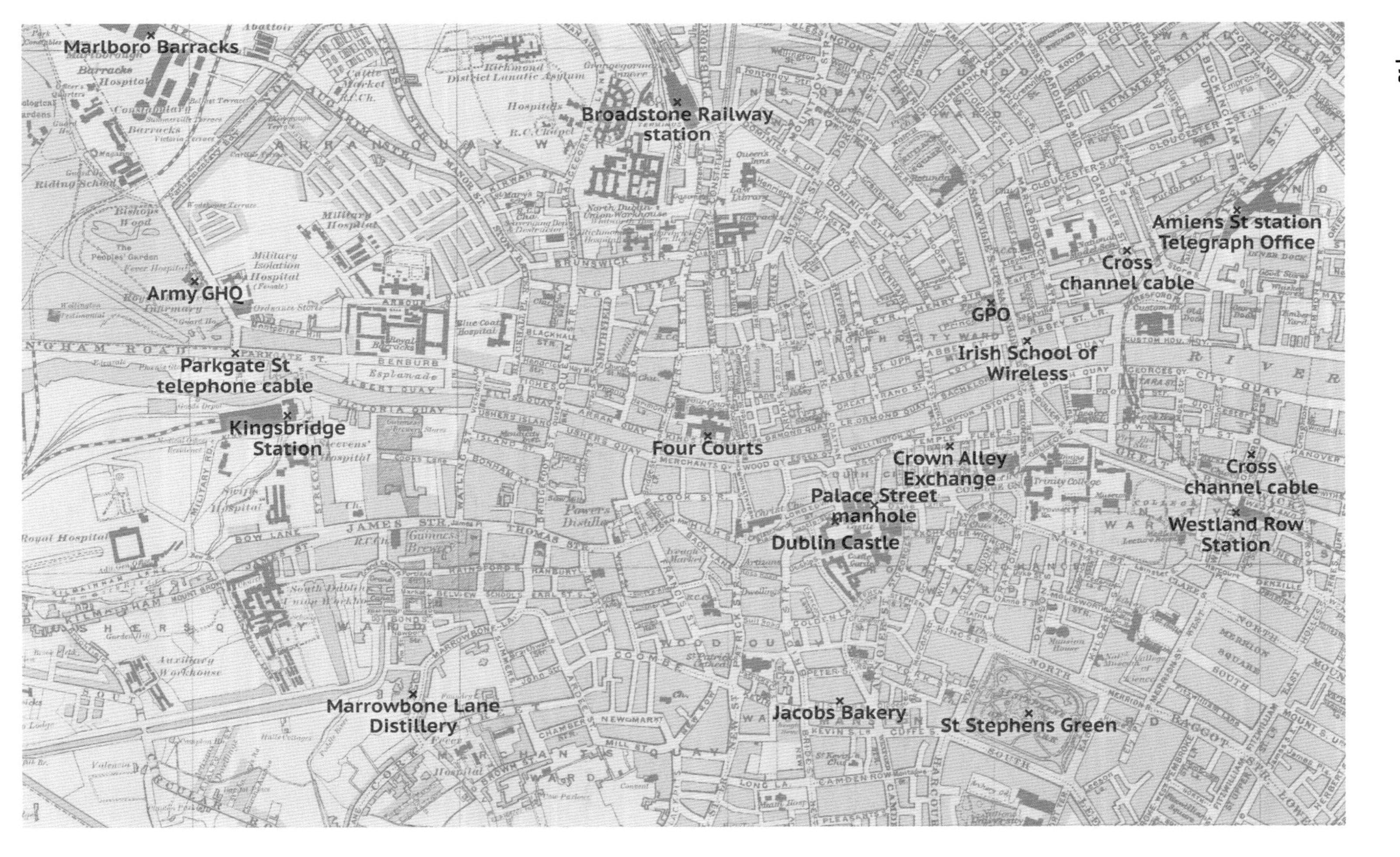

Figure 5.1: Locations of the key points in the telecommunications battle of 1916. *(Base map from the American Geographical Society Library, University of Wisconsin-Milwaukee Libraries)*

35

Dublin Metropolitan Police Telephone.

No. of Message. Station of Origin. Castle
Date 24-4-1916
Received at 12.26
Sent at 12.26 M.
From Supt G
To V.R. Lodge

The Sinn Feinn volunteers have attacked the Castle & have possession of the G.P.O They have Stephens Green Park in their hands and have turned out the people and locked the gates. Please take all necessary precautions (message taken by Secretary to His Excellency

81

Dublin Metropolitan Police Telephone.

No. of Message. Station of Origin.
Date 24-4-1916
Received at 4.5 P.M.
Sent at 4.8 P.M.
From S Sheehan
To Chief Supt

Mr. Pemberton senior Supt. G.P.O says communication may be had with London via North Wall and Amiens St Station please say can you let me have a car to take me to Newcastle Co. Wicklow

499

Dublin Metropolitan Police Telephone.

No. of Message. 3246 Station of Origin.
Date 28-4-1916
Received at 11-25 a.
Sent at 11-30 a.M.
From Dr. Keegan Jervis St. Hospl.
To Castle.

There are 28 or 30 bodies here, 6 of which are soldiers, and it is imperative that they be removed at once. They are here since the outbreak and it is a positive danger to keep them. Pse. say what can be done.

555

Dublin Metropolitan Police Telephone.

No. of Message. Station of Origin. Dublin 65.
Date 29.4.1916.
Received at 2.35 p.M.
Sent at 2.35 p.M.
From Caretaker, Eden Quay. Dublin Steam Packet Coy Office
To Police Office

My wife, three Children and myself are starving here and the Military will not allow us out Could you please do something for us, or ask the Military at the Custom House or Liberty Hall to relieve us in some way.

Figures 5.2 – 5.5: All calls through the Central Police Telephone Office in Dublin Castle were logged. The records from Easter 1916 form a detailed chronology of the Rising and its impact on the city. *(The Bodleian Libraries, University of Oxford, MS Nathan 476)*

Over at Army Headquarters on Infirmary Road the phone lines were also hopping. As well as taking a call from Dublin Castle, they also got word direct from the GPO about the attack there. Based on this report, Colonel H. V. Cowan, Assistant Adjutant-General, phoned four army barracks in the city to request reinforcements. At 12:25 Army Headquarters called the superintendent at nearby Kingsbridge (now Heuston) Station requesting the suspension of regular trains and the preparation of troop trains.[4] By 12:30 Col Cowan had used the army's own telecommunications network to phone the army barracks at the Curragh,[5] Athlone, Belfast and Templemore seeking reinforcements.[5] Col Cowan also tried to reach London but could not get through. Knowing that there was a wireless transmitter at the naval base at Kingstown, a junior officer offered to cycle there in plain clothes.[5] At the same time as Pádraig Pearse was proclaiming the establishment of an Irish Republic from the GPO entrance, this young officer was pedalling across the city to alert London of the insurrection. He made the 10 km journey in under 30 minutes and by 13:30 news of the rebellion had been flashed by wireless telegraph[6] to the Admiralty in London.

At Marlboro (now McKee) Barracks, near the Phoenix Park on the north side of the city, the telephone was ringing as well. By 12:30 reports of the attacks had been received from the GPO and from Army Headquarters. Based on these calls, the barracks hastily dispatched a troop of the 6th Reserve Cavalry which arrived in Sackville (now O'Connell) Street within minutes of Pearse's proclamation. By 16:00, thanks to Col Cowan's phone calls, troops had started to arrive from the Curragh. Within days the GPO had been destroyed by shelling by British forces, the rebellion had been crushed and the course of Irish history irrevocably altered.

The leaders of the Rising chose the General Post Office as the centre of the insurrection because it was the centre of most communications channels and had a commanding position in the heart of the city. It also had symbolic importance[2]: its flagpole carried the Union Flag, its portico the Royal Coat of Arms and its prominent public clock was surmounted by a crown. By March 1916, just weeks before the Rising, a major redevelopment of the GPO had been completed. The ground floor had been transformed into a light-filled public office, furnished with counters of colonial Burmese teak and a public telephone inside a 'silence cabinet' (Figure 5.6). Behind this public office lay the central postal sorting office. On the first floor, the entire Sackville Street façade had become a new telegraph office and trunk telephone exchange[7] that handled calls over the circuits linking Dublin with the rest of Ireland and with Britain. However, the main telephone exchange for local subscribers was 1 km away across the river in Crown Alley, a relic of the split in ownership for the 20 years between 1892 and

1912 when the Post Office controlled the trunk lines but most local subscribers were serviced by private companies. It was not enough just to command the GPO: the rebels would also have to control the Crown Alley exchange, for whoever controlled Crown Alley effectively controlled communications in the city.

Figure 5.6: The telephone 'silence cabinet' in the newly renovated public office of the GPO in March 1916 was destined to have a brief, if eventful life.

In insurrections around the world since the start of the twentieth century, control over the telephone network was considered vital: whoever controls the exchange can listen in on calls, vet all incoming calls, connect or disconnect calls as they see fit – in short, take complete control over who talks to whom and what is said. Eighteen months after the Easter Rising in Ireland, the Bolsheviks seized power in Russia. Taking control of the telephone exchange in Petrograd (now St Petersburg) was one of the first acts of the revolutionaries.[8] More recently, in Poland when martial law was imposed in 1981 to repress the Solidarity movement, telephone users heard a message before their call was connected: *Uwaga! Rozmowa kontrolowana!*[9] (Attention! This conversation is being monitored!). The leaders of the Easter Rising were well aware of the importance of telecommunications in any uprising and gathered intelligence about the telephone and telegraph networks and how best to disable them. However, unfortunately for the rebels, they were unable to use much of this knowledge on the day.

The intelligence gathering had begun at least a year before the Rising. Martin King, a member of the Irish Citizen Army and a cable joiner in the telephone section of the Post Office, recalled being asked in 1915 by Citizen Army leader James Connolly about how to cut off communications with England.[10] For weeks before Easter, members of the main rebel group, the Volunteers, had been gathering information about the network[11] aided by contacts within the Post Office Engineering Department. By such means the Volunteers gained knowledge of the telephone junction on Parkgate Street through which all the phone lines from the nearby Army Headquarters ran.[6]

In the week leading up to the Rising, the Volunteers devised a plan to destroy key links in the telecoms network. Two of them performed a reconnaissance around the city centre, noting manholes that contained important connections including one at Palace Street, near Dublin Castle, which gave access to many telephone lines used by the police, and one at Lombard Street with a special circuit between Dublin Castle and London.[10] The Volunteers' plan was to disable the cross-channel link from the GPO at points on Talbot Street and at Kingstown (now Dún Laoghaire), to sever the western trunk circuits serving Athlone, Galway and Sligo at Broombridge, to cut the telephone circuits for Dundalk/Belfast and cross-channel routes at Howth Junction and sever the circuits for Wexford/Waterford and the cross-channel telegraph lines at Westland Row and Shankill. Martin King himself was detailed to lead an attack on the main exchange in Crown Alley.

In the end, however, all this careful planning came to naught. Initially the Rising was planned for Easter Sunday, but the failure of the planned shipment of weapons from Germany on Good Friday called the viability of the Rising into question and Eoin MacNeill, Chief of Staff of the Volunteers, issued an order countermanding the planned Rising. The countermand was passed on by couriers around the country and printed in the *Sunday Independent* on Easter Sunday. But more radical voices insisted that the Rising go ahead, even without the planned shipment of arms, and fresh orders were issued by Pearse that the planned military plan be activated on Easter Monday. In the resultant confusion, only about a thousand rebels showed up.

Despite their depleted numbers, the rebels achieved partial success in disrupting the networks, damaging lines or equipment at 60 places over the course of the Rising.[7] All the telegraph circuits serving the GPO were cut around noon on Easter Monday. Assisted by Sean O'Keefe, Martin King severed cross-channel lines in a manhole at Lombard Street.[12] Seán Byrne cut the western trunk lines at Broombridge[13] and Richard Mulcahy disabled the northern trunk lines at Raheny.[14] A bomb planted in a manhole at Palace Street cut most lines from Dublin Castle, including the private circuit between the office of the chief under-secretary, Sir Matthew Nathan, and Military HQ on Infirmary Road.[6] Significantly, despite this damage, several lines from the Castle remained intact and were in constant use during the Rising to coordinate the government's response.

Amidst all this carefully planned takeover of the telecoms system, the humble telephone booth also came into play: in the first few minutes of the Rising in the GPO, Michael Collins imprisoned an off-duty British soldier in the telephone 'silence cabinet' that newly adorned the public office, using the telephone cord as a restraint. The solider, Second Lieutenant A. D. Chalmers of the 14th Royal Fusiliers, was buying stamps when the rebels stormed the building[6] and was eventually released as the British forces closed in on the GPO.

Crucially, however, the most important target – the main exchange at Crown Alley – was missed. Martin King was unable to carry out the planned attack on it because the men who were to have assisted him failed to show up. He reported this absence to rebel leaders Pearse and Connolly and was instructed to fall in with other men at the College of Surgeons.[2]

The staff on duty at Crown Alley were immediately aware that a rebellion was underway thanks to the flurry of calls being exchanged between the GPO, the Castle and the various military barracks. Shortly after noon, conscious that the exchange was completely undefended, one of the engineering staff on duty, Dubliner Sydney Verschoyle, phoned for military assistance.[15] The British Army responded by dispatching about 20 soldiers from the 3rd Royal Irish Rifles at Portobello (now Cathal Brugha) Barracks to the exchange.[16] Under the command of Kildare native, John Kearns, the soldiers had a difficult journey from Rathmines through a city descending into turmoil, and did not reach Crown Alley until 15:20.[15]

Aware that the main telephone exchange was still in operation, a small party of Volunteers led by Frank Thornton was belatedly dispatched from the GPO to Crown Alley exchange.[17] As they approached Crown Alley from Fleet Street, an old woman rushed forward shouting 'Get back, son, get back, the British are in the Telephone Exchange!'[6] Coming under machine gun fire from the recently arrived soldiers, the small party of Volunteers retreated[17,15]. The British military maintained control of the exchange throughout the Rising, allowing the government and military to communicate and, within a few days, crush the rebels.

As fighting continued through Easter week, the telephone network was also used to compile lists of casualties, arrange disposal of bodies and coordinate humanitarian action (Figures 5.4 and 5.5).[1] The twenty switchboard operators and other staff in the exchange at Crown Alley were trapped for five days by the rebellion occurring outside. Mattresses were set up and food purloined.[15] Rebel snipers took positions in buildings surrounding the exchange, occasionally peppering the switchroom with bullets. To aid evacuation and protect against shrapnel, the telephonists were told at one stage to wear their outdoor clothes, a cause of unexpected humour when one of the operators 'emerged from the retiring room not only fully hatted and cloaked but with her umbrella up'.[15]

Mary Louise Norway, wife of the Post Office secretary Arthur Hamilton Norway (and mother of future novelist Nevil Shute), wrote:

> The girls … have been on duty since Tuesday, sleeping when possible in a cellar and with indifferent food, and have cheerfully and devotedly stuck to their post, doing the work of forty. Only those on duty on the outbreak of the rebellion could remain; those in their homes could never get back, so with the aid of the men who take the night duty these girls have kept the whole service going. All telegrams have had to be sent by 'phone as far as the railway termini, and they have simply saved the situation. It has been magnificent![18]

Within a few hours of the start of the Rising, service had been restored at the telegraph office at Amiens Street (Figure 5.3). With the telephone service largely intact, officials could phone Amiens Street and dictate a telegram for onward transmission. The trunk telephone lines were diverted from the occupied GPO to the main exchange at Crown Alley[3] (which became the trunk exchange until 1944) allowing communications to be maintained with the rest of the country. Meanwhile a gunboat was deployed at Howth to guard the vital cross-channel cables.[1]

In the days following the Rising the rebels made sporadic attacks on telecommunications infrastructure surrounding Dublin and further afield. Telegraph lines were severed in Co Clare near Ennistymon, in Co Limerick at Bruff and in Co Wexford at Enniscorthy[1], while the recently erected trunk telephone line to Galway was cut due to 'trouble in Athenry'.[3] The Post Office's Superintending Engineer for Ireland, Edward Gomersall, took command of restoring service, becoming one of the first recipients of the newly created honour of OBE for his efforts.[19] With the assistance of colleagues who had arrived from Belfast, Post Office engineering staff travelled around the outskirts of the city and restored service, commandeering still intact subscriber lines to bypass sections of the circuits cut down by rebel forces.[18] With the rebellion still in progress such work was hazardous with Gomersall recording that 'Every day the engineering officers experienced the uncomfortable nearness of bullets, and, either when they were at work or were making their way to places where work had to be done.'[7] By Wednesday, 26 April, just 48 hours after the Rising had commenced, Gomersall was able to report that:

> I have opened a temporary telegraph office at Amiens Street station. I have 4 working lines to London and am arranging 7 more. I have also a circuit to Belfast. I have also got a Trunk telephone circuit to the Curragh, and one to Maryborough [now Portlaoise]. Communications with places W. & S. W. of Maryborough may be obtained by calling upon the telephone. Two circuits are available to Belfast, and communication may be obtained via the Dublin Exchange with London and Great Britain generally.[3]

Telecommunications services were back under government control and the Rising – from a military perspective – was doomed.

Within two weeks a new Telegraph Office had been equipped on the upper floor of the Post Office parcel sorting office on Amiens Street.[20] And by that stage, 12 of the Rebel leaders had been executed and public opinion across most of Ireland had turned hostile to British rule. It was the beginning of seven years of turmoil.

An order banning non-official telephone calls was enforced soon after the commencement of the Rising, ensuring that the rebels could not use the phone for their own purposes and leaving the network free for the military. Some claims were made that a Sinn Féin sympathiser at the Dublin exchange had telephoned his counterpart in Cork to inform him that 'Dublin has risen; let Cork rise' but that the intended recipient was away at lunch so that the call was answered by a unionist sympathiser[21] and as result Cork 'did not rise'. However there is little other evidence that such a call took place. There was also a report that early on Easter Tuesday someone in the Rebel-held GPO trunk exchange called Portarlington over a railway circuit to enquire about troop movements. The call was intercepted by an operator at Kingsbridge station who told Portarlington not to give answers to such questions until the identity of the caller could be verified.[2] Uncertain of the security of his own telephone line, while the Rising continued the Lord Lieutenant resorted to using French while speaking on the phone to his officials as he considered, possibly naively, that this would reduce the possibility of his conversation being understood by any rebels listening in to the call.[18]

While the rebels failed to take complete control of the telecommunications networks, they did, however, successfully exploit telecommunications to spread news of the

Rising to the wider world. The rebel leaders hoped that the news of a Rising in Ireland would garner public sympathy in the US and thus encourage official American support for an alternative government in Ireland. The leaders therefore wanted news of the Rising to reach the US as quickly as possible.[22] The Great War had badly affected the mail service with the US so other forms of communication were required. Despite the confusion over the timing of the Rising and their limited numbers, the rebels successfully used both the transatlantic cable service and radio transmissions to spread news of the Rising to the wider world.

Down in Kerry, brothers Eugene and Tim Ring worked in the cable station in Valentia, just as their father had done. The brothers were members of the Irish Republican Brotherhood (IRB) and were tasked with acting as liaison between Pádraig Pearse and John Devoy, leader of the republican group Clan na Gael in the United States. Tim Ring had instructions from the IRB to send a coded telegram to the household of John Devoy in New York to advise him when the Rising was about to take place. With wartime censorship in place, the pair decided to send a test message in advance. Accordingly in mid-April Eugene Ring sent a message to the station on the other side of the Atlantic to enquire if the operator at the other end wished to buy a bicycle. Despite the apparently innocent content, the telegram was detected by the Post Office checker. It was clear that any message, no matter how innocuous it sounded, which originated at the cable station would be discovered.[23] Thus Tim Ring's cousin, 18-year-old Cumann na mBan member Rosalie Rice who worked in the post office in Kenmare, was recruited to the scheme. The coded message to be sent to advise of the Rising was to be along the lines of 'Mother operated on successfully today.'[24]

By Easter Saturday it was concluded that a Rising was pending and Rosalie Rice sent the arranged coded message from Kenmare post office that afternoon. The telegram was duly delivered to the home of John Devoy at about 7 a.m. on Easter Monday, which was about noon in Ireland. By coincidence, with the Rising delayed by a day, this was just as events were indeed kicking off in Dublin[25]. In New York, John Devoy had scheduled a press conference for the morning of Easter Monday with the assumption that he would by then have received a telegram from Ireland and could announce that a Rising had taken place. Of course, all the message had said was that someone had been 'operated on successfully', not that the Rising had been successful or even taken place. But this detail was glossed over and pre-printed press releases with stories of an insurrection in Ireland were rushed onto the streets of New York. The British Consul, whose offices were opposite the press conference venue, taking an interest in the fuss outside, got their hands on the news and relayed it to the Foreign Office in London via Washington.[22] By the time the message reached London, the Rising had in fact just commenced, so the reality matched the hype.

A major investigation was launched by the British as to how news of the Rising had apparently reached sympathisers in the US before it had reached the government in London. Unfortunately for Tim Ring, the Gaelic American newspaper decided to mention that the news had reached Clan na Gael in New York from a remote fishing village on the west coast. This narrowed the line of inquiry for the British who examined all telegrams that had been dispatched to America from places that could be so described. They honed in on the Valentia cable station, resulting in the arrest of the Ring brothers and their cousin Rosalie Rice. Tim was jailed with his 1916 colleagues in the Welsh prison camp at Frongoch, his brother Eugene was incarcerated closer to home in Cahirciveen barracks[26] while Rosalie Rice was briefly jailed in Tralee. After Tim's release, the British government would not let him return to Ireland, so Western Union reassigned him to the Accounts Department of their London office until 1920 when he was allowed return to Ireland and his job at Valentia cable station.[22] Eugene, however, lost his job, not for his role in sending the message about the start of the Rising but for using the telegraph to send a personal message.[27]

The rebels did not just use the telegraph network to inform the world of the rising. The IRB leader, Joseph Plunkett, took a deep interest in the emerging technology of radio and was keen to harness its potential use for communicating to the wider world. Back in Dublin, across the road from the GPO, at the junction with Abbey Street, lay the Irish School of Wireless. Due to security restrictions associated with the Great War, its wireless equipment had been dismantled and the aerials on the roof had been taken down. In any event, the use of radio was still largely confined to point-to-point telegraphy transmissions such as between a ship and the station at Malin Head, or between the transatlantic station at Clifden and its counterpart in Canada. The ability to transmit voice by radio was in its infancy and broadcast radio was limited to a few experimental stations in the USA.[14] Nonetheless, on Easter Monday, a detachment of rebels led by Fergus O'Kelly left the GPO, crossed Sackville Street, took possession of the Wireless School and reassembled the radio apparatus. Under heavy sniper fire they erected a new aerial and readied the station for transmission. Hour after hour, they tapped out a message prepared by James Connolly: 'Irish Republic declared in Dublin today. Irish troops have captured city and are in full possession. Enemy cannot move in city. The whole country rising.'[28] This brief message, under very different circumstances to Marconi's transmissions from the Kingstown regatta 18 years earlier, is regarded as the first radio news bulletin transmission in the world.[29]

In addition to its effects on telecommunications infrastructure, the Rising also had a profound impact on the people who worked in the service, dividing them on political grounds. While a number of those involved on the rebel side came from within the telecoms service, most staff initially remained loyal to the Post Office of the United Kingdom.[30] The loyalty of the workforce was commended by Arthur Hamilton Norway, Secretary to the Post Office in Ireland: 'the conduct of the staff during this crisis and throughout this war has been quite excellent. In the restoration of the public service after the insurrection they acted with zeal and public spirit.'[31] While several staff members were granted extra leave for the risks they had taken during the Rising[2], the 'zeal' so admired by Norway was not matched by improved wages or conditions. There were claims that the Post Office planned to deduct the cost incurred in providing food and mattresses to the operators at Crown Alley who worked through the Rising from their overtime payments,[32] though in the end this did not happen. In the light of this it was unsurprising that postal workers began to identify with the cause of Irish independence.[30]

Of course, this growing support for independence was also a reflection of attitudes amongst the population as a whole, giving rise to the War of Independence that raged nationwide from 1919 to 1921. Doubtless learning from their failure to capture Crown Alley in 1916, the independence movement targeted the telecommunications service during the War of Independence, intercepting communications between arms of the British apparatus and destroying infrastructure. Kevin O'Reilly, a volunteer in the Intelligence division under Michael Collins, took a job as a night telephonist in Crown Alley exchange from 1919 to 1920. There he tapped the lines used by the British military and sent weekly reports to senior figures in the IRA.[33]

The campaign to damage telephone and telegraph infrastructure caused considerable, if largely short-tern, disruption. In July 1920 an attempt was made to burn down the main Belfast exchange on Queen Street[34]; in the early part of 1921, the trunk lines from Belfast were cut on several occasions.[35] In April and May of 1921, the IRA destroyed telephone and telegraph apparatus throughout South County Dublin.[36] By mid-1921 the IRA and British forces had reached an impasse and commenced negotiations. A truce was declared in July 1921, though peace proved short-lived. The Anglo-Irish Treaty was signed on 6 December 1921, giving Dominion status to 26 of the 32 counties. As part of the transition of power, the British authorities took stock of strategic infrastructure located in the future Irish Free State, and closed down some facilities, such as the Admiralty's Direction Finding station at Carnsore Point.[37]

A Provisional Government was established in 1922, with Corkman J. J. Walsh assigned the non-cabinet position of Postmaster General. Becoming boss of Post Office was an ironic career twist for a former Post Office clerk who had trained in wireless telegraphy,[20] before joining the Volunteers and participating in the Rising at the GPO.

At midnight on 31 March 1922, control of the Post Office in Ireland passed to the Provisional Government of the nascent Irish Free State. The leaders of the new State, most of whom like Walsh had fought in 1916, had learned a hard lesson during the Rising about the importance of telecommunications in times of war. The question remained, however, as to whether they understood the importance of telecommunications in times of peace.

CHAPTER SIX

AN ROINN PUIST AGUS TELEGRAFA

On the morning of 11 July 1929, a group of government ministers, senior civil servants and members of the judiciary gathered under the portico of the rebuilt General Post Office in Dublin. This new GPO was quite different to the version that had been largely destroyed during the Rising. The British royal coat of arms that had adorned the portico for a 100 years had been chipped off. And in contrast to the woodwork from colonial Burma used in the short lived 1916 remodelling, the interior now incorporated materials from all four provinces of Ireland: sandstone from Donegal and marble from Kilkenny, Cork and Connemara.[1] This was now an Irish, not a British, building – or rather would be once the formal opening took place.

This formative event for the new country was led by W. T. Cosgrave, head of the government of the Irish Free State. At noon on that day Cosgrave ascended the platform on the portico and reminded the invited guests and thousands of onlookers of the symbolism of the time and place: 'On Easter Monday of 1916 at this very hour they came here and took possession of this building – a handful of men against the might of an Empire but strong in spirit and faithful unto death.'[1] On Cosgrave's order the tricolour was raised on the top of the new building. The flag hoisting was performed by Gearóid O'Sullivan, just as he had done during the Rising 13 years before.[1]

Cosgrave opened the door of the Post Office and entered the public office where he purchased a stamp and sent a telegram to the International Telegraph Bureau in Berne.[1] He then made a telephone call to the Phoenix Park residence of the Governor General, the British Monarch's representative in the Irish Free State. In spite of the precise timing of the event, Governor General James McNeill was not at home to take the call so Cosgrave had to leave a message with the Governor General's son.[2] No matter that the telephone call failed; the flags and emblems were all in place.

This 1929 version of the GPO was a symbol of the new nation and its form and functions revealed the concerns of the new order. The new wing along Henry Street housed purpose-built studios for the government-owned radio station, which would soon be audible across the country.[3] However, the new Central Telegraph Office and associated accommodation for motorcycles on the Princes Street wing were not completed in time for the opening, and no public telephone exchange was incorporated within the new building. These omissions suggested something of the attitude towards telecommunications infrastructure by the new state.

The government of the new State did not seem to understand the potential role of the telephone in aiding economic development. By contrast it quickly grasped the potential of electricity to transform lives, implementing a radical new form of organisation to spread power across the country. The government also understood the potential of radio to be a unifying voice available across the land. Developing the telephone network, however, was not viewed as an integral part of building a new nation based on agrarian self-sufficiency, and as a result it was starved of capital and stultified by the structures of the civil service. Like the flags and emblems at the GPO, the limited attempts at development of the telecommunications network were often more symbolic than useful.

The destruction of the GPO and countless other buildings in central Dublin in the Rising and the years that followed presented the new state with an opportunity to implement real transformations, not just symbolic ones, to the infrastructure of the capital and the functions of its public buildings. In particular, the destruction allowed the authorities to implement the radical proposals for a 'Dublin of the Future'. This new city plan, devised before the Rising, proposed, amongst many other changes, a new 'Union Station' in the markets area, the relocation of the GPO to the Custom House so as to be closer to rail and sea connections, and the repurposing of the former GPO site as a new City Hall.[4]

The new city plan won a prize in the 1913 Civics Exhibition, but few of its proposals were acted upon even after the destruction of the years between 1916 and 1922. Many of them, especially the extensive new road network and underground rail links to the new Union Station, were deemed to be beyond the resources of the authorities of the fledgling Free State. More importantly, the GPO building had attained mythological status due to its pivotal role in the Rising and remained steadfastly where it had been. Hence the shattered city was reconstructed in a largely identical layout to its pre-1916 form with just some of the street names altered: Sackville Street, home of the GPO, became O'Connell Street. The new state seemed keener on cosmetic changes than major infrastructural improvements.

There was a hint that some in government grasped the potential role that infrastructure could play in aiding economic development when the Postmaster General of the Free State, J. J. Walsh, told the Dáil in 1923:

> Our country is competing with up-to-date countries such as Sweden and Norway and England, and if the people are placed in the position of dray-horses as against race-horses, so to speak, they are going to get left [*sic*]. We are anxious as a result of some of those economies to sacrifice something in the extension of the telephone.[5]

These proved to be mere empty words. There were many pressing issues for the government of the Free State, there was little money for capital investment, and most of the electorate were more concerned with basics like food and shelter than the luxury of a home phone. Official policy was one of 'small government': Ernest Blythe, the first Minister for Finance, almost made good on his promise to run the country on £20m (*c.* €25m) a year.[6]

Lack of finance, however, does not fully explain the dismal state of the country's telecommunications network in the decades following independence. Whatever J. J. Walsh might have told the Dáil in 1923, the prevailing view in government circles and beyond was that improving telecommunications was of little benefit in an Ireland based on agricultural self-sufficiency. Suggestions that the telephone might be a useful means for farmers to obtain up-to-date market information gained little traction, partly because it was considered that a nationwide radio service could fulfil this need more economically. Scant attention was given to promoting the telephone, or to curb the mounting losses from the telegraph service. Equally, little thought was given to the control and structure of the telecommunications services which remained a division of the Department of Posts and Telegraphs (*An Roinn Puist agus Telegrafa* in Irish) under direct control of the Minister.

A change in government in 1931 barely altered the situation: the policy of the new Fianna Fáil administration placed a greater emphasis on industrial development but also on protective tariff barriers. There was little need for an Irish industrialist, protected from competition by a tariff wall, to use the telephone to access up-to-date market information or strike a deal with a foreign customer. In Éamon de Valera's (in) famous radio broadcast to the nation in 1943, he spoke of how:

> the Ireland that we dreamed of, would be the home of a people who valued material wealth only as a basis for right living, of a people who, satisfied with frugal comfort, devoted their leisure to the things of the spirit.[7]

There was no place for the telephone in this vision.

To be fair, the Irish Free State had been born into turmoil, and the initial attention of the new government was focussed on simply keeping the country running, not planning how telecommunications might aid the economic development of the new state. Twelve weeks after the Provisional Government took over responsibility for the telephone, telegraph and postal services in the 26 counties, the embryonic country was plunged into a civil war. Telecommunications infrastructure was badly affected: in August 1922, anti-Treaty forces occupied the main telephone exchange in Cork city at 91B South Mall and smashed up the equipment with sledgehammers as they left[3] (Figure 6.1). Interviewed in 1998, Post Office technician Maurice 'Mossie' O'Sullivan recalled arriving in the exchange after it was attacked:

> ... when a fellow came into me. He was one of Dev's crowd and he was looking for a phone call to Macroom. And says I to him 'in the name of so-and-so, how could you get on to Macroom – would you look at the bloody place. You wrecked it yesterday'.[8]

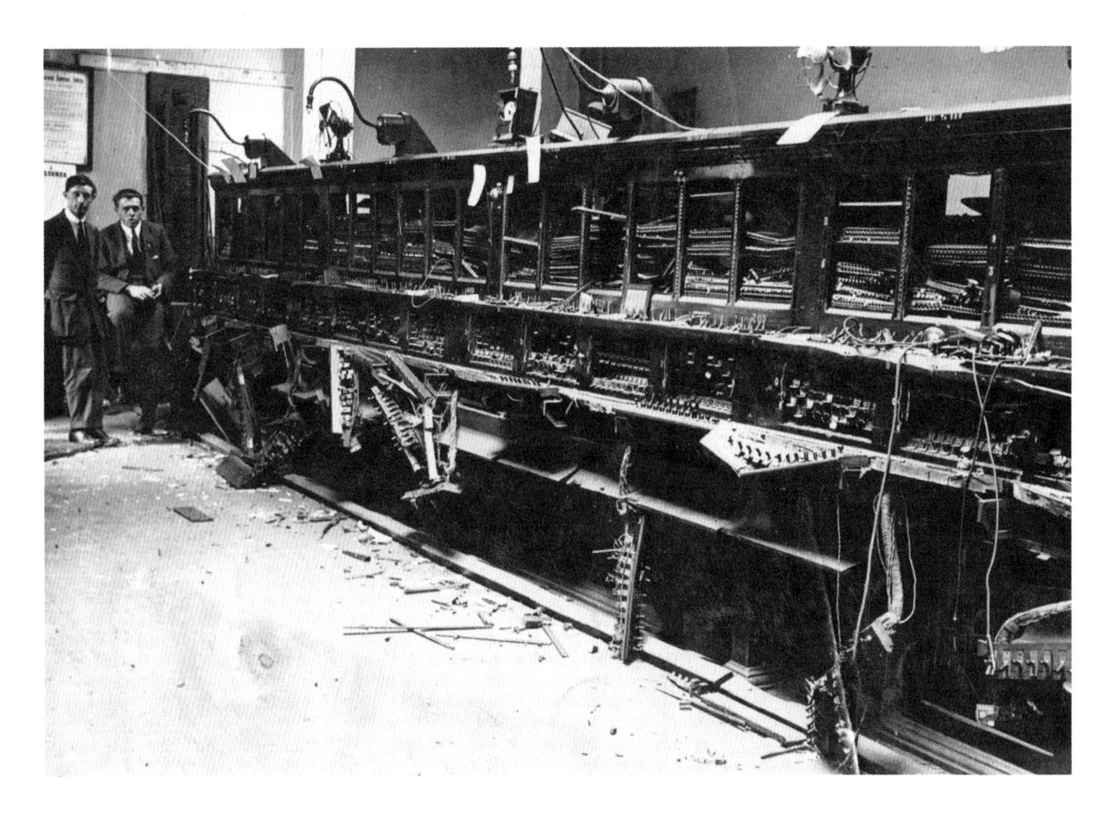

Figure 6.1: The telephone exchange in Cork following damage by anti-Treaty forces in 1922. *(© An Post/The Post Office 2021, courtesy of the An Post Museum & Archive)*

Early in 1923, the telegraph line connecting Blacksod in Co Mayo and its weather station to the outside world was cut by anti-Treaty forces, provoking the Colonial Office in London to write to the Governor General in the Phoenix Park requesting him to intervene with the Free State government on the grounds that 'the meteorological reports received from this station are of considerable importance ... They are also of much interest to the Meteorological Services of other European countries, which have hitherto been accustomed to receive them four times daily.' The line was duly repaired by the Post Office on 2 March only to be cut down on the same evening near Bangor Erris. Another attempt to repair the line had to be abandoned as the Post Office staff were accosted by armed men, again near Bangor Erris.[9]

Meanwhile in Dublin the telephone exchanges at Drumcondra[10] and Blackrock[11] were destroyed and there were several attempts to capture the central exchange at Crown Alley.[12] The Dublin home of J. J. Walsh was attacked as part of a campaign targeting Government ministers.[13] Sometimes the destruction of infrastructure had more of the character of a spectacularly drunken night out than a concerted campaign of terrorism. On 4 January 1923, about 20 armed men attacked Kingscourt station in Co Cavan, set fire to a train and kidnapped a driver:

> The 20 or so men involved remove a quantity of wine from the goods store before leaving. Some 4 km away at Kilmainham Wood station they break into the goods store and come away with a supply of porter and groceries which they load onto a lorry commandeered from the McEnteee Bros of Nobber. The men then make for the village of Kilmainham Wood where at 03:00 they break into the Post Office and smash the telephone apparatus.[12]

The turmoil caused by the Civil War on telecommunications had impacts beyond the borders of the new State. An annex to the Anglo-Irish Treaty, which paved the way for the establishment of the Free State, left the British government with continued control over submarine cables and wireless stations.[14] The anti-Treaty side used this as justification to attack the cable stations in Co Kerry. On 29 August 1922, Clarence Mackey, who had taken over the Commercial Cable Co. founded by his Dublin-born father, telegrammed W. T. Cosgrave to complain that the Waterville station had suffered intimidation of staff and damage to equipment so severe that it had been completely out of service for much of August. Mackey warned that unless the government agreed to guard the stations 'we shall be compelled at great regret to withdraw all our cables and equipment permanently from Waterville and transfer them to England'[15]. The Free State army's presence at the stations was duly stepped up.

After the Kerry cable stations were returned to service, the military sometimes made creative use of them to communicate. Telegrams between army units in different parts of Kerry were sent from Waterville to Valentia via the undersea cable to Weston-super-Mare, onto London, then across the Atlantic to Bay Roberts, across Newfoundland to Heart's Content and back to Kerry.[16] Such a telegram clocked up 8287 km to make a journey of 14 km as the crow flies. Similar creativity was deployed in Co Mayo. With no Post Office telephone system in the county[17] the army made such extensive use of the railway's private network that the stationmaster at Mulranny in Co Mayo complained that he and his family could not sleep 'on account of the Military using the telephone night and day'.[18]

A little further down the west coast, the Marconi transatlantic wireless telegraphy station near Clifden in Co Galway suffered a triple whammy as the new state took shape. The first issue was improved radio technology: by 1922 it was possible to transmit reliably to North America from Britain[19], rendering a station on the very periphery of Europe unnecessary. The other two issues were political.

The second issue occurred during the Civil War when the station was taken over by anti-Treaty forces on 24 July 1922. Unlike the drunken attack on Kilmainham Wood railway station and post office, this was a deliberate and politically motivated attack to silence the wireless telegraph station[20] by destroying the condenser and receiver[21] and expelling the staff. It took five months of conflict (Figure 6.2) before the station was permanently recaptured by government forces.

The third issue then kicked in. Despite the employment and income generated by the station's 200 full- and part-time staff, Minister Walsh viewed the Marconi company and its Empire-wide radio network as part of the *ancien regime*.[19] This was the final straw for the Marconi company and it decided to close its Clifden station.[22] The government ignored local appeals and did nothing to prevent the closure, which became permanent in 1925 when the station's equipment was sold for scrap.[23] Like at the GPO in 1929 where the removal of the royal coat of arms seemed to be more urgent than completing the telegraph office, the independent government could sometimes prioritise petty nationalism over economic welfare.

Incidentally, the Marconi company found a more sympathetic ear north of the border. In 1922 the company contacted the new Northern Ireland government with a view to taking over the telephone and telegraph service in the province. Its proposal was received favourably in Belfast but foundered as it became clear that control over the Post Office would remain in London.

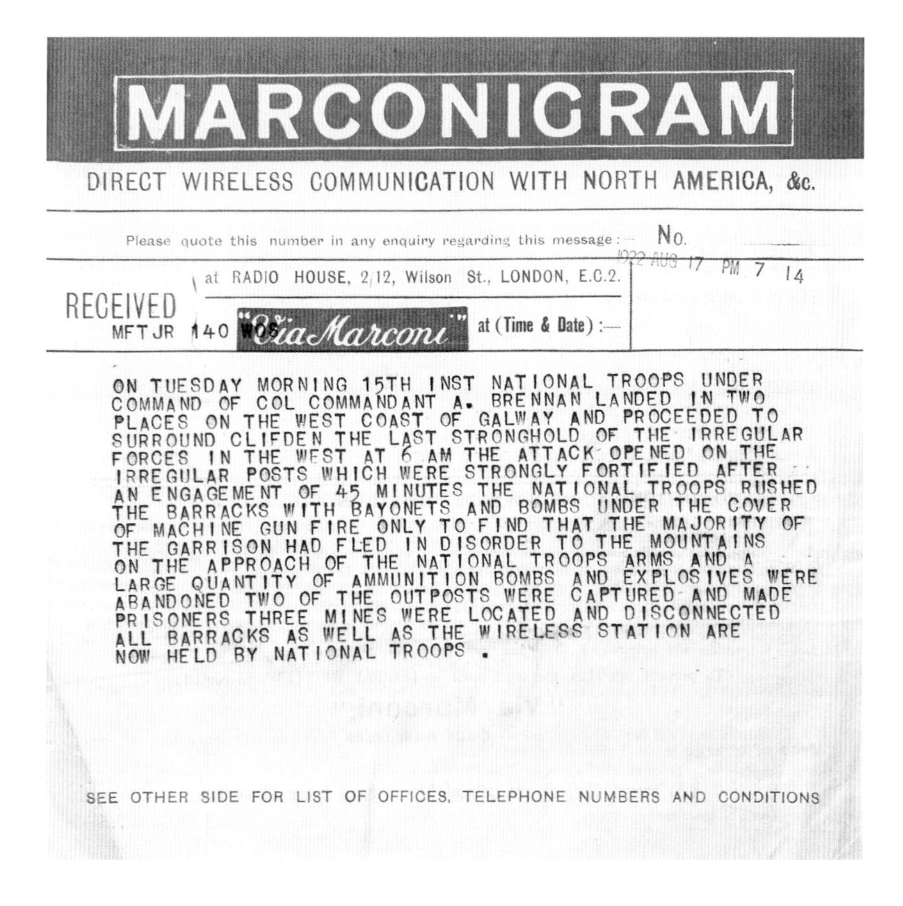

MARCONIGRAM

DIRECT WIRELESS COMMUNICATION WITH NORTH AMERICA, &c.

Please quote this number in any enquiry regarding this message :— No.

1922 AUG 17 PM 7 14

RECEIVED at RADIO HOUSE, 2/12, Wilson St., LONDON, E.C.2.

MFT JR 140 WOS "Via Marconi" at (Time & Date) :—

ON TUESDAY MORNING 15TH INST NATIONAL TROOPS UNDER COMMAND OF COL COMMANDANT A. BRENNAN LANDED IN TWO PLACES ON THE WEST COAST OF GALWAY AND PROCEEDED TO SURROUND CLIFDEN THE LAST STRONGHOLD OF THE IRREGULAR FORCES IN THE WEST AT 6 AM THE ATTACK OPENED ON THE IRREGULAR POSTS WHICH WERE STRONGLY FORTIFIED AFTER AN ENGAGEMENT OF 45 MINUTES THE NATIONAL TROOPS RUSHED THE BARRACKS WITH BAYONETS AND BOMBS UNDER THE COVER OF MACHINE GUN FIRE ONLY TO FIND THAT THE MAJORITY OF THE GARRISON HAD FLED IN DISORDER TO THE MOUNTAINS ON THE APPROACH OF THE NATIONAL TROOPS ARMS AND A LARGE QUANTITY OF AMMUNITION BOMBS AND EXPLOSIVES WERE ABANDONED TWO OF THE OUTPOSTS WERE CAPTURED AND MADE PRISONERS THREE MINES WERE LOCATED AND DISCONNECTED ALL BARRACKS AS WELL AS THE WIRELESS STATION ARE NOW HELD BY NATIONAL TROOPS .

SEE OTHER SIDE FOR LIST OF OFFICES, TELEPHONE NUMBERS AND CONDITIONS

Figure 6.2: This telegram, dated 17 August 1922, announced the recapture of the Clifden station by the Free State army, though in fact fighting continued until December. The telegram proved to be the last ever sent from Clifden. *(The Bodleian Libraries, University of Oxford, MS Marconi 193)*

The partition of the island into two separate administrations was the main trigger for the onset of the destruction of the Civil War but partition itself had direct consequences for the telecommunications network. The new border was crossed by many existing telegraph and telephone lines; for example, the phone circuits into Rosslea, Co Fermanagh and Middletown, Co Armagh came from the Co Monaghan side.[24] The trunk telephone circuit between Dublin and Sligo was routed via Dundalk and across what was now Northern Ireland.[25] With the end of the Civil War in 1923, initial security concerns about such arrangements abated[26] and, for several years, part of the internal communications network of each of the two parts of the island was looked after by the 'other side' (Figure 6.3).

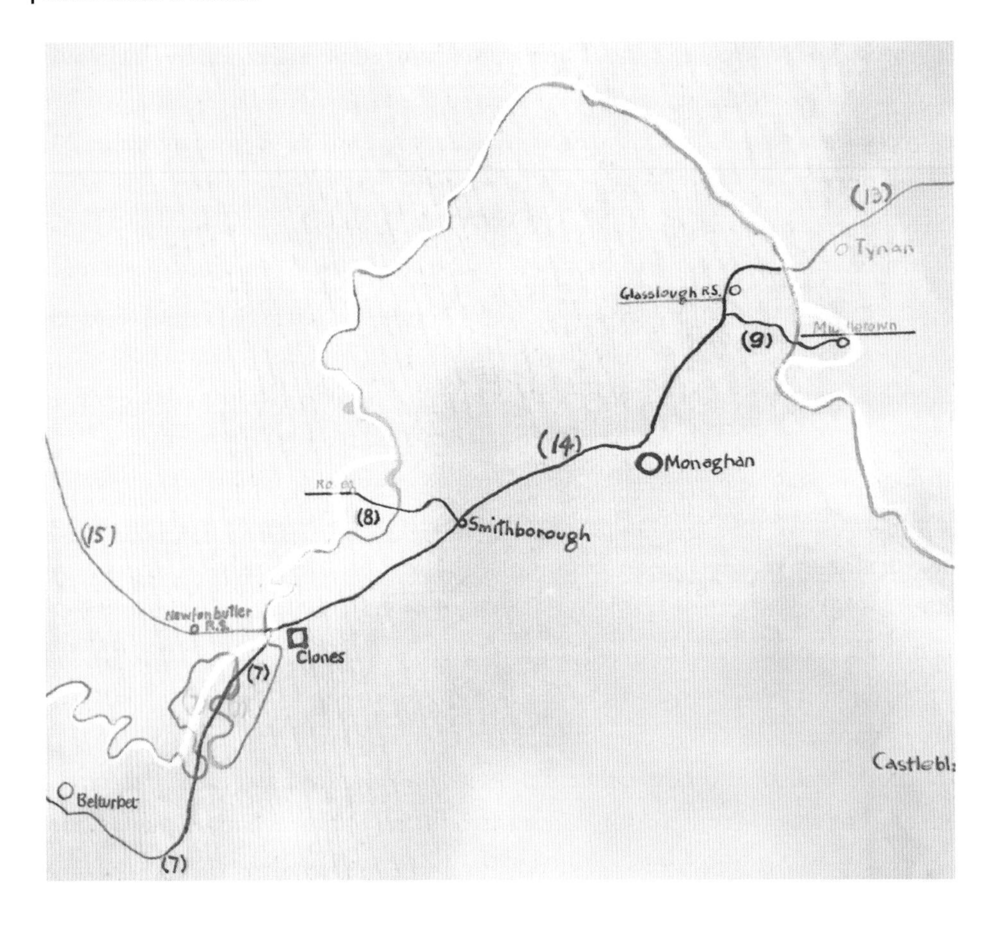

Figure 6.3: An extract from 'Diagram of Routes Crossing the Nor. Ireland – Free State Border'. The green line indicates the political border; the yellow line indicates the maintenance border initially agreed between the respective Post Offices. *(Courtesy of BT Heritage & Archives)*

By the end of the decade, mainly for financial reasons, new lines had been erected to allow the networks to be largely separated.[3] For example, new trunk circuits were erected along the main Dublin to Sligo railway line, so that phone calls between the two places avoided Northern Ireland.[25] This separation was not absolute so that as late as 1951, the telegraph line from Malin Head to Dublin ran through Northern Ireland, with the bill for maintenance footed by the UK Post Office.[27] While customs posts were erected along the frontier from 1923, a hard border of telecommunications was avoided with no extra charges applied to calls or telegrams to a destination on the other side of the border.

In the midst of all the exertions required to set up a new country, partition a telecoms network, repair wartime damage and cope with a civil war, Post Office services were paralysed by a divisive strike. The dispute centred on a decision by the Provisional Government to cut a cost-of-living bonus that had been paid to Post Office workers. At 6 p.m. on 10 September 1922, 19 days after the assassination of government leader Michael Collins, all operators at Dublin's Central Telephone Exchange at Crown Alley stopped work, leaving only two supervisors to handle calls. The Government argued that civil servants did not have the right to strike and, with the Civil War raging, used aggressive tactics of forcibly removing picketers and recruiting new staff in an attempt to keep services operating.[28] This was mirrored around the country, badly disrupting telephone, telegraph and postal services for almost three weeks. Despite its comparative brevity, the strike left a bitter legacy between staff in the Department and the government in power, with unions perceiving Minister Walsh as haughty and aloof.[29]

Telecommunications services were also affected by other manifestations of nation building. In May 1922 the new government asserted its autonomy by arranging with the Commercial Cable Company and the Marconi company to bypass London. Telegrams between the Free State and North America were to be sent directly from the stations in Waterville and Clifden respectively[30], though within weeks both stations were out of service thanks to the Civil War.

Another result of the new order was that placenames all over the country were altered and new government agencies introduced. These changes had significant impacts on telecommunications services. For instance, the Royal Irish Constabulary and Dublin Metropolitan Police were disbanded and policing taken over by the new Garda Síochána. Thus in the 1921 Telephone Directory[31] the number for the Royal Irish Constabulary in Maryborough, Queens County, is Maryborough 5. In the 1929 *Leabhar Seolta Telefón*[32], the number for the Garda Síochána in Port Laoighaise, Co Leix, is Port Laoighaise 5: a new name and a new force but the same phone line. Similarly, in 1921, the occupants of 32 Nassau Street in Dublin were listed as 'Taxes, H. M. Inspectors of - Dublin (1st District)'[31] who could be reached on Dublin 4895. By 1929, the occupants have become 'Revenue Commissioners Dublin (1st District)' but could still be contacted on Dublin 4895.[32]

The title 'Postmaster General' was changed in 1924 to 'Minister for Posts and Telegraphs' (*'an t-Aire Puist agus Telegrafa'* in Irish),[33] though the holder of the new position remained J. J. Walsh. This was another largely tokenistic measure as, despite the change of title, the position remained an 'extern' (non-cabinet) role until 1927. It was part of the pattern where symbolic gestures were favoured over deeper transformations.

Another largely symbolic gesture was a programme to extend the geographic reach of the telephone, which commenced in 1923. As illustrated by a 1926 map (see Frontispiece), the telephone service was extended to reach all counties, bringing the total number of exchanges from to 617 by 1928[32]; an impressive increase of 221 per cent in six years.

In theory it was transformative infrastructural investment but in reality the programme was more tokenistic nation-building. The expansion was performed on an increased but still meagre capital budget and, as we will see, the less visible infrastructure of trunk lines tended to be neglected. All the new exchanges were manually operated but many of them had only one of two subscribers so their operating costs were high compared to the income generated. It is difficult to avoid concluding that this geographic expansion of the network was an early example of spreading investment across a wide geographic area to avoid creating local political discontent. Unfortunately the result of dispersing limited resources so thinly was that the benefits everywhere were negligible. This scattergun approach to investment proved to be a recurring issue for successive Irish governments.

One of the reasons why many rural exchanges were so small was that the structure of charges for line rental and calls made telephones prohibitively expensive outside cities and towns. On top of the annual line rental of £7 10*s*. (€9.52), an extra 25*s*. (€1.59) in retail applied for every furlong (201 m) beyond the first mile from the exchange. A further charge was applied if the exchange serving them had fewer than eight subscribers, which was quite likely in a rural area. Rural dwellers also typically paid more for calls as the local call area covered only a 5-mile (8 km) radius, outside of which higher rates applied.[25]

In 1927 the Department made an attempt to woo rural dwellers with a rural party-line scheme where several houses shared the same pair of lines back to the manual exchange, effectively updating a similar programme from 1911. Despite their popularity in other countries, notably Canada, and an advertising campaign targeting farmers, the party-line scheme proved no more successful than its 1911 predecessor and two years later it was reported that there were only seven such lines in use. Michael Heffernan of the Farmers' Party told the Seanad in 1931: 'Senators may perhaps guess the reason. It may be that as a race we are nervous of letting our neighbours know what we are talking about to another subscriber.'[34]

Mr Heffernan's analysis was probably at least partly true but the price of £43 per year (about €262 per month at 2021 prices) for a location 5 miles (8 km) from an exchange[25] with two other subscribers sharing the line was probably a greater deterrent for many prospective customers. As a result, while the number of telephones did increase, the number of lines per capita remained low in comparison with other countries (Figure 6.4) and was only a fifth of the UK level in 1938.

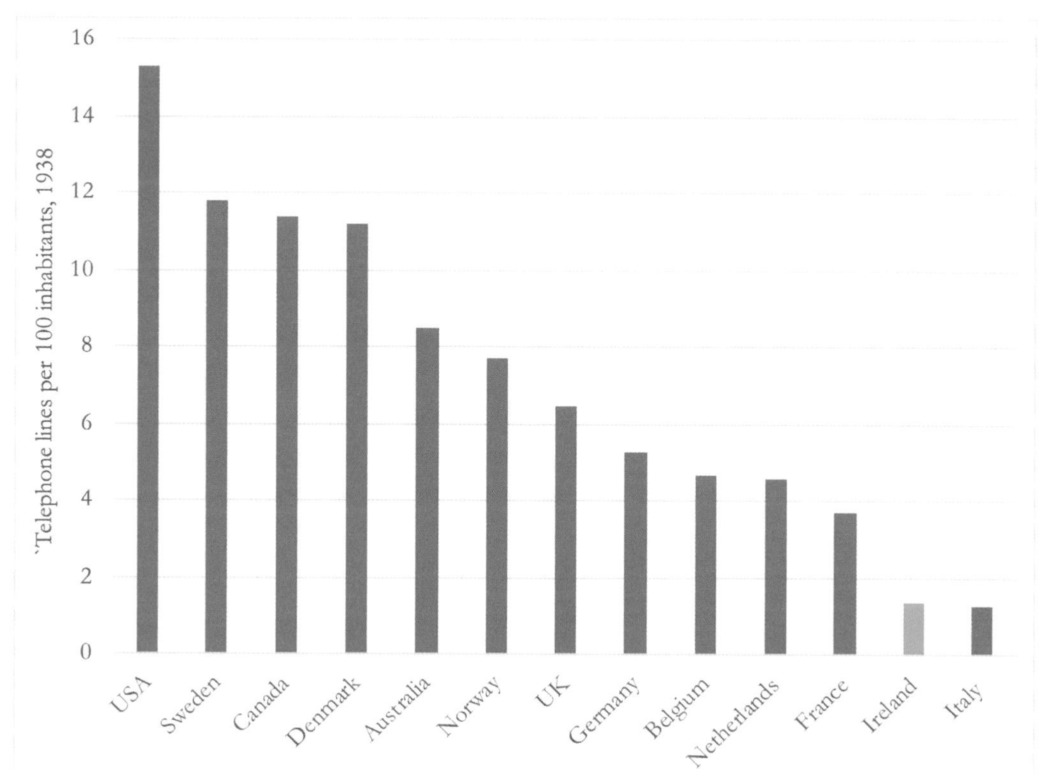

Figure 6.4: In 1938 the number of telephone lines per capita in Ireland remained low by international standards. Source: International Telecommunication Union, *Statistique générale de la téléphonie 1939* (Bern: 1940).

To add to the litany of woes facing the rural telephone user was the manual exchange. While the larger switchboards were staffed 24 hours a day, this could not be justified at smaller locations, which in 1929 amounted to 572 exchanges of a total of 654. Such locations typically had service between 8 a.m. and 8 p.m. on weekdays while on Sundays service was provided for a grand total of 90 minutes. Of course, most of the exchanges with restricted opening hours were in rural areas, further reducing the value of having a telephone in such areas.

Some suburban exchanges lacked 24-hour service as well and the drawbacks this posed in the event of an emergency were illustrated in a 1923 court case. On 15 November 1923, a Daniel T. Lambe was charged with breaking £150 (€190) worth of glass at the Oblate Retreat House in Dublin's Inchicore. Father O'Reilly had tried to summon the police by phone but could get no reply from Crumlin exchange, though it was supposed to be open until 1 a.m. Drawing attention to the fact that the Central Exchange in Dublin provided a 24-hour service, but branch exchanges such as Crumlin did not, Mr Lupton, the prosecuting barrister, declared, 'Everybody in the House of

Retreat could have been murdered that night, and nobody could be got to help them.' Making his preference for private enterprise clear, Mr Lupton mourned that in 'the old days of the National Telephone Company, before the phone came into the hands of others, one could get an answer at any hour of the day or night.'[35] Others blamed the new administration, with the *Freeman's Journal* lamenting that:

> The telephone service badly needs the improver's hand. It could hardly be worse than it is. Day and night its inefficiency is demonstrated. The Free State ought to insist upon getting as good and as disciplined a service as the foreigner was able to provide when he was in control.[36]

A solution to the problem of providing 24-hour service on an economic basis was to remove the need for an operator. The technology to do this was already in existence, though its origins lay in corporate rivalry rather than in saving operating costs. In 1878 a Kansas City undertaker called Almon Strowger learnt that his competitor's wife worked as a telephone operator and suspected that she was diverting business calls meant for him to her husband. With the help of his nephew and other business partners, he invented a telephone exchange that needed no operator because it automatically put the caller through to the number they wanted.

His system was refined by the invention of the telephone dial, which creates trains of on/off current pulses corresponding to the digits 1–9, and 0 (which sends 10 pulses). A Strowger exchange is based on the concept of selectors. In its simplest form, a selector consists of a semi-circular bank of contacts, each of which is connected to a different subscriber, and a wiper that can be moved to connect with any of the contacts (Figure 6.5). When a caller dials a number, the wiper steps around the bank of contacts in line with the number dialled. Thus, if the caller dials the digit 4, four pulses are sent to the exchange and the wiper is moved four times so that it makes contact with the fourth set of contacts. Through the use of selectors that move in both horizontal and vertical axes, and by connecting multiple selectors together, many thousands of subscribers can be supported.

Exchanges following Strowger's design, sometimes called step-by-step exchanges, were produced and installed throughout the world as the twentieth century progressed. In 1914 a Strowger private automatic branch exchange (PABX) was installed in the Sirocco Engineering Works in Belfast[37] and in 1921 a similar facility was installed at the Guinness brewery in Dublin[38] but no automatic exchanges had been installed by the Post Office for public use in Ireland by the time of independence.

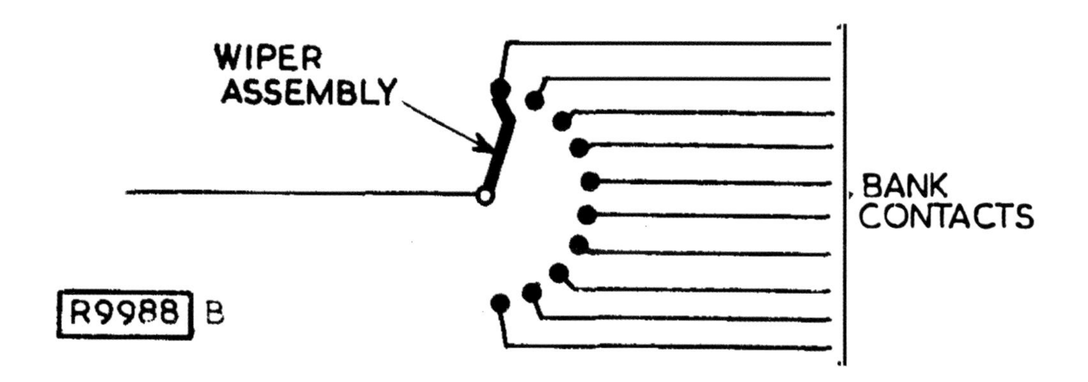

Figure 6.5: A simplified diagram of a selector, the key element of Strowger's automatic exchange. It consists of a movable arm known as a wiper, and a contact bank. The wiper steps around the bank of contacts according the number dialled. *(Courtesy of BT Heritage & Archives)*

On 24 July 1927, Ireland's first public automatic exchange was put into operation in part of the former Ship Street Barracks at the southern perimeter of Dublin Castle.[39] The Ship Street exchange came into service four months before London's first public automatic exchange opened at Holborn, a fact that must have provided some pride to staff in the still-new Department. Costing £20,000 10*s.* 8*d.* (€25,395), the new exchange served the south-west of the city, west to Inchicore and east to Stephens Green. This area included the Oblate Fathers Retreat House in Inchicore – in the event of another break-in at the Oblate Retreat House, Father O'Reilly would now be able to summon the Gardaí even at 1 a.m.

Subscribers on the new automatic system in Dublin were allocated five-digit numbers. Allowing for the use of the number 0 to dial the operator and reserving numbers beginning with 1 for other special services, a 5-digit scheme allowed for a theoretical maximum of 80,000 lines. At the time, it was believed that telephone users would be unable to remember numbers more than 5 digits long which is why many cities adopted an alphanumeric pattern where the exchange 'name' was dialled before the numeric part of the number. In Dublin, however, the Department correctly forecast that a 5-digit pattern would be sufficient for many years to come. Sadly, this meant that Dublin never had numbers like PEnsylvania 6-5000, made famous in the song, or WHItehall 1212, the number for Scotland Yard that the British public was often urged on news bulletins to phone with information about crimes. Despite the relatively simple numbering system in Dublin, the Department added detailed instruction to the telephone directory about how to dial a number (Figure 6.6).

3. **WHAT TO DO WHEN ORIGINATING A CALL FROM AN AUTOMATIC TELEPHONE.**

(a) **Before lifting your receiver, refer to the current issue of the Telephone Directory and ascertain the correct number of the subscriber you require.** If the number is not in the Directory, dial " O " and ask for **" Enquiries."** When you are told the number, hang up your receiver for a few seconds before making your call.

HOW TO DIAL.

(b) **Before dialling, always place the receiver to your ear and listen for the " Dial Tone."** This is a continuous low-pitched buzz, and indicates that dialling may be commenced. The " Dial Tone " is usually heard at once ; if it is not heard within a few seconds, hang up your receiver for a second or two and try again. If the tone is not then heard, your line or instrument is out of order, and you should advise the Supervisor from another telephone.

(c) **After you have heard the " Dial Tone,"** place a finger firmly in the opening in the dial through which the first figure—reading from left to right—of the required number is seen.

(d) Pull the dial round to the stop, remove your finger from the opening, and without touching the dial, let it return to rest.

(e) Proceed in the same way with the other figures of the number.

(f) If the number dialled is available, you will then hear the " Ringing Tone." This is a double " burr-burr "-ing sound heard at intervals, and indicates that the distant subscriber's bell is being rung.

(g) **The " Number Engaged " signal is an interrupted high-pitched buzz—equal periods of tone and silence.** If this is heard, hang up your receiver for a few minutes and dial again.

(h) **A continuous high-pitched buzz heard after dialling** indicates that the number is unobtainable for some reason other than " number engaged " or " no reply." On hearing this, hang up your receiver, verify the number and dial it again. If the " unobtainable " signal is again heard, hang up your receiver for a few seconds, then dial " O " and ask for **" Enquiries."**

Continued on next page.

Figure 6.6: The 1928 telephone directory included detailed instructions about how to use an 'automatic telephone'. Point (f) refers to the double 'burr-burr' ringing tone, still used today in Ireland and in most countries that were once part of the British Empire.

Dublin was also the home of Ireland's second automatic exchange which was inaugurated in 1928 on Fitzwilliam Lane, off Merrion Street, to serve the south-east of Dublin city including much of the city's administrative and commercial premises. After a hiatus of a few years partly caused by the 'Economic War' with the UK, automatic exchanges were opened at Rathmines (1936), Terenure (1936), Clontarf (1938) and Dún Laoghaire (1939).[17] It is noteworthy that these were (and indeed still are) located in largely affluent suburban residential areas. By contrast the poorer north city and inner suburbs were served by exchanges on the south side; presumably there was insufficient demand to warrant constructing a local facility.

The difficulties of providing a 24-hour service to subscribers served by small exchanges were not unique to Ireland and, as the technology became cheaper, automation was applied to rural exchanges. Sweden, with an even lower population density than Ireland, began a programme to roll out small, robust, low-maintenance automatic exchanges in rural areas.[40] Closer to home, the UK Post Office embarked on a similar programme with a modular exchange system called a UAX (Unit Automatic Exchange) that could be installed in blocks of 25 lines to provide small automatic exchanges at the minimum possible capital outlay.[41] As a result, and in contrast to the situation south of the border, the first places to get automatic service in Northern Ireland were small towns and villages, the programme starting with Aughnacloy, Co Tyrone in 1929.[42] The choice of a town on the main Dublin to Derry road 500 m from the border for the first such facility may not have been an accidental one on behalf of the authorities in Northern Ireland. Many more rural areas in Northern Ireland were upgraded in the following years.[43]

Nine years later, a semi-automatic rural system was implemented on 24 September 1938 serving the North County Dublin villages of Malahide, Donabate, Rush and Lusk using American equipment[38] but this trial was not extended further. Despite the official view of agriculture as the backbone of the economy, there was no attempt to bring automatic service and attendant 24-hour service to the rural areas that comprised the majority of the Free State until after World War II. Southerners would, in the words of de Valera, have to remain 'satisfied with frugal comfort'.

While Northern Ireland started automating exchanges in small towns, its cities were not ignored. Belfast was converted in 1935 when all 18,000 subscribers were moved over to five new automatic exchanges on 2 November. The cutover was quite the event, with a special BBC radio programme broadcast from the new Telephone House at Cromac Street on the night before.[44] Derry followed in 1937, with connections to 15 surrounding exchanges, including Bridgend, Lifford, Moville, Buncrana and Letterkenny in the Free State. The pro-nationalist *Derry Journal* reported approvingly that:

> dialing gave little trouble, as most had become familiarised with it through seeing it on the films … Free State operators … are now able to ring Derry subscribers direct or contact distant Six Counties exchanges as far as Coleraine and Omagh without manual assistance from Derry.

With perhaps just a hint of superiority, the *Derry Journal* report further noted that 'This is the first time that this part of the Saorstat has had experience of the automatic system.'[45]

The availability of automated service conferred a certain status, most especially in the USSR. In 1922, on Lenin's initiative, a special automated network was installed for the Kremlin bureaucracy, parallel to the public telephone network. This official party telephone was called *vertushka* – the Russian word for 'pinwheel', or 'whirligig', because it used dial telephones at a time when the ordinary telephone system relied on manual switchboards. This separate official telephone system was retained even after the public system in the USSR started to use automatic exchanges and, still known as *vertushka*, existed right through until the demise of the USSR.[46]

In Ireland the official emphasis on agricultural and protected industries posed another problem for the development of telecommunications as all the equipment required had to be imported, affecting the country's balance of payments. Only 14 per cent of the workforce of the 26 counties was engaged in industry, compared with 35 per cent in Northern Ireland[47] and the Free State lacked any tradition in the advanced engineering skills required to manufacture telecommunications equipment. As a result the Strowger switching equipment for the new automatic exchanges was supplied by Standard Telephones and Cables, a British subsidiary of US-based International Telephone & Telegraph (ITT).[48] The purchase of British-made equipment was unsurprising in view of the shared history and common language, though contrasted with the large part the German firm Siemens played in developing Ireland's electricity supply. The small domestic market and lack of a technological manufacturing tradition in the 26 counties meant that it took many decades before any manufacturing of telecommunications equipment took place in the State.

In the meantime, domestic capabilities were limited to the repair and testing of telegraph and telephone apparatus at the Post Office factory near Kingsbridge (now Heuston) station in Dublin.[3] Its competence in even those limited activities was criticised by the Comptroller and Auditor General in 1924, who reported that the cost of repairing some articles was greater than that of buying them new on the open market.[3] This was another example of a policy, sometimes explicit, to use the Department as an employment creation agency, regardless of the effect on the economics of the services it provided. In 1931 the Department noted that with over 12,000 staff it was probably the largest employer in the State.[49] Retaining manual switchboards kept operators in jobs, and the labour costs involved could simply be passed on to the customer.

With private telephones too expensive for most people to afford, public phones provided a vital amenity. But like so many other facilities, public phones were a victim of the policy of 'too little, too late' that was endemic in the Department of Posts and Telegraphs. To be fair, the Department had inherited a poor legacy from the days of British rule. On-street kiosks, which had first appeared in Berlin in 1881,[50] spread to many other cities – Stockholm boasted 500 by 1925. However, kiosks did not initially find favour in the United Kingdom as the then privately owned telephone companies were not allowed to install them on publicly owned streets.[51] Instead public call offices were provided, typically within commercial premises. For example, Switzer's department store (now Brown Thomas) on Dublin's Grafton Street installed a 'Public Telegraph Call Office' in 1900.[52] Use of a call office phone was limited to the opening hours of the establishment in which it was located, restricting its utility. The establishment with the call office would advertise the service by means of a sign with the legend 'You May Telephone From Here.' Following independence, these signs, now green in colour, also bore the Irish text '*Tá Telefón ar Fáil Anso* [*sic*].'

Following the nationalisation of the telephone service in 1912, attitudes to on-street kiosks relaxed but rollout was slow. Sometimes this was due to opposition from local authorities. Dublin Corporation initially opposed the first kiosk on College Green which was eventually erected in May 1925. Waterford Corporation spent many months in 1935 adjudicating on an application from the Department of Posts and Telegraphs for a phone box on the Quay.[53] By 1927 telephone boxes had reached Cork and Limerick[54] and by 1931, the three million citizens of the Saorstát had 14 kiosks to choose from.

Additional phone booths were installed in Dublin in 1932 in advance of the Eucharistic Congress (Figure 6.7)[55] but provision remained inadequate, to judge by an exasperated *Irish Times* columnist in 1935 who lamented, 'Who has not waited for minute after precious minute outside the one lonely kiosk in a large district, hoping

Figure 6.7: Installation of two new telephone kiosks on Dublin's Aston Quay for the Eucharistic Congress in 1932. *(Image Courtesy of the National Library of Ireland)*

Four generations of telephone kiosks. Clockwise from the top left: (Figure 6.8) The first telephone kiosks in the Free State installed between 1925 and about 1930 were of the UK Post Office K1 design, such as this survivor in Foxrock in Dublin; (Figure 6.9) This P&T design was used across the country between about 1930 and 1980, its ubiquity making it a symbol of the nation; (Figure 6.10) Eircom-branded boxes replaced most earlier designs in the 2000s; (Figure 6.11) Most of the few surviving phone boxes were replaced with this design by eir in 2021 against hope that the young thing inside eventually would run short of coppers?'[56] While the language might seem a little sexist to the modern reader, the *Irish Times* was making a fair point about the number of kiosks. In 1939 in the UK there was one phone box for every 1,242 people[57]. In Ireland, with a grand total of 159 kiosks[25], the equivalent ratio was one per 18,239 people. De Valera's vision of Irish people devoting 'their leisure to the things of the spirit' did not include chatting on a public telephone.

The first telephone kiosks in Ireland were of the UK Post Office's K1 design (Figure 6.8), recognisable by the small orb on the roof, and were manufactured by a firm in London as tenders from local firms proved much more expensive.[58] A different model, unique to Ireland and made from concrete with the word *Telefón* in the old *Cló Gaelach* script at the top of the box, was adopted a few years later and became the standard model for the following decades.

The Irish department followed the lead of the UK Post Office in regard to the equipment inside the boxes. Introduced in Britain in 1925, the A/B coinbox telephone was so-called because when the called party answered, Button 'A' had to be pressed before a two-way conversation was possible. In the event of no reply, engaged tone, or a wrong number, the caller could, in theory at least, have the coins returned by pressing Button 'B' (Figure 6.12). When connected to an automatic exchange, the mechanism was wired so that a caller could talk to the operator by dialling 0 without inserting any coins. The operator would ask the caller for the number required, determine the charge and ask the caller to insert the fee before connecting the call.[59]

Figure 6.12: An A/B coinbox telephone. Public phones like this were in use from the late 1920s until the early 1990s. From the 1930s, the phones were enhanced to accept three different denominations of coins (initially 1d., 3d. and 6d.) each of which made a different sound as the caller inserted it; the operator could determine how much money had been inserted by means of a microphone inside the mechanism.

As local calls were untimed there was no pressure on the caller to be brief which, as noted above, was a cause of considerable friction between the caller and anyone waiting to use the same public phone. During the 1940s coin box phones in Dublin were modified to cut off a call after five minutes, warning of the impending termination being given by 'pips'. Discussing the development in 1945, the then Minister for Posts and Telegraphs, Patrick J. Little, noted, 'It is a source of satisfaction to record that the complicated equipment required was designed by our own engineering staff.'[60]

Every innovation brings new opportunities and it was soon discovered that, by obstructing the coin return chute, the coins released when button B was pressed could be prevented from being returned to the caller. In 1934 the *Irish Times* reported on a case at Dublin District Court:

> Hubert Ginochy (18), with an address in Bride street, was charged with obstructing the services of the Post Office by inserting tramcar tickets in a slot of the telephone kiosk at Burgh quay, attempting to steal money from a receptacle in the telephone, and loitering with intent to commit a felony ... The defendant, who denied going near the kiosk, was sentenced to fourteen days' imprisonment.[61]

The more dextrous user could also make calls for free by tapping the receiver cradle to mimic the pulses generated by the dial. Despite these shortcomings the A/B coinbox remained in use for over 60 years, making it a reluctant design classic.

The paucity of telephone boxes was a visible manifestation of a wider problem of inadequate investment. The geographical expansion of the telephone network undertaken in the early years of the state was performed on an increased but still meagre capital budget. As a result, the less visible infrastructure of trunk lines tended to be neglected: there were only three circuits between Dublin and Cork[62] and two between Dublin and Waterford.[63] Thus, while in May 1926 the Minister could declare in the Dáil, 'We have made the service practically universal. Every town and village, I think I can truly say, will have an installation by the time the next report is presented here',[58] he did not refer to how long a caller might have to wait to make a call over the single trunk line shared by the new exchanges set up in Roscommon and Mayo.[64]

For a short while around 1930 it seemed that Co Mayo might turn from a telecommunications backwater into a hub to rival Co Kerry. The reason was a request from the giant US telephone company AT&T and the UK Post Office to land a transatlantic telephone cable on the Belmullet peninsula.[65] The Irish government readily consented to the proposal – it had little option but to do so as the UK retained the right to land cables on the shores of the Free State under the terms of the Anglo-Irish Treaty of 1921. AT&T purchased 40 acres (*c*. 16 ha) of land at Frenchport and employed about 12 men for a period on preparatory works. The proposed cable was considerably longer, far more expensive and much more technically challenging than any other telephone cable of the time and the Great Depression caused the plans to be abandoned.[38] It was 1956 before the Atlantic was spanned by telephone cable and 1989 before Ireland was linked directly with the US by a cable capable of carrying telephone calls.[66]

In the intervening decades, transatlantic telephony became possible thanks to the medium of radio. The UK Post Office established a radio telephone station at Rugby, part-designed by talented Omagh-born engineer Charles Eric Strong[67], which could communicate with North America. On 26 August 1929 the service became available to subscribers in Dublin. For a mere £9 6*s*. for 3 minutes[68], equivalent to about €220 per minute at 2021 prices, Dubliners could gossip with New Yorkers. Telephone users in the rest of the state were denied this service as the speech quality over the primitive trunk network was too poor.[69]

Innovations from the world of radio also helped improve Ireland's primitive trunk landline network. From the 1920s, thermionic valves, invented by Lee De Forest for use in radio, began to be used to amplify speech on trunk telephone circuits so that much thinner and cheaper cables could be utilised. Further improvements occurred from 1927 when Bell Telephone Laboratories in the US perfected what was called the carrier system which carried multiple conversations over a single circuit, effectively using radio transmission over a set of wires. This early form of frequency division multiplexing could initially squeeze three conversation channels onto a single circuit but this was later extended to 12 and then 24 channels.[70] Such enhancements improved the quality of speech and provided a cost-effective way to increase network capacity, making it possible in many places to eliminate 'delay' working and connect trunk calls while the caller remained on the line.

The first carrier system in Ireland was opened between Dublin and Belfast in May 1932.[17] The impetus for this upgrade was the Eucharistic Congress, which as we saw also provoked the installation of the Athlone radio transmitter and additional public phones – clearly it required religious intervention to improve the communications network. Circuits between Dublin and all the main centres were eventually converted to carrier systems later in the 1930s[17] but, in what was to become a pattern for decades to come, the extra capacity became fully utilised almost as soon as it had been provided.

This was also the case when a pair of new undersea telephone cables between Howth and Nevin in Wales were laid in 1937–8. These cables were early examples of the coaxial type,[71] providing a total of 16 cross-channel circuits, compared with the two previously available. (Coaxial cables will turn up again several times in this story). On the day after the cables opened for calls, in a short article optimistically titled 'Telephone Delays Abolished', the *Irish Times* commented that 'it is anticipated that there will be no delay in telephone calls from Dublin to London and Liverpool.'[72] Within two weeks, the Irishman's Diary column in the same paper reported that it took over ten minutes to get a call connected to London.[73]

Other promises by successive ministers that trunk calls would be available on a 'no delay' basis remained merely aspirational. In 1931 the Fianna Fáil TD for Longford-Westmeath, Michael J. Kennedy, gave the example of a constituent in Delvin who asked for a call to Mullingar, 20 km away. The call was booked and five or ten minutes later:

> he went out, entered his motor car and drove to Collinstown, a distance of seven miles ... He called the Mullingar subscriber from Collinstown, got through, and returned by motor car to Delvin. Even when he reached Delvin the original call had not been put through and he cancelled it. That is no exaggeration.[74]

Such complaints were to continue into the 1970s.

If Mr Kennedy had actually been able to get through, he would have been charged *6d.* (about €2.00 in 2021 prices) for just three minutes of conversation. Call prices were reduced a little in 1936 but, even so, a daytime call to London cost *4s.* for three minutes, about the price of four and a half pints of beer at the time.[75] The Department expected the price reductions to lead to a fall in revenue but their pessimism was unfounded and an increase in subscriber numbers and call traffic led to higher overall revenue.[76] Of course, this imposed further strains on an overburdened network. The Department of Posts and Telegraphs was caught in a catch-22. Promoting use of the telephone, such as by reducing the charges, would increase demand for capital investment but, with other, higher-priority projects in other departments fighting for the same limited capital budget, such funding was not available. The safer policy was not to encourage demand and thus avoid any heavy capital requirements. Ireland was the dray horse competing against the race-horses, as warned of by J. J. Walsh in 1923.

Other financial and policy issues hampered development of the network. One of these was an emphasis on short-term gain at the expense of long-term stability and value. This was highlighted when the need to provide extra capacity as a result of the 1936 price reductions required the employment of an extra 60 engineering staff – the same skills that had been declared redundant in a cost-saving measure just three years previously.[38] There was also a related problem of fitful funding. A building to house a new exchange in the Dublin suburb of Clontarf had been completed in 1929[38] but it took a further nine years for the Department to equip it.

A further issue was the tendency to treat telephones as a cash cow. The telephone section of the Department made a profit almost every year from 1931–2 until the 1970s[25] but such profits were milked to shore up the loss-making telegram service rather than reinvested in telephone infrastructure. In 1936–7, income from the telegraph services covered only 65 per cent of expenditure. The deficit of £104,263[76] (€132,387) represented a subsidy of 4c per telegram. Efforts to improve the efficiency of the telegram service did little to curb these losses. Morse code was replaced by Baudot and teleprinters, reducing the manual effort required. In Dublin the pneumatic tube system from the 1870s was partly reinstated to connect the busy post office at College Green with the telegraph office in the rebuilt GPO.[38] Another innovation in the capital was the introduction in 1929 of motorbikes to deliver telegrams. The *Irish Times* commented approvingly that 'day in and day out these messengers give, probably, the fastest telegram service in the world, and they never fail in their job.'[77]

The interest by the *Irish Times* in the speed of the telegram messengers was understandable, for the newspaper industry was a heavy user of the service and continued to benefit from the concession rates granted at nationalisation in 1870. Perhaps this also explains the lack of attention to the consistent losses racked up by the telegraph service. The main newspapers had telegraph offices on their own premises so they could receive stories directly from agencies and correspondents. The concessionary rates afforded to press telegrams meant that, for all the traffic generated, they had little impact on the mounting losses of the telegraph service.

The problems of weak financial management, underinvestment and poor marketing were not, however, universal across all public services. On 29 July 1929, 11 days after opening the rebuilt GPO in Dublin, W. T. Cosgrave performed another opening ceremony, one that was of even greater national significance. This time he was at Ardnacrusha hydro-electric power station, which, together with a new state-owned Electricity Supply Board, was set to transform Ireland's electricity supply. Following a special blessing by the Bishop of Killaloe, Dr Fogarty, Mr Cosgrave declared that 'Henceforth the Shannon will be harnessed in the service of the nation, distributing light, heat and power throughout the Saorstát, increasing at once the comfort of our homes and the productive capacity of our farms and factories.'[78]

The Shannon Scheme was an ambitious project, quite out of character for the financially cautious government. In addition to the massive engineering project at

Ardnacrusha, the scheme also established a national utility that would take over all the existing electricity supply entities, public and private, and build a national grid to connect them up to power from the Shannon.

Unlike the telecommunications service, which was provided directly by a government department, the Electricity Supply Board (ESB) would be a semi-state body owned by the State but intended to be independent of government control and financially self-sufficient in the medium term. The £5.8m (€7.4m) cost of the Shannon Scheme, equivalent to almost 20 per cent of the Government's annual revenue budget for 1925,[79] was criticised as being unnecessary and expensive by many, while the proposed publicly owned Electricity Supply Board was described in some quarters as 'socialist', a dirty word for a fiscally and socially conservative government. Nevertheless, the electricity grid spread across the country during the 1920s and 30s, connecting up cities, towns and villages – but few rural homes. This was because, as with the telephone service, the cost of installing cables to isolated houses was often impossible to justify on short-term commercial grounds. In 1939 the Minister for Industry and Commence, Seán Lemass, asked the ESB to prepare a report on rural electrification, paving the way for the successful Rural Electrification Scheme that began after the war.[80] The problems of providing a telephone service in rural areas drew much less attention and took far longer to resolve. The Agricultural Credit Corporation (founded in 1927) and Aer Lingus (1936) were other examples of semi-state bodies established in this period, owned by the State but operating largely on a commercial basis.

A common factor with all these semi-state companies is that they were new enterprises, or involved the new semi-state body taking over functions from private companies or local authorities. None involved devolving power from central government. By contrast, control over the telephone network was a valuable form of political patronage, a fact that would become increasingly important as use of the telephone expanded. Successive ministers and senior civil servants were unwilling to devolve decisions about the telephone service to an independent enterprise. Perhaps this was why the government spurned an approach to take over the telephone service made in 1930 by the US-based International Telephone & Telegraph (ITT), which operated the telephone networks in Spain and several Latin American countries.[81] Financially it must have looked attractive to a cash-strapped government to receive a lump sum in exchange for offloading a service that was not considered core, but the offer was not pursued. It was not until the 1970s that serious consideration was given to reforming how the telecommunications network was owned and governed.

The desire by government for tight control was also apparent in deliberations about the country's new radio service. Radio broadcasting was viewed by government as an instrument to build a nation in line with its own viewpoint. During the debates preceding establishment of the station, the Minister, J. J. Walsh, argued that without a government-controlled station 'nothing but British music hall dope and propaganda would be received by those listening-in throughout the Saorstat'.[82] Thus responsibility for radio came under the direct control of the Minister for Posts and Telegraphs and the staff involved were employees of the Department,[83] like postmen and telephone operators.

The new radio station commenced broadcasting on New Years' Day 1926. The Dublin transmitter's wavelength, initially 390 m (770 khz), and station callsign, 2RN, were allocated by the British Post Office, whose consent for wireless services was required under the terms of the Anglo-Irish Treaty. The Minister, who had been so hostile to the Marconi company over the Clifden telegraph station, had apparently changed his views as the studio equipment and the transmitter at McKee Barracks were supplied by the company with Guglielmo Marconi cabling his congratulations from Rome on the opening night.[19] A radio station in Cork followed in 1927 and further plans were made to extend the reach of the government's voice across the rest of the country.

In line with these plans, a high-powered transmitter was installed in 1932 at Moydrum outside Athlone, near the geographical centre of the island. As we've already seen, like many other pieces of infrastructure it was pressed into service for the Eucharistic Congress and formally opened afterwards on 6 February 1933 by Éamon de Valera, who had become leader of the government in the year before. The now national broadcaster initially called itself Radio Athlone, though the programming was fed over an open wire telephone circuit from the GPO in Dublin[84] running alongside the railway line via Mullingar. The insecurity of this communications link was exploited on St Patrick's Day in 1936 when the circuit was tapped somewhere between Dublin and Athlone. Instead of Éamon de Valera's broadcast to the nation, listeners heard an unidentified voice repeatedly saying '*a chairde*, this is the IRA.'[85]

The station became Radio Éireann in 1938, its diet of agricultural prices and farming techniques fostering the concept of Ireland as an agrarian economy while the government's view of Irish culture was represented by the emphasis on programmes in Irish, coverage of Gaelic games and concerts of céilí music.[86]

The Marconi-built transmitter at Athlone had an initial power of 60 kW, making it one of the most powerful transmitters in Europe at that time.[87] Curiously, the most powerful other transmitters in Europe at the time belonged not to the established great European powers, but to Czechoslovakia, Poland and Hungary[67], which like Ireland, were all newly created states. Their governments understood the importance of the new medium of broadcast radio in nation building. The Athlone station's power was increased to 100 kW from 1937, providing good reception to all of Ireland and, after dark, much of western Europe, a feature which was to cause some issues during the war. The converse was also true: in addition to 'British music hall dope and propaganda' from the BBC, after dark broadcasts from all over Europe could be picked up by most wireless sets. The daily *On The Wireless* column in the *Irish Times* highlighted offerings from afar, such as concerts from Breslau (now Wrocław) and opera from Heilsberg (now Lidzbark Warmiński), as alternatives to the often stolid fare from Dublin's GPO.

In 1940 R. M. Smyllie, editor of the *Irish Times* and no fan of Radio Éireann's view of Irish identity, considered that 'No matter how bad the Irish programmes may become, 170,000 will continue to take out licences every year in order that they may listen to the BBC.'[88] Evidently, at least in Mr Smyllie's circles, Radio Éireann's efforts at nation building were failing as potential listeners quite literally tuned out. Similar discussions about British television were to ensue in the 1960s and 70s. Despite such criticism the fact that in 1932 citizens from Donegal to Kerry could listen to an Irish radio station was a genuine transformation for the still-new country.

The Government wanted to extend Ireland's voice to North America and further afield, and, to help achieve this, a shortwave transmitter was added to the facilities at Athlone in February 1939. The first shortwave broadcast was made on St Patrick's Day 1940 but the 1.5 kW transmitter was totally inadequate.[89] There was a further attempt at shortwave broadcasts just after the war but this was also quickly abandoned.[89] It was not until the arrival of internet streaming around the turn of the century that Irish broadcasting became available to a global audience.

The government understood the potential of radio as means of building a nation just as it appreciated the role of a national electricity grid in developing an economy based on agriculture. On the other hand developing the telephone network was not viewed as an integral part of building a new nation based on agrarian self-sufficiency, whose new central post office lacked a telephone exchange. Any initial indications that the government grasped the potential role of the telephone in aiding economic development had been quickly extinguished and the Department's limited attempts at development of the telecommunications network were largely tokenistic and scattergun.

The national radio station, even if its output was constricted by the government's narrow view of Irish identify, was a least a sign of modernity and a unifying voice available across the land from 1932. Ireland had, by the magic of Marconi's wireless, taken its place amongst the nations of the world and, on the dials of millions of wireless sets across Europe, the name Athlone glowed proudly alongside places like Breslau and Heilsberg. Within seven years the world would be turned upside down and the names of those two cities would lose their place on the radio dial – and on the map of Europe.

CHAPTER SEVEN

NEUTRAL AND ALLIED

Maureen Flavin was at work in the post office in the remote village of Blacksod on the very west coast of Co Mayo. It was the morning of 3 June 1944. Maureen remembers the day well. Not only was it the day of her 21st birthday, it was also the day she received two intriguing phone calls from the lady with the English accent.

Maureen worked alongside the postmaster Ted Sweeney and his sister Frances Ashforth.[1] It was a busy job for a small village post office, for, in addition to handling the post and operating the newly installed telephone switchboard, they also had to record meteorological readings (Figure 7.1) and phone them through to the Meteorological Service in Dublin.

Figure 7.1: The Post Office in Blacksod, Co Mayo, in the 1940s. The Stevenson screen which housed some of the weather instruments is visible in front. *(Met Éireann)*

Blacksod served this meteorological role because of its location. Ireland is often the first place in Europe to receive weather systems rolling in from the Atlantic to the west. As the most westerly weather station in Ireland, the readings taken at Blacksod at the time were the best predictors of the weather across not only Ireland but also Britain and further into Europe. The weather-reporting job had become even busier over the previous week following a new decree from Dublin: they were now required to take the meteorological readings on the hour every hour,[1] rather than the usual four times per day.

At 11:00 a.m. that morning, Maureen's colleague Frances had phoned the report through to Dublin as usual. Shortly afterwards, the bell on the switchboard rang with a trunk call. Maureen answered.[2] The caller, a woman with an English accent, asked Maureen to confirm the weather report Frances had just sent through.[3] It was a surprising call: it was unusual to be asked to check the readings, and a female English voice was a rare commodity in the ranks of the civil servants who worked in the Meteorological Service.

Maureen, no doubt intrigued, requested the caller to hold on while she asked her more experienced colleague Ted Sweeney to check and repeat the readings on the instruments at the post office.[4] Ted, who was also the Blacksod Point lighthouse keeper, knew the meteorological instruments and the weather as well as anyone in the west of Ireland. The anemometer dial showed a Force 5 from the south-southwest and the barometer pressure was continuing to fall.[3] After a year of taking weather readings alongside Ted, Maureen knew these were all portents of an approaching storm.

At noon, it was Ted's turn to consult the instruments and phone the readings to Dublin. Soon after, the switchboard rang again, and again Maureen answered it. 'It was the same lady,' Maureen recounted some 70 years later. 'The lady with the English accent and she asked if we could please check and repeat the very latest weather observations we had sent from Blacksod.'[2] Maureen, perplexed at the sudden interest in the weather in this remote part of Co Mayo, checked again with Ted and confirmed the readings to the caller.

What Maureen did not know at the time was that the lives of the 150,000 troops involved in the planned D-Day invasion of Nazi-occupied Europe depended on those weather reports. After months of planning and preparation under General Dwight D. Eisenhower, a date for the Allied invasion had been selected. D-Day was planned for Monday, 5 June, when moonlight and tidal conditions were considered optimal.[2] On 3 June, with two days to go, the weather in England looked fine. But that morning's 11:00 weather report from Blacksod, sent to Dublin and onto the UK Met Office's wartime home at Dunstable, indicated a storm arriving from the Atlantic.

If the reports from Blacksod were correct then this storm would be over the English Channel on the day selected for the landings, potentially a disaster for the Allies. Postponing the invasion was risky as tidal conditions would not be suitable after the 6th. The window of opportunity was narrow. The forecasters advising Eisenhower had to be absolutely certain that the intimations of an Atlantic storm were correct, hence the phone calls to repeat the readings on Maureen's birthday. With the poor weather report from Blacksod checked and double-checked at Blacksod, Eisenhower decided to postpone the landings by 24 hours.[2] It was a leap of faith.

The following day, her modest birthday celebrations over, Maureen was back on duty. Having taken the readings at 15:00, Maureen once again plugged her headset into the post office's switchboard, and as usual selected the trunk line to the nearby town of Belmullet and asked for Dublin. Once the Belmullet and Ballina operators had put her through[1] Maureen asked the Dublin operator for 22251, the number of the Meteorological Service's head office in O'Connell Street, a number she now knew off by heart. The meteorological assistant on duty recorded the readings; Maureen released her plug to end the call.

Once again, but unbeknownst to Maureen, Ted and Frances, the assistant in the O'Connell Street office punched the readings from Blacksod and five other weather stations across Ireland up on a paper tape encoded with an American cipher and fed this tape into a teleprinter. On the other side of Irish Sea, a teleprinter in the UK Met Office tapped out the latest readings from the six stations in Ireland,[5] just as it had done each time the hourly report came in.

The 15:00 report was instantly sent to Eisenhower's team. The readings from Blacksod that Sunday showed, at last, an increase in the barometric pressure (Figure 7.2). The clearance so anxiously awaited by Eisenhower was approaching. Later that day, armed with Maureen's latest report, Eisenhower gave the order that the D-Day landings were to proceed. The largest seaborne invasion in history took place on Tuesday 6 June 1944, its success due to the arrival of calm weather from the west and the efforts of Maureen Flavin and her colleagues in Blacksod in a supposedly 'neutral' Ireland.

This sharing of weather data from Blacksod was remarkable for several reasons. The phone line that Maureen Flavin used to pass on the reports had been installed only four years previously when the Belmullet peninsula gained a telephone service at long last, primarily to serve the nearby army Look Out Post.[6] More remarkable still was that the Irish government was sharing meteorological information with the British thanks to a secret agreement for the 'mutual exchange of reports between our Meteorological Services and the United Kingdom in time of war',[7] in spite of the

Form 7441 DEPARTMENT OF INDUSTRY AND COMMERCE, METEOROLOGICAL SERVICE
STATION WEATHER RECORD

(B889).A.8746.Wt.4436—P.793.3.150Pads.1/44.A.T.&Co.,Ltd. G.9*

INSTRUMENT	TIME MARKS G.M.T.			SCREEN TEMPERATURES READ AT G.M.T.
Barograph	1300			
Thermograph				
Hygrograph				
Rain Recorder				
Anemograph				

Wet bulb muslin last changed on 3rd June 1944

Read at G.M.T.	Max. (T_xT_x)	Min. (T_nT_n)	Grass Min. (T_gT_g)	After setting of Max., Min.,* and Grass Min.: Dry Bulb	Max.	Min.	Gra. Min.	RAINFALL mm	RR	SUNSHINE Total for Day (SSS)
0700							—			
1800			—							

STATION Blacksod Pt.
Sun day 4th June 1944
Sheet No. 4

Hour G.M.T. / Observer's Initials	CLOUD Forms Low C_L	Medium C_M	High C_H	Amounts Low N_h	Total N	Height of base in feet h	WEATHER Since last report W	At time ww	VISIBILITY Land V	Sea V_s	WIND Direction DD	Speed m.p.h. F	PRESSURE mb Barometer as read / Aneroid	Attached Thermometer °A / M.S.L. Pressure (PPP)	Change of Pressure / app	TEMPERATURE °F Dry Bulb (TT)	Wet Bulb / Dry—Wet ($t_1t_1t_1$)	Humidity % By Hygrograph / Calculated	State of Sea (S) / State of ground (E)	NEPHOSCOPE Form of Cloud (C) / Direction (dd)	Time in secs. (tt) / Speed V_rV_r	Form of Cloud (C) / Direction (dd)	Time in secs. (tt) / Speed V_rV_r	REMARKS
1300	L	—	—	10	10	900	c	c	J	J	WSW	X	995.5	287	+2	54.6	50.0	X	6	—		—		
M.F.	6			8	8	3	2	02	7	7	22	7		97.0	201		4.6	70	2					
1400	Cu	—	—	9x	9x	1100	c	c	J	J	WSW	X	995.6	287	−2	53.5	49.2	X	6	—		—		
M.F.	8			7	7	4	2	02	7	7	22	7		97.1	701		4.6	70	2					
1500	L	—	—	9X	9X	1100	cPRc	cPR	J	J	WSW	X	996.1	287	+8	54.0	52.1	X	6	—		—		
M.F.	6			7	7	4	8	25	7	7	22	7		97.6	704		1.9	87	2					
1600	CL	—	—	7	7	1300	b	bc	J	J	WSW	X	996.5	288	+8	55.0	51.0	X	6	—		—		
[illegible]	9			5	5	4	x8	01	7	7	22	7		97.8	804		4.0	74	1					

Hour		III C_LC_M	$wwVhN_h$	DDFWN	PPPTT	UC_Happ	$t_1t_1t_1SV_s$	RRT_xT_xE	$T_gT_gT_nT_n$	SSS			$CddV_rV_r$	$CddV_rV_r$	Checked by	Time Sent
1300	ADmiralty	1046-	02738	22768	97055	7-201	04667					Neph.				13 01
1400	cloudy	1045-	02747	22727	97154	7-701	04667					Neph.				14 01
1500	CAPRE	1046-	25747	22787	97654	5-704	01967					Neph.				15 01
1600	Cape	10490	01745	22785	97853	7-204	04067					Neph.				16 01

FOR USE AT METEOROLOGICAL SERVICE HEADQUARTERS ONLY

1	2	3	4	5

*I certify that a legible carbon copy of this sheet has been filed at the station.
*I certify that the entries on this sheet are true copies of the corresponding entries on the copies of Form 7440 filed at the station.
(Signature) [illegible]
*Delete whichever does not apply.

Figure 7.2: The weather observations from Blacksod on 4 June 1944. The reading for 15:00, initialled 'M.F.' for Maureen Flavin, showed a pressure of 996.1mb, the first reading to show an increase of any significance and indicating that a clearance was approaching. *(Met Éireann)*

official implacable policy of neutrality. Of course, this official neutral stance was in fact an elaborate mirage and the telecommunications service was intimately involved in maintaining this deception.

When war broke out in 1939, Éamon de Valera, who had become leader in 1932, 'kept Ireland benevolently neutral for Britain'.[8] De Valera's government used the telecommunications networks to maintain this mirage of neutrality with strict censorship imposed on radio broadcasting, telephone and telegram traffic. Furthermore during 'the Emergency', as it was dubbed in Ireland, a propaganda battle was played out on the radio waves, and the telecommunications network was stretched to the limits by changed transport patterns.

While neutrality kept the country's limited armed forces out of an active role in the conflict, new defence facilities were established all around the country. The government built a network of Look Out Posts (LOPs) at strategic points around the Irish coast to report every military activity observed in the air and sea, especially submarine. These LOPs were manned around the clock by volunteer members of the Local Defence Forces with typically two men working on each shift.

Ultimately, 83 LOPs were established between 1939 and 1942. The locations chosen were often the sites of the Napoleonic War era signal stations, though, unlike in 1804, the east coast was included in the network of LOPs. In a few posts, such as Bray Head on Valentia island in Co Kerry (Figure 1.3), the original Napoleonic signal tower buildings were reused[9] but in most cases huts comprised of prefabricated concrete components were erected. Optical signalling using semaphore or other means had insufficient bandwidth and speed for a twentieth-century war so instead the telecommunications network was to be used. This posed a problem, however, as most posts were in remote locations, many miles from existing telephone lines so that at first only a few were equipped with telephones.

The initial solution was for a coastwatcher to cycle to the nearest Garda station with a phone connection to send a report. Since the nearest connected Garda station could be over 12 km from the LOP, this imposed a delay of up to an hour in the transmission of a report to headquarters[10] during which time a bomber aircraft could have crossed the country, dropped its explosives, and been well on the way home. In December 1939, an opposition senator scoffed:

> All round our coast, numbers of untrained country lads ... are being paid at the rate of 30*s.* or £2 a week, and planted like palm trees on the tops of hills overlooking the sea – scanning the horizon with the naked eye for, I presume, enemies of our neutrality.

> When they imagine they see an aeroplane or seaplane they walk five or six miles to the nearest telephone box to communicate the news to headquarters.[11]

In fact the situation was even worse than Senator MacLoughlin outlined because the telephone box was a rare commodity in the cities – and unknown in rural Ireland. His scepticism was understandable: Ireland's defence capabilities were very limited, so if an invading force was actually detected little could be done.

There was initial opposition from various Government agencies to installing telephones in all of the LOPs due to the high cost.[10] In March 1940 the Naval Attaché to the British ambassador visited LOPs and reported back to London that of the 12 stations in Co Donegal none had a telephone and it took up to 90 minutes to dispatch a message.[12] The fact that a British Naval Attaché could potter around Ireland visiting military installations was an indication of the type of neutrality being pursued.

With increased security concerns after the fall of France in June 1940, when there were real fears of an invasion from either the Axis or Allied powers, the policy shifted and the Department of Posts and Telegraphs was asked to urgently provide telephones at the posts that hitherto had no telephone lines.[13] Within months almost all LOPs were connected, the programme including, at long last, a trunk circuit from Ballina to the Belmullet peninsula that proved its value four years later.

Under the same defence programme, 24-hour service was provided at all exchanges with lines serving LOPs and telephones were installed in those remaining Garda Stations that hitherto lacked them.[14] Just two LOPs remained unconnected to the telephone network, Parkmore and Foileye, both in Co Kerry. In the latter case, reports were carried 7 km by bicycle to Kells railway station, telephoned from there to Cahersiveen railway station where they were transcribed and carried on foot to the local Garda station and thence telephoned to the Command Intelligence Officer in Cork.[10] Ironically, this telecommunications blackspot was located 15 km as the crow files from the cable station at Valentia, which continued to be a global communications hub throughout the war. The information received by the regional Command Intelligence Officers was compiled[15] and quietly forwarded to the British, just like the weather reports.

While information about the weather and German naval movements flowed freely along the telecommunications network to the British, communications by ordinary citizens along the same telecommunications network were strictly controlled. Both the Irish and British governments had their different reasons for applying controls, which meant that a telephone call between Dublin and London might be monitored by two separate censors.

The British started to listen in to telephone calls between Britain and Ireland from the outset of the war. The telephone was of particular concern since the instant communication it provided was the obvious route for the most up-to-date and thus valuable intelligence. For example, when it was proposed during the opening weeks of the war that the main British naval fleet be moved to the Clyde, Churchill warned that the Germans would soon know all about it, thanks to the Irish and the telephone:

> There are plenty of Irish traitors in the Glasgow area, telephone communication with Ireland is, I believe, unrestricted, there is a German ambassador in Dublin. I should expect that within a few hours of the arrival of these ships it would be known in Berlin.[16]

Churchill's concern about the German legation (embassy) in Dublin was shared across the British defence and intelligence community. It was widely held that once any sensitive information reached Ireland, it could easily fall into the hands of German operatives and be relayed to Germany. The legation in Dublin was able to telegram Berlin; such messages were routed through London where the British could delay them but, until they broke the cipher in 1943, they could not read them. The British also suspected, correctly as it turned out, that the German legation held a clandestine radio transmitter[17] which could be used to transmit intelligence, including that gleaned from Britain, back to the fatherland.

The British regarded the border as too porous to police effectively and thus MI5 applied censorship to communications with Northern Ireland as well. While Churchill was concerned about 'Irish traitors' deliberately passing on intelligence they had gleaned in Britain, MI5 censors were more worried about the potential security breaches of apparently innocuous calls from Britain to loved ones in Ireland such as 'by seamen telephoning to their girls saying they cannot visit them because they are going with a convoy, etc'.[18] As the war intensified, further restrictions were imposed. From June 1940 only telephone conversations in English were allowed by the British authorities and the censors had the ability to immediately interrupt any call considered suspicious.[19] The British suspended all non-official cross-channel calls for five months from 5 April 1944 ostensibly for security concerns. Though it affected Northern Ireland too, the *Belfast News-letter* described it as 'another step in Britain's plan to isolate Eire and thus prevent vital information from reaching the enemy'.[20] Of course, the British and Irish governments continued to allow vital information, such as Maureen Flavin's weather reports, to flow under the Irish Sea, unimpeded and unknown to most.

On the Irish side, censorship was applied in order to maintain the illusion of neutrality and from August 1941, all of the 31 trunk telephone circuits between Dublin and the UK, including Northern Ireland, were monitored by four censors based at Dublin Castle. Calls for official business made by certain 'listed' persons were exempt from censorship, with a secret memo informing such persons that 'the person making the call, whether from his Office or his private residence, should notify the operator that the call is one on official business and that he is a person who is entitled to exemption from censorship.'[21] The authorities evidently did not consider that a woman would need to make a call on official business.

Since about 100 censors would have been required to monitor all lines on a 24-hour basis, only random checks could be undertaken[22] – presumably the possibility of being monitored by a censor was considered sufficient to curtail what was discussed. Like their British counterparts, Irish censors could also terminate a call if banned material was discussed. This power was open to abuse with telephone conversations of journalists who were unsupportive of government sometimes being cut off apparently out of spite.[23]

The Emergency provided the government with an excuse to tighten surveillance on domestic communications too. A large number of telephones were tapped, particularly the lines of people regarded as sympathetic to Nazi or IRA causes. The telephones of all foreign diplomats and legations in Dublin were monitored and relevant transcripts passed to the Department of External Affairs.[24] Of course, surveillance was not applied evenly: the British representation in Dublin was allowed to operate private wires (leased lines in today's parlance) to Belfast and London.[22] The German legation in Dublin could not avail of a similar facility as all circuits to mainland Europe were routed via London and service over them had been suspended at the outbreak of war due to British policy. The German legation lost another possible communications channel in 1943 when, under mounting pressure from the British and Americans, the government badgered the German minister to place the legation's radio transmitter in a bank vault.[23] The bank branch selected was the Munster and Leinster (now AIB Bank) on Dame Street, where de Valera also kept his bank accounts.

Another one of those to find themselves under scrutiny due to this tightening of domestic security was J. J. Walsh, the country's first Minister for Posts and Telegraphs. He had retired from politics in 1927 but had come to the attention of the intelligence arm of the Irish army, G2, during the war for his attempts 'surreptitiously to gain favour with certain Germans by passing on … information of a military nature concerning Northern Ireland'.[23] Despite such concerns, the Minister for Justice, Gerald Boland, refused a request from G2 to tap Walsh's phone. IRA sympathisers were monitored as, in line with its tradition of 'England's difficulty being Ireland's opportunity', the IRA cooperated with Nazi Germany during the early part of war. The government's opposition to the IRA was resolute. When an IRA pirate radio station began broadcasting from the Dublin suburb of Rathmines, its location was quickly identified and raided while on air. One of the station's operators, Mayoman Jack McNeela, was sentenced to two years imprisonment for the wartime crime of 'conspiracy to usurp a function of Government'; he died in detention while on hunger strike.[22]

Rathmines was a surprising hotbed of Republican activity during the early years of the war. In October 1941, during a raid on a house in Rathmines, Gardaí found a document called 'Comprehensive Military Report on Belfast' in the handbag of an IRA courier called Helena Kelly. The chilling 24-page report produced by the IRA's Northern Command included a detailed section about Belfast's Telephone House, with descriptions of each of its six floors, two comprehensive sketch maps and details down to the level of the location of spanners that could be used to dismantle generators. It is likely that the IRA intended to pass the report to Dr Hempel of the German Legation in Dublin for onward communication to Berlin where it would assist with Nazi attacks on Belfast. While there was close cooperation between G2 and their British counterparts in MI5 it seems that the existence of this document was not shared with the British.[25] Perhaps the existence of such detailed information in Dublin, clearly intended to help the Nazis further destroy Belfast, was too embarrassing for the Irish government to admit to.

Government concern about the IRA also drove some of the controls applied to telegrams; indeed the letters 'IRA' could not be included in a message.[22] There were telegraph censors in the GPOs in Dublin and Cork and watching staff at the cable and wireless telegraphy stations in Kerry and Donegal. All telegrams in and out of the State were checked, as were many inland telegrams. Copies of all inward telegrams were retained at the GPO in Dublin. Only messages in English or Irish were permitted and messages in cipher were prohibited. A message deemed suspicious could be altered or simply not dispatched to the recipient. Alternatively, the sender or the recipient might be asked to prove their identity or explain the reason for the message. Traffic over private wires used by newspapers was not generally subject to checks but the authorities could terminate such a service if they suspected that it was misused. In any case, the newspapers' content was subject to strict censorship

A different censorship regime applied to the transatlantic telegraph stations in Co Kerry, which remained in operation throughout the war. With automated repeaters in use, there was no easy way to read the messages in transit passing through the stations and, in any case, applying censorship would have provoked the ire of the countries whose messages were being intercepted, probably leading to the stations being bypassed. A military guard and watching staff were installed to ensure that no messages were originated locally thereby evading the censors in Dublin and London.[22] In 1943 the Commercial Cable Co. asked for government consent to devote one of the cables landed at Waterville to US Office of War Information messages.[26] The government of neutral Ireland agreed and thus sensitive Allied information was able to flow between Washington and London via Co Kerry.

Despite wartime restrictions on telegraph messages, the telegram continued to be used for communications and even enjoyed a brief renaissance, temporarily reversing the steady decline that had started from about 1902 (Figure 7.3). The enforced brevity of the telegram gave rise to succinct communications, even when the sender was probably little concerned about cost. This was exploited by Winston Churchill in his late-night telegram from Chequers to Éamon de Valera in December 1941 containing

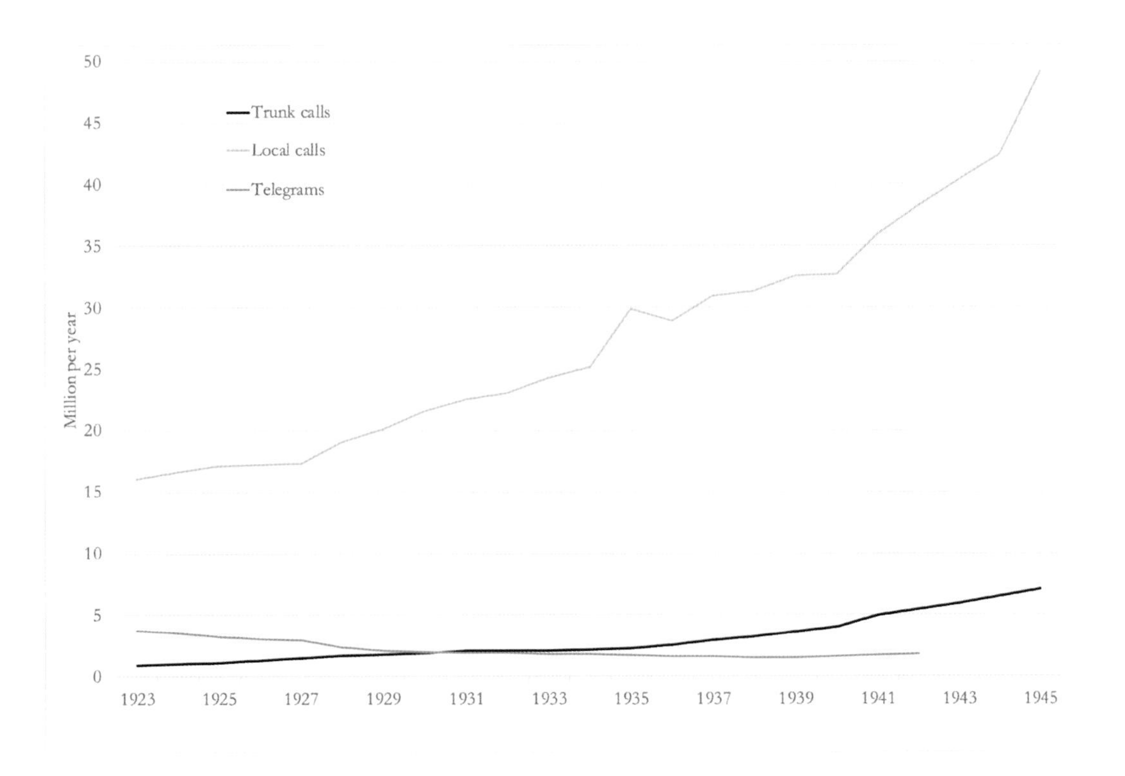

Figure 7.3: Telephone and Telegram Traffic 1922–45. Sources: Flynn, Roddy, *The Development of Universal Telephone Service in Ireland 1880 - 1993* (1998) DCU, Doctor of Arts thesis; Dáil Debates, various dates.

just 20 words: 'Now is your chance. Now or never. "A Nation once again". Am very ready to meet you at any time.'[27] To an Irish recipient, the phrase 'A Nation once again' implied a promise of Irish unity but the ambiguity was probably intentional. De Valera did not reply by telegram and did not seriously pursue the offer.

The temporary renaissance of the telegram was the result of severe restrictions on physical transport. With few native sources of fuel in Ireland, petrol for private motoring became unavailable and train services curtailed. Many lines had only one train per day and journeys were slow and erratic. During the winter of 1942 a train from Killarney to Dublin took 23 hours to make the journey.[28] The use of bicycles and horse-drawn transport surged. Similar to the effects of the Covid-19 pandemic beginning in early 2020, the telephone also substituted for travel to the extent that the number of trunk calls made in 1945 was virtually double the pre-war level.[29]

The strain these extra calls placed on an already inadequate telephone network was partly alleviated by infrastructural improvements that took place despite the wartime conditions. The ability to undertake this was thanks to some unusually prescient thinking from the Department of Posts and Telegraphs. As most telecommunications equipment was imported, in view of the deteriorating political situation in Europe in 1938, the Department had purchased an extra £100,000 (about €7.2m at 2021 prices) worth of telephones and equipment.[30] This reserve represented three years' supply of normal demand. From this the Department had to install not only extra lines at the aforementioned coastal LOPs, but also extra equipment related to censorship.

A programme to improve telecommunications with the islands off the west coast continued during the Emergency so that, by 1942, 19 islands off the west coast had been provided with telegraph service – six by cable, one by overhead line and 12 by radio. Bere Island and Arranmore gained telephone as well as telegraph communication with the mainland.[31] The larger of Aran Islands, Inishmore, had its telephone cable connection with the mainland restored after a gap of several years when only a temporary wireless telegraph service was available.[32]

The Department was also able to install a new 45-operator switchboard early in 1944 in a building on Dublin's Exchequer Street (Figure 7.4) (now the home of the Fallon and Byrne food store and restaurant) as a part of a programme to improve the trunk service.[33] Operators dealing with inland trunk calls, who had been moved in 1916 from the GPO, finally vacated their temporary and congested home of 28 years at Crown Alley. It was through this new Exchequer Street switchboard that Maureen and her colleagues phoned in their weather reports in June 1944. The Department also rustled up some more trunk carrier systems to add a little extra capacity on routes from

Figures 7.4 – 7.5: The new trunk switchboard that opened in 1944 on Dublin's Exchequer Street. Operators working here in June 1944 answered those calls from Maureen Flavin in Blacksod to the Met Service. *(Figure 7.4 (top) courtesy of the Institution of Engineering and Technology Archives)*

Dublin to Clonmel, Tralee and Limerick.[30] However, supplies of more sophisticated equipment dried up almost immediately at the onset of the war. Approval had been granted in September 1938 for the installation of an automatic exchange in Cork city[34] but due to the war it took ten years to complete.

As the war dragged on, the telecommunications service had to operate increasingly on a 'make do and mend' basis. In May 1945, just after the war had ended in Europe, the Minister Patrick J. Little explained to the Dáil that:

> owing to shortage of instruments, of underground cable and of wire for overhead circuits, it has recently been necessary to impose further restrictions on the provision of service. A list of priorities has been drawn up, preference being given to essential and emergency services, hospitals, doctors, etc., and this must be rigidly adhered to until the supply position improves.[30]

Of course, the privations suffered in Ireland were nothing compared to the devastation the war had wreaked on much of the rest of the world. At war's end Berlin had only 750 working telephone lines[35], its long-distance exchange equipment seized by the Red Army[36] and carted off to the USSR.

The last years of the war had other effects on Ireland's infrastructure, though not of the destructive kind. A massive increase in air traffic brought to an end the previously tranquil life of the aviation radio stations. These had been established by the Department of Posts and Telegraphs in 1936 at Baldonnel in Co Dublin, Ballygireen, Urlanmore and Rineanna (now Shannon Airport) in Co Clare and Foynes in Co Limerick[37] to communicate with aircraft. The life of the air traffic control staff was transformed as the Foynes flying boat station became a civil aviation hub with the number of flights soaring from to 33 in 1937 to 1,098 in 1943[37], all communicating with the ground via Morse code. A veritable Casablanca on the Shannon, in addition to passengers in transit between destinations like Lisbon, Botwood in Newfoundland and Poole in England, Foynes was also home to airline staff, US consular officials, British immigration officers and MI5 agents – all of whom were closely monitored by G2.[10] While Foynes quickly slid back into relative obscurity as land planes took over from flying boats, the aviation radio service established in 1936 was retained and expanded, continuing to play a vital role for transatlantic air traffic to the present.

Radio in its broadcast form became another battleground of the war. Radio had uses both as a navigation aid and as a propaganda weapon. At the outset of the war the UK government was concerned that the Luftwaffe could use broadcast radio signals as direction beacons for their bombing raids. To make this more difficult the BBC synchronised all of their radio transmitters onto two frequencies.

The problem for the British was that the powerful 100 kW transmitter at Athlone could also be used by the Luftwaffe to guide air raids on the UK. Pressure from the British government to synchronise Athlone with the other Radio Éireann transmitters in Dublin and Cork was initially ineffective as the Irish authorities lacked the equipment required to make the changes. To reduce the utility of the Athlone transmitter for German bombers, in September 1939 the BBC took three low-powered transmitters at Redmoss outside Aberdeen, Penmon on Anglesey and Clevedon near Bristol out of regular service. In a measure to confuse Luftwaffe aircraft, receivers picked up Radio Éireann and relayed it by landline for retransmission at the three BBC sites.[38] In October 1940, the Irish authorities acquired the equipment required to synchronise the three Radio Éireann transmitters at Dublin, Cork and Athlone thus reducing their value as navigation aids, and the BBC relays ceased. There is no record of whether British listeners missed their regular dose of hurling and céilí music after the retransmissions ended.

Irish listeners too lost their dose of hurling and céilí music, intermittently. Even with the three Radio Éireann transmitters synchronised, the authorities in neutral Ireland agreed to silence the Athlone station during Nazi bombing raids of UK cities in 1940–1.[10]

The ability of radio waves to cross borders unimpeded by censorship meant that the content of radio broadcasts had huge propaganda potential. Those BBC relays of Radio Éireann were not the only Irish voices to be heard from stations in Britain during the war. From 1941 the BBC started its *Irish Half Hour* programme, commissioning the popular Dublin comedian Jimmy O'Dea. Ostensibly intended for Irish people serving in the British armed forces, it was really intended to encourage listeners in Ireland to listen to the BBC and thus hear the British view of the war.[39] The Germans used radio more overtly, employing William Joyce (aka Lord Haw-Haw), whose father was from Ballinrobe in Co Mayo, to make the *Germany Calling* programme. Intended for both British and Irish audiences, the programme attracted a large audience of Irish listeners[39] bored with the dull offerings of Radio Éireann. The Nazi broadcaster also made a programme specifically for Ireland which included a large amount of Irish language content, on the grounds that *gaeilgeoirí* were more likely to hold anti-British views and thus be sympathetic to the German cause.[39] However, with few radios in Gaeltacht areas, the Irish language broadcasts had little impact and were dropped after two years.[39]

It was understandable that Irish listeners might want to tune in to foreign stations, as the government imposed strict censorship over all media content, including Radio Éireann broadcasts. As a neutral state Ireland could not be seen to be helping a belligerent power by revealing prospective weather conditions, so there was a complete prohibition on meteorological reports. This was enforced so strictly that sports commentators could not give any hint about the weather conditions on the pitch during a match. Of course, behind the scenes, the State continued to send meteorological reports to Britain[2] but the censorship helped maintain the illusion of neutrality. The strict censorship had more serious impacts: news of the atrocities of the Holocaust and Nazi death camps was not permitted to be printed until after the end of the war.[40]

The end of the war in Europe came 11 months after those weather reports from Blacksod were phoned in by Maureen Flavin. Peacetime brought a restoration of international cooperation, with weather reports from Blacksod now openly shared with meteorological services across the world. Maureen continued to share responsibility for taking the readings with Ted (Figure 7.6) but he was no longer just a colleague – they married in 1946.[4]

Ireland was left relatively unscathed by a war that had shattered the infrastructure of much of two continents, killed millions and redrawn borders. The *Irish Times* and the rest of post-war Europe no longer enjoyed those broadcasts of opera from the radio station at Heilsberg: the remnants of its transmitter now lay in Poland while the bombed-out opera house was now in the USSR. True, there were plenty of problems with telecommunications infrastructure in Ireland. But these were of the home-grown variety, as the next 35 years would clearly show.

Figure 7.6: Ted and Maureen Sweeney (née Flavin) after their marriage in 1946. *(Courtesy of Vincent Sweeney)*

CHAPTER EIGHT

A PAINFUL CASE

One day in the late summer of 1972 a man arrived at my father's dental practice in central Dublin clutching his jaw in agony. The receptionist, Eithne, recognised him from a previous appointment. She sat him down in the waiting room and headed into the surgery. My dad put down the drill; Eithne gave the patient's name and asked if he could be fitted in. My dad agreed to see him and asked Eithne to bring in the patient's dental chart.

This was not an unusual event. Toothache can be excruciating and my father would always try to slot in anyone suffering with pain despite his schedule. Being a dentist was not a strictly nine-to-five occupation and my dad sometimes had to contact my mother to let her know that he would be delayed owing to an emergency. However, at that time he had no easy means to do this. We had recently moved house and were on the waiting list for a phone – along with 30,000 other Irish homes and businesses.

So, a little while later, Eithne ushered the poor man into the surgery as my dad studied his dental chart. One detail jumped off the oblong pink and white card: 'Employer: P&T'. My dad suddenly remembered that, on a previous visit, the patient, now prone and writhing in the dental chair, had told him he was a telephone technician with the Department of Posts and Telegraphs.

'Open wide', my father said as he picked up the mirror to identify the offending tooth. Casually, he asked, 'Are you still with P&T?'

'Ye-e-s', came the strangulated reply.

My dad was a firm believer in the personal approach – if a tap was leaking, why look up the Golden Pages when a patient who was a plumber was due for a check-up next month? Here was an opportunity to fix a problem bigger than a leaky tap. Before he explained to the P&T man about how he would treat his toothache, my dad asked,

'Any chance you could push us up the waiting list for a phone?' Even with his aching jaw, the poor P&T man paused before consenting: yes, he would try to get a phone installed in our house so long as my father cured his damned toothache. The abscess was drained, a prescription was written and the toothache was alleviated. On leaving the surgery, the relieved technician said, only half-jokingly, 'I hope I don't get a ministerial sanction.'

The following week, a marigold and white P&T van pulled up outside our house. Within hours, a shiny black model 746 phone was sitting resplendent on our hall table, with the number 384596 carefully written in black ink in the centre of the dial. There was just one small problem – the phone did not work. Somewhere between our house and Blanchardstown exchange there was a break in the cable. For weeks, every morning I would race down the stairs in my school uniform and lift the receiver, desperately hoping that somehow a team of nocturnal technicians had fixed the cable and connected us to the outside world. Quite who six-year-old-me was going to phone is a mystery, but this did not stop my longing to hear the comforting purr of a dial tone emanate from the shiny black device.

Eventually one day, three months later, I came home from school to be greeted by my mother with joyous news. Our shiny black phone had rung! We were thrilled: many people waited three years, not three months, for a working phone. 1972 was a painful year to look for a new phone line in Ireland. Indeed the whole period from 1945 to 1980 was an excruciating combination of obsolete equipment, overloaded lines and outdated attitudes leading to anguish, heartache and frustration for all involved.

While Ireland emerged relatively unscathed from World War II, the country's telecommunications network was far from healthy. It contained antiquated equipment – recycled, cobbled together and kept working for the want of anything else. In Cork city the exchange building at 91B South Mall that had been considered unfit in 1914[1] was still in use 35 years later.[2] The switchboard it housed was the same one, patched-up and extended, that had been smashed by the axes of anti-Treaty forces in 1922.

A proposal by the cabinet in 1945 to modernise the telephone network had quickly run to ground. The Department of Posts and Telegraphs poured cold water over the bid to expand the network to 100,000 lines. Its opinion was that a total of 80,000 telephones would represent saturation and could only be reached if:

> everyone slightly above the artisan class had a residence telephone and if fairly small shops as well as every business supplying labour were to have a business line. To effect such development there would have to be radical change in the outlook of the

> population towards the telephone which most of them at present, rightly or wrongly, regard as a luxury.[3]

The plan actually adopted was so watered down that the *Irish Independent* acerbically commented:

> The Ten-Year Plan certainly does not err on the side of rashness ... the prospects of seeing telephones in farm-houses need hardly be considered this side of the year 2000 A.D.[4]

But even these limited objectives were not met. By 1951 delivery targets were being missed due to staff shortages and increased public demand.[5]

This miserly attitude to extending the phone system contrasted with the progressive approach to the electricity network. The ESB initially faced the same problems expanding its network beyond the State's cities and towns because the high capital cost of connecting isolated rural dwellings would not be met by the revenue generated. In 1943 the government agreed to subsidise a scheme to electrify rural households, allowing for a rural electrification programme that commenced three years later. The programme aimed to connect 280,000 new consumers over ten years[6] – a far more ambitious target than that set for telephones. Innovative marketing techniques were deployed, with canvassers calling door-to-door and community organisations like the GAA and Irish Countrywomen's Association called into action to persuade rural dwellers to sign up for electricity.[6] In the 30 years before the programme's completion in 1976, over 420,000 customers in rural Ireland had been connected,[6] transforming life in the countryside.

The success of rural electrification in the 1940s and 1950s was particularly remarkable in view of the dismal economic backdrop of the era. The Irish economy stagnated while the rest of Western Europe enjoyed an unprecedented post-war boom. Emigration increased further with the keys of council houses being handed back as families fled unemployment and misery.[7] Public expenditure was limited by an insistence on balancing the books while economic policy continued to promote agriculture and protectionism. The perception persisted that there was little need for telecommunications in an Ireland based on agricultural self-sufficiency.

A new government from 1948 did little to improve the State's infrastructure, cancelling a transatlantic air service after the aircraft had been delivered and axing a short-wave radio service after the 100 kW transmitter had been installed in Athlone by the Marconi company.[8] Political scientist Tom Garvin suggested that:

> In the eyes of Irish democracy and its servants, transatlantic airlines, telephones and motor cars were all the toys of the rich rather than the essential instruments of economic and social advance: the machineries of the future were upper-class and therefore to be distrusted.[9]

In this zero-sum culture the perception was that the national cake was of a fixed size and the only way to get a larger slice was by reducing the size of the other slices. Thus investment in telecommunications could only take place at the expense of investment in other, more visible, realms such as housing. A different recipe, like investing in the infrastructure that could allow the economic cake to rise, was inconceivable. As the secretary of the Department of Posts of Telegraphs during this period, León Ó Broin, concluded in his 1986 autobiography: 'The problem was not of our making. There were no votes in telephones.'[8]

Even when the money was free, or at least cheap, the state seemed unable to envisage how it might be spent to develop modern infrastructure. Such was this case with Marshall Aid, which provided the Irish government with $149m in soft loans and grant-aid.[10] In keeping with the philosophy that had pervaded Ireland since independence, the majority of the aid was spent on agriculture, by assisting schemes to drain and reclaim land for farming, though there was no obvious or immediate increase in agricultural output[11] as a result.

In keeping with the prevailing culture, the Department of Posts and Telegraphs continued the same policies of short-termism and parsimony after the war. Sometimes it was case of penny-wise and pound-foolish. In August 1941 telephone lines in and around Ballsbridge in Dublin were transferred to Merrion automatic exchange over new copper cables, allowing the Ballsbridge manual switchboard to be closed and thus saving staff costs. Within seven years, with capacity of the latter filling up and increases in the price of copper, a new exchange had to be opened in Ballsbridge and the lines rerouted again.[12]

That new exchange in Ballsbridge was followed in 1949 by an automatic exchange for Cork city,[2] promised since 1938. Some smaller towns were equipped with Rural Automatic Exchanges (RAXs) similar to the Unit Automatic Exchanges (UAXs) used in the UK since the 1920s.[13] In keeping with the general miserly policy, one of the first of these exchanges to be installed, at Charleville Co Cork, used some components recycled from the UK.[14]

Installation of these RAXs was hampered not only by the limitations of the P&T budget but also by problems in other parts of the public sector, such as the Office of

Public Works (OPW), which was responsible for the buildings to house telecommunications infrastructure. While P&T would have liked to use a standardised building, the OPW continued to design each exchange building individually with the Department describing the building process as 'painfully slow'.[15] Eventually in 1962 the OPW produced a design for a Standard Rural Automatic Exchange building.[16] With its bare concrete walls and clerestory roof, it slowly became a familiar sight across Ireland.

The Department of Posts and Telegraphs also fell victim to projects instigated by other departments for their own political purposes. Towards the end of the Emergency unemployment relief works were undertaken to provide 70 miles (110 km) of concrete pipe ducts for trunk telephone cables, though the Department's internal reports of the time make clear that such a scheme was not required for engineering reasons.[17] A few years later, when the Department started to develop a trunk network of underground cables, it did not bother using this ducting but instead contracted the building contractor Sisk[5] to dig new trenches for the cables. Of much greater value to the Department was the 64 km network of ducting under the streets of Dublin formerly used to house power cables for the tramway system that it had acquired in the late 1940s to use for telephone cables.[16]

Those underground trunk cables forming an initial backbone serving Dublin, Athlone, Limerick, Cork and Waterford had a capacity of 600 channels, a vast increase over the existing routes. This huge jump was due to the use of coaxial cable. 'Coax' is a multi-layered cable with a solid or stranded copper wire at its core, surrounded by an insulator and wrapped with a metallic shield of foil or braided wire, with a plastic jacket on the outside (Figure 8.1). The dimensions of the cable are controlled to give a precise, constant conductor spacing. As a result, a single coax can carry multiple signals at different frequencies with low losses.

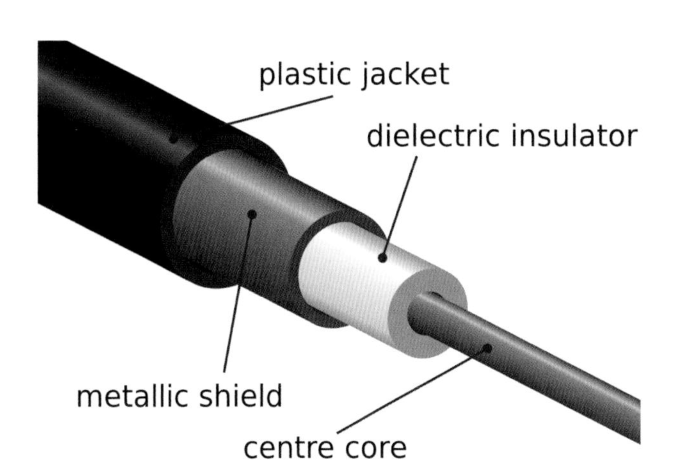

Figure 8.1: A cross-section of a coaxial cable. *(Tkgd2007, CC BY 3.0, via Wikimedia Commons)*

An early use of coaxial cable technology was for the undersea cables laid between Howth and Nevin in Wales in 1937–8 which provided 16 cross-channel circuits, each telephone conversation being effectively transmitted on a separate frequency. As technology improved the bandwidth offered by a coaxial cable increased further and, as we will see later in this story, coax is still widely used to carry not just telephone calls but also television pictures and broadband.

As well as making a start on improving the capacity of the network, the Department managed to introduce some new services, though sometimes the motive was partly one-upmanship. In September 1946 Dublin gained a special number to contact the emergency services. The number selected, 999, was the same as that already in use in many parts of the UK. There, an emergency number had been introduced in London after a fire disaster in 1935 in which five women died after calls from neighbours to the operator seeking the fire brigade went unanswered, swamped by other calls.

The UK Post Office had selected 999 as it was felt that repeating digits would be easier to dial, especially in the dark or thick smoke. 111 was rejected because it could be triggered by overhead lines rubbing together in the wind. 000 could not be used as 0 was then used as the number for the regular operator, while the remaining options, from 222 to 888, were already in use as the initial digits of regular phone numbers. The clincher for the UK Post Office in favour of 999 was that public telephones were already configured to allow 0 to be dialled without inserting coins. Only a simple change to this mechanism was required to also allow 9, or more specifically 999, to be dialled for free. Ironically the service was introduced in Dublin before the city's public phones had been so modified.[18] The reason may have been another case of the Department rushing to get ahead of the UK Post Office which was just about to extend the 999 service to Belfast.[19]

The 999 service was a clear success with the *Irish Independent* describing it in glowing terms in 1951:

> ... the first shot is the dialling of '999' on a telephone. The telephoned report is flashed from the 24-hour police shortwave radio system and intercepted by the patrol cars ... To-day this alarm system is one of the biggest deterrents to would-be criminals in the metropolitan area.[20]

Initially confined to Dublin, the 999 service was extended to most cities and towns with an automatic exchange. By 1962, telephone users from Charleville Co Cork to Carrickmacross Co Monaghan could dial 999 to summon help in an emergency. Whatever the initial motive, it was an enlightened move by P&T; by contrast the US did not introduce their equivalent 911 service until 1968. In 1991 the EU adopted the number 112 as an emergency number for use in all member states; by 1998 Ireland had introduced this in parallel with 999. In 2016 an average of 4,400 calls each day were received from operators handling calls to 112 and 999[21] – quite an increase from the 20 calls a day reported by the *Independent* in 1951. About 9 per cent of the calls were classified in 2016 as 'noisy', caused mainly by overhead lines pulsing the digits 112 inadvertently as they touched one another in the wind.[21] The decision all those decades ago not to use 111 was a wise one: the EU's decision to use 112 perhaps less so.

The 999 service reached Limerick in 1957 with the opening of the city's automatic exchange. This was the first crossbar exchange on the public network,[2] marking a radical technological shift by P&T away from the step-by-step technology developed by Almon Strowger in 1878. Exchanges based on the crossbar principle had fewer moving parts than their step-by-step predecessors, thus increasing their reliability and reducing maintenance. Another efficiency was gained from their use of the Common Control principle which meant that the control equipment used to select the dialled number was only required for the set-up of the call. Just like a human switchboard operator, once the call was established, the control equipment then became available to connect the next call.

The change in technology also marked the end of the dominance of the British firm STC as a supplier. The Limerick exchange was supplied by L. M. Ericsson, beginning a long relationship between the Swedish firm and Ireland. For smaller locations P&T also bought equipment from German firm T&N and French-made Pentaconta exchanges from CGCT.[2]

Initially automatic exchanges allowed users to dial other customers connected to the same exchange, and sometimes those in adjoining areas, but anything further afield still required the assistance of an operator. Efforts to overcome this started in the 1920s so that by the end of the 1930s it was possible to dial direct between many cities in Switzerland, Belgium and the Netherlands.[22] Development continued apace after the war: by 1951 residents of the New Jersey town of Englewood could reach 13 other areas across the USA including San Francisco, 4,130 km away. Meanwhile in Ireland a call to somewhere a few miles away continued to require an operator. The pain involved in this led Fine Gael TD, James Dillon of Monaghan, to tell the Dáil in 1953 that 'I would sooner get a tooth filled in this country than make a trunk call'.[23]

Finally, on 19 March 1957, Athlone got a taste of this shiny future as it became the first place in Ireland to be equipped for Subscriber Trunk Dialling (STD), which allowed customers to dial direct to Cork, Dublin and Waterford.[24] The system in Athlone was introduced with little media coverage. A short article in the *Westmeath Independent* instructed callers to use the prefix 7 to reach Dublin numbers, noting that 'Athlone is the first place in Ireland or Britain to have an automatic system for long distance trunk calls'.[24] The lack of fanfare and use of the code 7 rather than the now-familiar 01 is curious and may indicate a rush to get the system into operation. Perhaps, like the 1927 opening of Dublin's first automatic exchange five months ahead of London, there was a desire to beat the UK, where STD went into operation 18 months later.

The Department was not just taking its cues from the UK, however. In 1957–8 the United Nations Technical Assistance Programme provided fellowships to allow two P&T engineers to travel abroad for training in telephone switching, with Dr Henry Wroe, staff engineer in the Switching and Signalling Section, heading to the Netherlands.[25] As part of the same programme a Dutch telecommunications engineer, G. J. Kamerbeek, came to work with P&T.[26]

Exploiting Dutch experience made sense as they had devised a national numbering plan some 20 years earlier that allocated each area a code commencing with a zero. With this pedigree it was no surprise that the numbering plan devised was similar to that in place in the Netherlands, with the second digit of the code signifying the region (e.g. 2 for the south, 9 for the west) and the subsequent digits identifying the particular area.

However, it also borrowed a concept initially used in North America. On rotary dial phones, smaller numbers, such as 2, are dialled more rapidly than larger numbers, such as 9 where the dial has to rotate much further. Therefore, codes with lower numbers were assigned to more populated areas: hence Cork city was assigned the code 021 and the Aran Islands were allocated 099. Dublin, being the most populous area, was assigned the only two-digit area code, 01.

An unusual feature of the scheme was that it allocated codes for direct dialling to Britain and Northern Ireland within the national numbering plan. Thus for many years, the prefix for London was 031, tucked in between Kanturk Co Cork on 029 and Drogheda on 041. The national numbering plan devised back in 1958 stood the test of time and survives to this day with relatively few changes (Figure 8.2), though in 1991 the pseudo-national codes for Britain were retired so that calls to Britain had to be dialled using the international prefix.

By December 1958, STD had been extended to Cork,[27] using the now-familiar area codes. The rollout of STD also heralded a new charging system. In an automatic exchange, each subscriber had a meter that advanced once each time a dialled call was answered. Before STD, trunk calls were connected by an operator who looked up an index file unique to that exchange (Figure 8.3) to work out how to route the call and what rate to charge. The operator filled in the details of the calling and called number and the charging code on a paper ticket and, when the lights indicated that the call had ended, added the duration.

Figure 8.2: The current system of area codes, illustrated above, has been modified relatively little since its introduction in 1958. *(Comreg)*

	EXCHANGE	STD CODE	CH GP	C I	ROUTING
1272	Cloughmills	026 563	BQ/BE		Coleraine.63 NO
1273	Coachford		CFD		Cork. 11393
1274	Coagh	064 87	CK/BE		Magherafelt.7 NO
1275	Coalisland	086 87	DG/BE		Portadown. 7 NO
1277	Coleraine	026 5	CN/BE		See Section 1
1278	Comber	024 7	NA/BE		Belfast. 141 NO
1279	Cong**		BRE		Ballinrobe + X
1281	Cookstown	064 87	CK/BE		Magherafelt.7 NO
1282	Coolbawn (Tipp)**		NH		Limerick. 7766
1284	Cooraclare		KR		Limerick.1121+ X
1285	Cootehill		CTL		Dundalk. 16
1286	Cootehall		BY		CKN + X
[illegible]9	Cork	021	CK		See Section 1
1288	Cornamona **		BRE		Castlebar, 951+ X
1289	Creggagh**		ETN		Limerick.1129+ X
1290	Craigavon	076 2	P/BE		Portadown. NO
	CRAIGAVON				11/83

Figure 8.3: An index file from Cahersiveen exchange showing the charge groups and routing for operators to use.

For STD a completely different process had to be used. Instead of operators filling in paper tickets, the subscriber's meter (Figure 8.4) simply advanced when the call was answered and then periodically as the call continued, for example every 30 seconds on a call from Dublin to Drogheda or every 10 seconds on a call from Dublin to Cork. When STD was rolled out in some countries, including the UK, local calls became timed, but in Ireland they remained at a set charge, regardless of duration. At a press conference in June 1959 the Department's spokesperson explained that the new charging structure formed part of a plan to fully automate the whole system within 10 years.[28] He was being a little optimistic as, in the end, it took 23 years.

Figure 8.4: Subscribers' meters at Crown Alley exchange around the early 1950s.

This development of direct dialling was a bright light in a bleak landscape. In 1958, of 22 European countries, Ireland was ranked 17th in terms of the number of telephone lines per head, behind Spain, Czechoslovakia and East Germany (Figure 8.5). Only 76 per cent of those lucky enough to have a phone in Ireland were connected to an automatic exchange, compared to 99.9 per cent of those in Switzerland. Even in East Germany only seven per cent of lines remained on manual switchboards.[29]

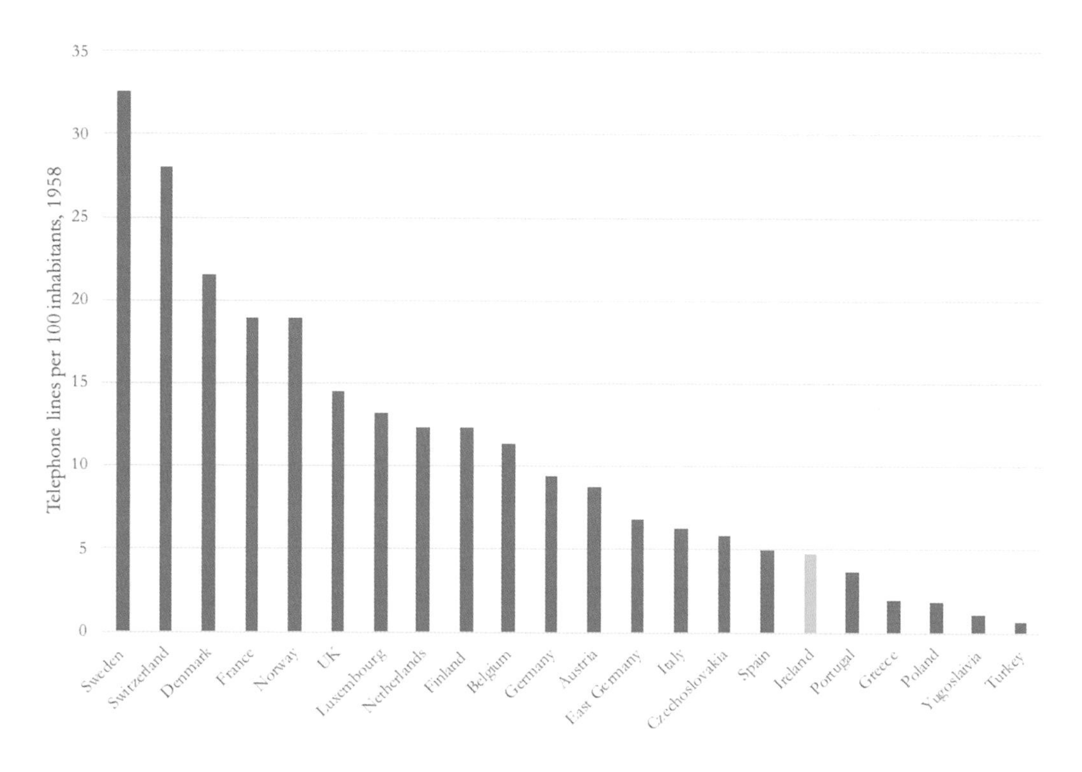

Figure 8.5: Telephone lines per 100 inhabitants, 1958. Of the 22 European countries included, Ireland was 17th. Source: International Telecommunication Union, *General Telephone Statistics* (Geneva: 1958).

For years, the underdevelopment of the country's telecommunications network had been largely a reflection of country's economic backwardness. But then the latter changed. The first *Programme for Economic Expansion*, largely attributed to T. K. Whitaker, Secretary of the Department of Finance, marked a turning point in Irish economic policy. The government pivoted from its previous policies of agriculture and protectionism towards industrial development and foreign investment.[30] In another radical step for an Irish government the programme actually acknowledged that 'a highly developed telephone system is an essential adjunct to competitive production and distribution and provides, in addition, a desirable social amenity without any net charge on the Exchequer.'[31]

The *Programme for Economic Expansion* had a dramatic effect on the economy. Living standards rose by 50 per cent between 1960 and 1970,[30] powered by an influx of foreign investment. Emigration slackened and the population started to increase for the first time since the Famine. The fine words about the telephone system, however, proved to be mere words. Two years after publication of the *Programme for Economic Expansion*, capital expenditure on the telephone network in 1960–1 was a paltry £2m, which, allowing for inflation, was the same as it had been in 1950.[32] The country's increasingly globalised, booming economy further increased the pressure on its ailing telephone system. The pain became ever more excruciating.

The pain was probably felt worst by those waiting for a phone to be installed. A waiting list had started to develop towards the end of the war, when supplies were in short supply, with 4,000 applicants looking for a phone by 1946. As the list grew, a priority scheme was devised giving precedence to categories such as doctors and clergymen.[33]

There followed a decades-long pattern where successive Ministers for Posts and Telegraphs would attribute the growth of the waiting list to the large number of applications. Sometimes the tone suggested that it was the public's fault, such as when Minister Joseph Brennan complained that 'the number of applicants each year seems to increase and this rate of progress makes the problem more and more difficult to catch up with.'[34] Warming to his theme, the Minister continued:

> once new telephones are installed in private houses, all the friends and neighbours of those new subscribers want one too. When a telephone is installed in a rural area in a private house, nearly invariably there are five or six new applicants from the same area.[34]

If only the public would stop looking for more phones, the Department would be able to keep up with demand!

Indeed, the response of P&T to growing demand was not to welcome it but to discourage it. By 1965 new applicants had to pay a connection fee of £10 and the initial term of the rental agreement (Figure 8.6) upfront. The initial term was normally a year's rental but could be as long as five years for those subscribers located some distance from the nearest exchange,[32] further fuelling the perception that the government did not care about the more remote parts of the country. The Department admitted that the requirement for upfront payment was 'designed to contain the enormous latent demand within manageable limits'.[35]

Cuir aon fhreagra chuig:—
Address any reply to:—
AN CEANN-MHÁISTIR POIST
The Head Postmaster
Do Thagairt } Your Ref. } ....
Ár dTagairt } Our Ref. } Paul Mulvey

AN ROINN POIST AGUS TELEGRAFA
Department of Posts and Telegraphs
OIFIG AN PHOIST
Post Office
....
..............5/5/..........19 82

A Chara

We will be arranging shortly to provide you with telephone service. You are invited to call to the Post Office at .... to sign the contract and make the necessary payment. Office hours are 9.30 a.m. to 5.00 p.m., lunch hour excepted, Monday to Friday.

The following amounts will be payable in advance:-

Rental for the initial term of the agreement (24 months) .... £ 201.84
Connection charge .... £ 180

TOTAL £381.84

Kindly bring this letter with you.

Should you be unable to call within fourteen days please advise me if you still propose to rent the telephone otherwise it will be presumed that you do not require service and your application will be regarded as having been cancelled.

Telephone service is usually provided within about two to three months of the agreement being signed.

Mise le meas

Ceann Mhaistir Poist

CM 58

Figure 8.6: A letter accepting a telephone application in 1982. The amount demanded upfront, €484.84, represented three weeks average industrial wage of the time, but there was no commitment about delivery dates. *(Courtesy of Paul Mulvey)*

As the numbers on the waiting list grew, the waiting time increased. In 1964, of the 13,198 applicants, 1,237 had been waiting for over two years, 379 for over three years, while 29 applicants had been waiting for over four years.[32] The waiting list was also unevenly spread. Applications from premises remote from an exchange were left on the long finger due to the high capital costs involved in installation. From the late 1960s, problems were also acute in the burgeoning Dublin suburbs, with the fast-growing suburb of Tallaght feeling especially neglected. Tallaght alone accounted for three per cent of the entire waiting list in 1973,[32] while not a single one of the 1,000 houses in the Ailesbury estate had a phone in 1980,[36] three years after their construction.

Hours of Dáil time were spent by the Minister answering umpteen questions from TDs, all along the lines of when a phone would be installed in the home of Mrs Murphy in Ballygobackwards. The answer was invariably a non-committal aspiration to a date sometime in the future.

Waiting for a telephone was not a problem confined to residents of the Republic. In 1968 there were 2,000 waiting for a telephone in Northern Ireland. In Communist-era Poland, the average waiting time for a residential line was 13 years.[37] What was exceptional in Ireland, however, was that waiting lists spiralled during the 1970s while those in most other West European countries diminished.[38] In 1980, in spite of the best efforts of the Department to deter applicants, the waiting list reached a record 94,000,[32] representing 22 per cent of the installed lines.

The frustration felt by citizens waiting for a phone drove many of them to try desperate measures. In 1978 a Miss Kay Tracy from Garrafrauns in Tuam Co Galway wrote to the President to plead for his assistance in getting a phone line installed, as her 68-year-old mother was in poor health:

> If she had the phone it would be a great help for her. I am her only daughter and cannot leave her in the house alone. I hope that you will help to get the phone for us. We will be very greatful [*sic*] to you. Again I am very sorry to bother you.[39]

With continued problems delivering private phones, a public telephone was the only connection with the telecommunications network for many people. By the mid-1960s most post offices contained a public phone for use during post office hours but P&T policy was that kiosks could only be provided where they would make money. Complaints abounded from both small villages and large suburban housing estates clamouring for a phone box to be installed, eventually leading these financial rules to be relaxed.[32]

This did not eliminate the pain, however. P&T continued to use the 1920s-designed A/B public phone long after it had been replaced in its native UK by more advanced models. The A/B phone permitted only local calls to be dialled direct, adding to the pressure on operators. It was also designed for more innocent times and fell victim to the increasing vandalism that emerged from the late 1960s. In 1975 P&T's own figures showed that 23 per cent of the 610 telephone kiosks in Dublin were out of order.[40] A popular guide to Dublin published around the same time had a more jaundiced, or perhaps more accurate, view:

> Out of ten attempts to call from a public phone box, you are likely to find four of the booths smashed by vandals, three simply out of order, one where the coinbox has not been emptied ... With luck on your side your tenth attempt may be successful, provided of course that the other person's phone is not out of order as well.[41]

For those lucky enough to have their own phone, there was still suffering. Customers connected to smaller manual exchanges still lacked 24-hour service – indeed, on Sundays they had just 3.5 hours in which to make calls. Most countries had solved this problem by eliminating manual exchanges – but not Ireland. P&T thus had to resort to improvisation to provide 24-hour service to rural customers on small switchboards. The process adopted was that the local postmistress or postmaster, just before they clocked off, connected phones on their exchange to the trunk lines to the nearest main exchange that was staffed around the clock. For example, in the Co Mayo village of Tourmakeady, the lines were switched through to Ballinrobe at 8 p.m. each night. Since there were far more phones in Tourmakeady than trunk lines to Ballinrobe, each trunk line was connected to perhaps four or five phones, with each number assigned its own ringing signal. The system persisted into the 1980s; the priest in Tourmakeady at that time, Father Joe Kearney, remembers that he answered the phone at night only when it gave three rings followed by a pause: 'two rings was for the doctor and five rings was for the neighbour.'[42] Everyone sharing the same trunk line could hear everyone else's conversation. Conducting affairs, whether of a business or extra-marital nature, over the telephone was fraught with danger.

Even when the local exchange was in operation, and each phone had its own dedicated line and number, it was widely assumed that the local postmistress or postmaster was eavesdropping. Listening in on calls was considered a serious misdemeanour by the Department and was probably far less common than imagined. Whether the fears of eavesdropping were valid or not, they led to such levels of suspicion that priests in the west conversed over the phone in Latin in order to thwart any unwanted listeners.[42]

Of course, the personal service offered by a manual exchange could often be a useful one. In 1974, as Fermoy prepared to go 'auto', a writer in the *Cork Examiner* reminisced about the personal service offered to her by a telephone operator in Fermoy when she went into labour early:

> I lifted the receiver and asked for the doctor's number, only to be informed in a concerned tone of voice: 'You won't get her at home, for her car passed here going out on a call only five minutes ago. But I'll tell you now, there's no need to worry,' the blessedly reassuring masculine voice continued. 'I've an idea where she's going, for I put that call through, and I'll call that number in a few minutes when she's had time to get there; and then you can talk to her. Will I call the hospital number for you meantime? Just sit quietly on the chair by the telephone, and don't worry everything will be fine.' And so it was.[43]

Michelle Ward, who grew up in Glenties, Co Donegal, also had reason to be grateful to a telephone operator:

> One day, when I was about five, Dad was outside and heard the phone ringing. When he came in he asked who answered the phone. I had. When he asked who it was, I replied 'Somebody died and you have to tell somebody.' He did manage to figure out who it was by ringing the exchange and asking Nancy if she could remember who she had just connected to us![44]

Postmistress Nancy McLoone and her two assistants in Glenties[45] could provide a more personal service than was possible at the larger exchanges, through sometimes this anonymity could be useful. For example in the late 1950s, before becoming famous as a singer, Ronnie Drew worked as a night telephone operator in Dublin. He recalled one night when:

> ... a posh lady came on and demanded to be put through immediately to a number in England. I thought I'd let her cool her heels and I told her there would be a delay of an hour and twenty minutes. She became very impatient and I more or less told her what she could do with herself. She then said, 'Do you know who you are talking to?' I said I didn't. She said she was the wife of the Minister of Posts and Telegraphs and she would have me sacked in the morning. I asked her did she know who she was talking to. She said no. I said, 'Thanks be to Jaysus' and pulled out the cord.[46]

In any case the personal touch offered by a telephone operator provided merely a salve, instead of the surgery required to treat the chronic pain. For decades, the letters columns of newspapers and chambers of county councils up and down the country were filled with stories about the difficulties in making a call even over the shortest of distances. At a meeting of Leitrim County Council, Councillor Dick Ellis described waiting one hour and ten minutes to get through from Ballinamore to Drumshanbo, a distance of 16 km as the crow flies. 'I would be there and back on a bicycle',[47] he declared. Things were no better on the east coast: it took up to five hours to get a call from Butlin's holiday camp at Mosney to Dublin,[48] 26 miles (42 km) away, even slower than Councillor Dick Ellis on his bicycle. In Dublin, callers complained of waiting for up to an hour just to get an answer from the operator.[49] An OECD report published in 1963 noted that the telephone service was 'undergoing severe stress in relation to economic expansion and it is barely meeting demand'.[50]

Slowly, this particular pain was eased as capacity on the main routes was gradually increased. The midlands town of Athlone was involved in another first, with the inauguration of the State's first microwave telephone link in 1961 to connect the town with Galway (Figure 8.7). This microwave system was an updated version of the first radio system used between Belfast and Scotland in 1934 but using a far higher frequency of 7.4 GHz and providing a much greater capacity of 240 calls.[51] The microwave network was quickly expanded including a Dublin–Arklow link in 1962. With the Wicklow mountains in the way, the signal from the 4.5 m diameter dish atop Andrew Street exchange[52] was bounced off a reflector on the Hill of Howth, down the coast to Wicklow town[53] and then to Arklow.

Over the following years, a mast with microwave dishes became a familiar site in towns across the state as the microwave network was expanded to become the backbone of the telecommunications network. The tall mast erected on top of Dame Court exchange in February 1968[12] became a feature of central Dublin and remains a critical element in the infrastructure of various mobile phone networks, having long outlasted the exchange equipment in the building beneath it.

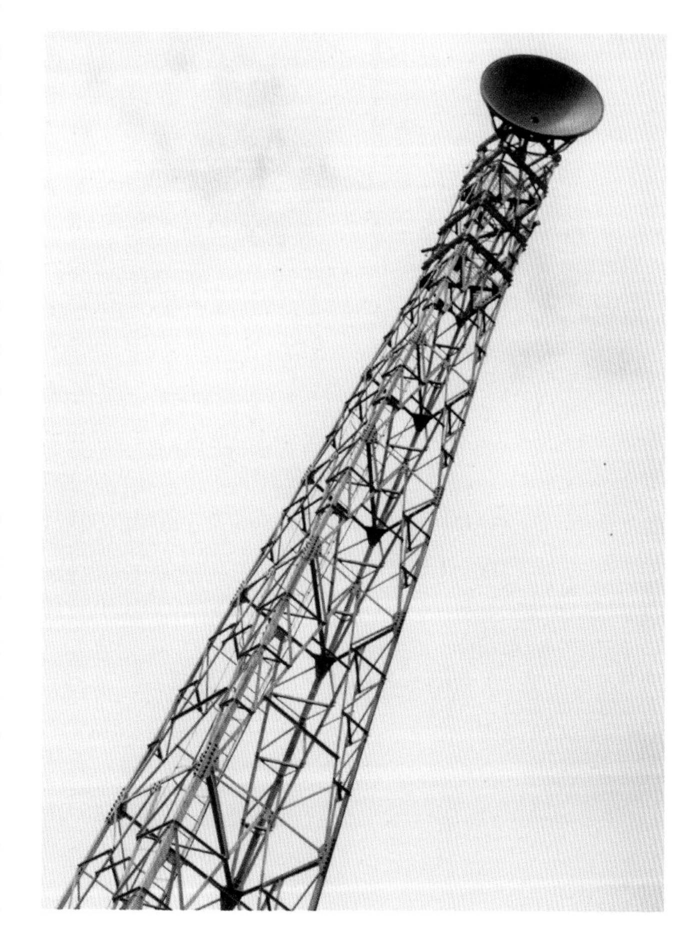

Figure 8.7: Microwave tower at Galway, photographed in 1960 shortly before it went into service. *(Courtesy of the Institution of Engineering and Technology Archives)*

The limitations of the telephone service had a positive effect, however, on another P&T service. This was the telex system, which was launched in Ireland in 1956. Telex was the child of the global telegraph network which, since the 1920s, had made increasing use of teleprinters, initially to transmit telegrams sent by the public and to send private messages, such as news from press agencies to newspaper offices. In the 1930s several countries created a switched network, where one teleprinter could be connected to any other on the network through an exchange, a network that became known as telex. Similar, private networks were created by the aviation industry for use in reservations and air traffic control.

The initial telex network in Ireland had 24 users[54] who, in a neat coincidence, could communicate with 24 countries. It was aimed at large companies conducting international business, reflected in the annual charge of £160 a year before a single message had been sent.[55] At such prices, the *Longford Leader* considered that:

> until the initial costs come down a good deal, as they inevitably will, a telex will not be seen in private residence – city or country. So no matter how unsatisfactory the telephone service is, the ordinary man will have to depend on the telephone or telegram for some time to come.[55]

Despite the high rental charge, telex proved popular with many business users offering a more reliable and, from 1964[56], fully automated service. It was particularly favoured by the travel and banking sectors, with the latter taking advantage of an inbuilt security feature know as answerback. At the beginning and end of the call, the recipient machine would automatically send a response that included its number, thus verifying that the message had been sent to the intended recipient in its entirety.

By 1973 overseas firms employed almost one-third of Irish manufacturing workers,[57] providing another impetus to use telex. It was cheaper for international messages and, since a message could be received even when the telex machine was unattended, there were no problems with different time-zones. To serve this growing need, in 1974 a new Stored Program Control (SPC) exchange built by Hasler was opened, the first computerised public telex exchange in Europe.[3] While the number of telephones per capita in Ireland consistently lagged below global averages, by 1981 the ratio of telex lines per capita was tenth in the world[58] and a new 15,000 line capacity exchange from Fujitsu was switched on at Mill Street[58] in Dublin's Liberties. The high point of the telex service was 1988 when 7,800 machines were in use. Over the following years the numbers dwindled as telex customers migrated to fax, email and other data services and in 2002 the Irish telex service ceased operating.[59]

The telegraph service also made increasing use of teleprinters with the last message in Morse code tapped out from Limerick Junction in 1958.[2] As the telephone network continued to spread, the telegram became largely confined to purposes such as wedding greetings

(Figure 8.8). With speed of delivery for such messages less critical, in 1987 telegrams were replaced by Telemessages where the message was dispatched to the recipient by regular post.[60] Telemessages survived a little longer than telexes, but were scrapped in 2008.[61]

B. or C. | Taille le n-íoc | **Telegrafa an Phoist** | No. C 221

Recd. at 10.13 M. | s. d. | Ba chóir an fhoirm seo a chur le aon fhiafraí mar gheall ar an telegram seo. | Sent at ... M. | Date Stamp

From ... By ... | (This form should accompany any enquiry respecting this telegram.) | To ... By ...

Prefix | Sineadh | Office of Origin and Service Instructions | No. of Words

Z48 9.26 BRAMHALL MR 15

MR AND MRS FAY GRESHAM HOTEL DUBLIN =

CONGRATULATIONS AND BEST WISHES =

BETTY AND FRED MANCHESTER +

K2555.P.40244/B.3.500,000.8.6/64. MUINNTIR DOLLARD. Spl.

Figure 8.8: For many years, the most high-profile use of the telegram was to send greetings to the bride and groom at a wedding reception.

Another descendent of the public telegraph network was the private communications network used by air traffic control. In 1966 an agreement was reached between the UK and Irish governments to share responsibility for air traffic control for the sections of the North Atlantic controlled by the two countries. Prestwick in Scotland assumed the air traffic control function and Ballygirreen in Co Clare took responsibility for communications, with the joint operation called Shanwick as a conflation of *Shannon* and Prest*wick*.[62] Ballygirreen was linked with Prestwick and the other stations controlling aircraft over the North Atlantic by teleprinter, some over landlines and some using radioteletype (RTTY).[63] While telephone users in Ireland put up with long waiting lists and shoddy service, it seemed that the Department could deliver the goods when it was under scrutiny from abroad.

In the purely domestic sphere, however, most of what little innovation there was in the world of telecommunications seemed to occur beyond the reaches of the Department of Posts and Telegraphs. An example of this was Ireland's first dedicated telephone-order business which started in the 1960s. Dubliners could phone in orders, not for a hot pizza, but for an ice-cream cake from HB Dairies for delivery to their home by refrigerated van. With freezers a rarity, the telesales business was soon taking 40 orders a day for birthday parties and even weddings.[64]

Also beyond the stultifying grasp of P&T, the ESB established its own private telecommunications network, installing one of the largest private exchanges in Europe[65] and using its electricity distribution network to carry calls between its various premises.[66] Its telecoms supplier, Technico, was founded by Austrian-born former P&T engineer Otto Glaser. In 1960 Glaser established another company, Telectron, to manufacture telecoms and electronic equipment, a company that eventually grew to employ 800 staff at sites from Dublin to the Aran Islands.[66] Two years later, north of the border, a large factory was opened at Monkstown in the north Belfast suburbs by STC, which had supplied much P&T equipment over the decades. In 1972 this nascent sector expanded further when Ericsson set up a manufacturing plant in Athlone.[67] Ireland was beginning to develop its own skills in telecommunications.

Ireland's developing industrial sector was assisted by the Kilkenny Design Workshops, established by government in 1963 to improve design standards in Ireland. One of their designers, Damien Harrington, produced a new P+T logo in 1969.[68] The logo's simple, modern form (Figure 8.9) presented a strikingly progressive brand image, unfortunately largely at odds with the service actually delivered.

Figure 8.9: The 1969 P+T logo, designed by Damien Harrington of the Kilkenny Design Workshops. *(IE/NIVAL KDW/GD/162/01: Collection National Irish Visual Arts Library (NIVAL), NCAD, Dublin.)*

In 1970, facing a deteriorating economy, the already inadequate P&T capital budget was pruned[32] and construction staff laid off.[69] The Minister plaintively outlined the problems faced by the service due to continued starvation of capital:

> My Department have had the experience of repeatedly being pulled up short in their construction programme because of lack of funds and having to devote engineering resources to make-shift and uneconomic expedients which multiply difficulties for the future.'[70]

In response, opposition TD and future Taoiseach Garret FitzGerald suggested that the service be reorganised as a semi-state body and he railed against the cutbacks and the State's treatment of telecommunications in general: 'If it were under the control of anybody other than the State there would be enormous expansion and the demand would be met by the supply.'[71]

Being part of government did not mean that P&T had an unbounded social remit – indeed in most cases it applied financial criteria as stringently as the most capitalistic of private sector corporations. A case in point was the remote Black Valley in Kerry, whose 23 households lacked both electricity and telephones. A campaign for a telephone connection for the valley started in earnest in 1968. With no progress on the issue, residents boycotted the 1973 election, with local man John Joe Donoghue telling the *Kerryman* newspaper of the problems residents endured:

> Only today I had to drive the eight miles to the Gap of Dunloe to phone Killarney Hospital to discover if my wife and our new son, John Gerard, were ready to come home. It is more than time this sort of thing came to an end.[72]

The boycott seemed to achieve initial success[72] but once the election was out of the way, hopes were dashed. The Minister for Posts and Telegraphs, Conor Cruise O'Brien, told the Dáil that 'there are no developments which at present would enable us to bring a telephone service to the people of the valley without inordinate cost.'[73] Electricity arrived in 1977 with the aid of an EEC grant,[74] but there was no sign of telephones until 1983, when residents were offered service – for an advance payment of a mere £4,400,[75] over €13,000 at 2021 prices. There were no takers. A single public phone using a VHF radio link was installed in 1985 but it was not until 1990 that a deal was struck to offer private phones at a rate residents could afford.[76]

León Ó Broin, who worked with several ministers during his long reign as Secretary of the Department of Posts and Telegraphs, suggested that the low priority afforded to telecoms was related to the personality of successive Ministers:

> The Post Office always suffered from a shortage of capital for telephone development, and with Ministers allowing themselves to be treated as small boys there was never any real hope of catching up with the demand.[8]

Conor Cruise O'Brien became Minister for Posts and Telegraphs in the coalition government that came to power in 1973. While his focus sometimes strayed from his portfolio, O'Brien was not a Minister who would allow himself to be treated as a small boy. The new government announced a £175m (€222m) investment plan, dwarfing all previous plans.[32] However, this belated heavy investment added to costs and by 1977 the traditionally profitable telephone service was losing almost £500,000 a week. The Minister blamed the loss on 'culpable underinvestment in the Sixties and early Seventies' but justified the continued capital investment as 'any further delay would not merely be an inconvenience but an intolerable handicap to business and industry'.[77] Despite the extra investment there was no attempt to reform control of telecommunications. Garret FitzGerald had railed against mismanagement of telecommunications when in opposition in 1970, but the government he was part of did nothing to reform it.

The financial concerns of the telecommunications service in the Republic during the 70s seemed trivial compared the problems in Northern Ireland, where a targeted campaign of damage to telecoms infrastructure was carried out during the Troubles. In Belfast, with frequent bomb scares and night staff reluctant to brave uncertain streets on their way to work, there were often delays getting through to the operator. In August 1970 the Post Office placed newspapers adverts to apologise for the standard of operator service[78] (Figure 8.10), though arguably calls to the operator in Belfast in the midst of the Troubles were still answered more quickly than similar calls in relatively peaceful Dublin.

TELEPHONE SERVICE

BELFAST EXCHANGE

The Post Office regrets that it is having difficulty in maintaining a satisfactory standard of telephone service to its customers in the Belfast area due to the present civil disorder.

The many incidents which occur within the City produce a large number of calls to the operator. Service is also being disrupted from time to time by the partial evacuation of the telephone exchange on receipt of bomb warnings. Operating staff are working under great pressure to maintain a reasonable time to answer on all calls.

Telephone users are asked to co-operate:

1. **By dialling as many calls as possible direct from their own telephones, both local and trunk calls. Dialling codes are available in the booklet issued to each subscriber. Additional copies can be obtained on request by dialling 191.**
2. **By making full use of the Telephone Directory.**

The exchange staff will continue their dedicated efforts to provide a good service and the co-operation of all customers would be appreciated.

T. S. WYLIE,
Telephone Manager

Figure 8.10: Advertisement in the *Belfast Telegraph* of 9 August 1970. *(Courtesy of BT Heritage & Archives)*

In the eight months to April 1972, 25 telephone exchanges in Northern Ireland were bombed, affecting 1,800 lines and requiring £1m worth of repairs.[79] In the same month a bomb destroyed the exchange in Newtownbutler Co Fermanagh, six months later the 180 lines it served remained dead because the authorities would not repair it in the face of ongoing bombings.[80] The IRA, which had prepared a report about how to destroy Belfast's main exchange at Telephone House for the Nazis in 1941, made many further attempts to bomb the facility during the Troubles. In 1974 the British Army managed to defuse a bomb left in a hijacked van outside the exchange. If the 300 kg bomb, the biggest in the history of the Troubles to that point, had exploded it would almost certainly have brought down the entire building.

Remarkably, in spite of this hostile environment, the phone network in Northern Ireland was fully automated by 1970 when Rathlin island's 15-line manual switchboard was replaced with an automatic exchange.[81] The island had gained a telephone link to the mainland in 1935 using a short-wave radio link.[82] Marconi would undoubtedly have been pleased at the continued use of the radio technology he had pioneered on the island.

While the South, by and large, did not have to contend with damage to infrastructure, the service south of the border was intermittently affected by the Troubles. During the early 1970s telephone exchanges were frequently evacuated due to bomb warnings. While these were almost always hoax calls, they disrupted operator services, sometimes for hours, while buildings were searched. Telephonists on both sides of the border were often at the frontline in another respect, as bomb warnings were frequently phoned in to operator assistance numbers, not always in the locality of the bomb. Many lives were saved as a result of operators accurately noting the details of the call, determining the location of the appropriate emergency services and swiftly passing on the information.[83]

The telex service also saw use in the Troubles, being employed by Sinn Féin to send communiqués to press agencies, including statements from the IRA claiming responsibility for terrorist attacks. Sinn Féin's newspaper, *An Phoblacht*, reported gleefully of a raid on their West Belfast press office during which an over-eager British soldier threw the telex machine out of an upstairs window only to later discover that, as Post Office property, it had to be replaced.[84]

In 1972 an explosion at the Standing Stones relay station, the northern terminus of the Dublin-Belfast microwave link, caused severe disruption not just to cross-border calls but also to services between the Republic, Great Britain and beyond.[85] Coincidently the two hilltop locations on the southern side of this microwave link at Cappagh Co Kildare and Mount Oriel Co Louth (Figure 8.11) had been used by Edgeworth for his optical telegraph systems over 170 years before, since both optical and microwave communications are dependent on line-of-sight. However, while the optical telegraph relay at Cappagh was considered 'defensible against everything but cannon' in 1804, this was not sufficient against possible bombs from 1970s paramilitary organisations. As a result the army was deployed to guard these two repeater stations.[86]

Thanks to the Troubles, even the routing of telephone traffic sometimes became a political issue. In 1978 due to capacity problems on the direct routes from Northern

Ireland to Britain, the UK Post Office leased capacity for phone calls between Northern Ireland and Britain via Dublin. When this emerged in 1982, there was controversy that telecoms traffic related to military and security services in Northern Ireland was being routed via the Republic.[87] An additional route from Northern Ireland via the Isle of Man added in the 1980s removed the need to use the link via Dublin.[87]

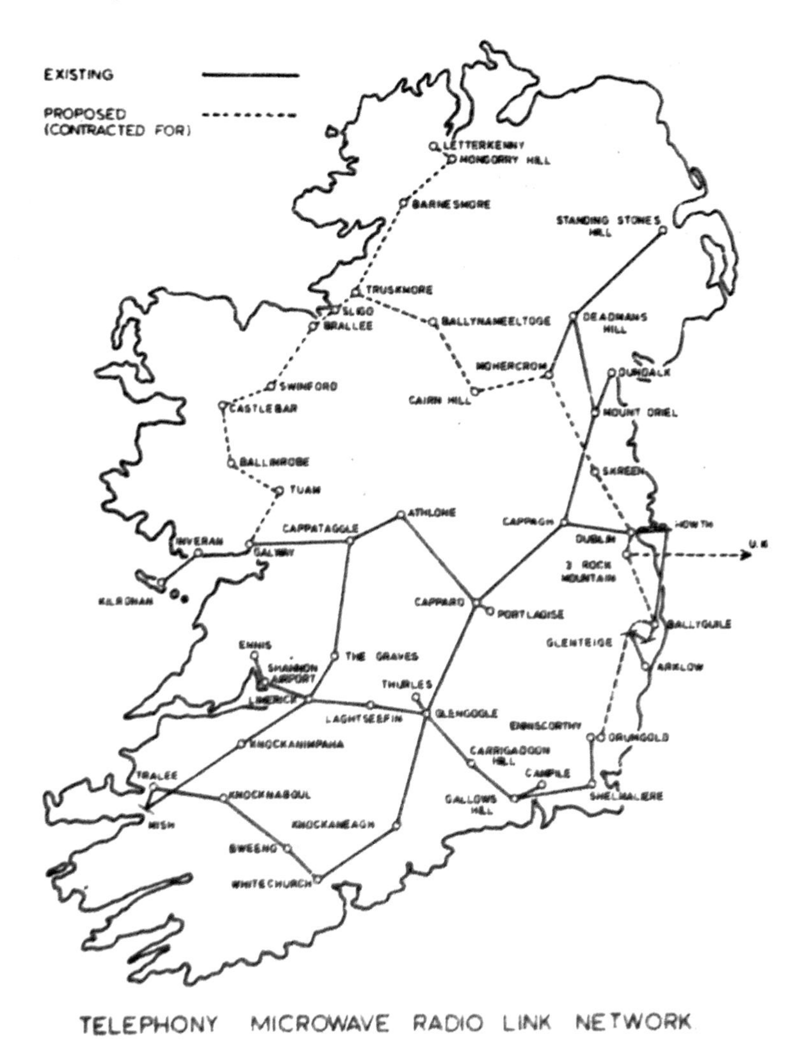

Figure 8.11: The microwave network in 1978. Cappagh in Co Kildare and Mount Oriel in Co Louth were locations used by Richard Lovell Edgeworth for his optical telegraph system.

The severe impact of the bombing of the Standing Stones relay station above Belfast showed the continued reliance of the Republic on the UK for its external telecommunications links. This dependency had continued even after 1962 when two telephone circuits came into operation between Ireland and the US. An operator in Dublin could now talk directly to her counterpart in AT&T's international hub in White Plains, New York instead of first having to talk to an operator in London.[88] A call cost £3 (€3.81) for three minutes.[89] However, the two circuits were routed via the UK as the new transatlantic telephone cables bypassed Ireland and landed instead in Scotland and France.[90]

International telecommunications grew even more rapidly in the early 1970s as the economy continued to expand and Ireland joined the EEC. By 1974 operators at the international switchboard in a new eight-storey telecommunications building on Dublin's Marlborough Street had access to 200 international circuits linking Dublin with New York, Frankfurt, Paris, Madrid and Amsterdam. On New Year's Day of the following year, International Subscriber Dialling (ISD) was introduced. Initially it enabled subscribers connected to exchanges in central Dublin to dial numbers in Belgium,[91] undoubtedly chosen to facilitate calls between civil servants in Dublin and their colleagues in Brussels. By March the service had been extended to France and Luxembourg.[92]

As with many other developments, Ireland was late to the ISD party. Residents of Brussels had been able to dial direct to Paris by 1956,[93] 19 years before they could reach Dublin. By 1964 the growth in international calling led to the allocation of country codes. Apart from the single-digit codes 1 – shared between several countries in North America – and 7 – allocated to the USSR – the initial digit of the code signified the region: 2 for Africa, 3 and 4 for Europe, 5 for South and Central America, etc. The next one or two digits indicated the specific country with smaller countries such as Ireland allocated a three-digit code. The objective was to minimise the length of the number that had to be dialled. Ireland was allocated 353 – but it would be 11 more years before anyone could actually dial it.

There was a particular impetus for this belated introduction of ISD. In 1975 Ireland hosted the EEC Presidency for the first time, including a summit of the leaders of then nine member states. The painful reputation of the Irish phone system was such that some British reporters expressed doubts in advance that Ireland could host such a summit of leaders with its attendant entourage of officials and reporters.[94] In the end the special telephone exchange set up in the Press Centre at Dublin Castle staffed by 'girl telephonists from the International Exchange' handled 1,300 calls over the two days[95] without a hitch.

By 1977 ISD was available in Shannon as well as all Dublin exchanges, and 21 countries could be dialled direct. This led to curious anomalies: the Dublin phone book carried dialling instructions for Wagga Wagga in Australia but not Enfield Co Meath.

The Gardaí at Kilcock, Co Kildare could, with a little bit of luck, reach the New South Wales police within seconds, but to call their colleagues 12 km away in Enfield meant dialling 10 and waiting, sometime interminably, for an operator to answer. In the same way that the Eucharistic Congress had led to the provision of extra telephone kiosks in Dublin, ISD was introduced to Galway in 1979, in time for the Papal visit that year.[96]

International Subscriber Dialling was not the only service to be introduced in Ireland long after many other countries. In most countries it was possible to call the operator and ask for a reverse charge call; a particularly useful facility for anyone who needed to make a long-distance call from a payphone without carrying a sack of coins. The Irish public had to wait until 1976, decades after most other places, for such a facility.[97] The service died with the withdrawal of operator services on 1 October 2007.[98]

It was a similar story with the Speaking Clock. The world's first such service had been introduced in 1933 in Paris[99] but Irish phone users had to wait until 1970 before they could enjoy a similar service. From that year, by dialling 1191 it was possible to hear the time read by an automated voice belonging to telephonist Frances Donegan (later McGrath) from Co Mayo. She received a cheque for £50 for her services from the Minister, Gerry Collins.[100] The Speaking Clock remained a popular service for decades until the ubiquity of mobile phones and internet connected devices from the late 1990s rendered it increasingly redundant. The service was turned off in 2018.[101]

The limited services and poor condition of the telephone network became recognised as a barrier to economic development. In 1975 a third of Irish industrialists interviewed by the Confederation of Irish Industries cited the parlous state of telecommunications as a primary inhibitor of economic development[102] while in 1978 the Industrial Development Authority (IDA) reported that the deficiencies of the telecoms services 'is now the most serious deficiency in our industrial infrastructure, and is hurting the IDA effort daily'.[35] Efforts by the IDA to raise such issues were not always welcomed. Padraic White, later head of the IDA, describes a 'chilly' meeting in the 1970s between the agency and the Secretary of the Department of Posts and Telegraphs, Proinsias Ó Colmáin. The latter questioned the right of the IDA 'to voice the slightest criticism or advance the mildest suggestion'.[103]

Even when the IDA did manage to attract foreign businesses the problems continued. Swedish businessman Lars Edman wrote to P&T in 1980 excoriating the telephone service available at his factory near Macroom and the Department's response to his criticism:

> Before I got established in Ireland, I was enticed by I.D.A.'s advertisements in Swedish newspapers saying that the Irish government will look after you even after you get

> established. It is then very astonishing to find about your attitude … Would you like to buy rotten sausages at the full price?[39]

Management attitudes, structures and policies also led to discontent and frustration within the Department's own staff. Corkman Louis O'Halloran, who joined P&T after graduating in 1958 as an electrical engineer, found that working within the civil service meant that 'you couldn't be an engineer, you were a slave to a system … Frustration really drove me out of it'.[104] His college classmates who had joined the ESB had 'a lot more scope to be engineers', outside the confines of the 'cumbersome' civil service.[104]

Most of the Department's technical staff came through a different route, joining the Youth-in-Training programme at age 16 or 17. Trainees were recruited each autumn into a three-year programme, with on-the-job training interleaved with three-month terms spent in colleges such as Kevin Street in Dublin. The programme began in 1955[105] (Figure 8.12) and by 1970 intake had expanded to 200 or more each year with the name changed to Technician Trainee (TT),[106] though as late as 1974 entry was limited to 'young men' (Figure 8.13). Dubliner Ben Jones, who joined P&T as a TT in 1978, has happy memories of his years in training, describing them as 'amazing, I got to try everything from climbing poles to working underground.'[107]

The rigidity in management attitudes extended further than fixed views on gender roles. A 1977 article in *Magill* magazine, beginning with the assertion 'Our telephone service is a national sick joke'[108], described human resources policies within P&T that sounded as though they had not changed since Agnes Duggan had started her job with the United Telephone Company in 1881. Night operators in Andrew Street exchange walked out in protest against 'oppressive supervision', including a requirement to sign a form requesting special permission from the supervisor to visit the toilet.[109]

Industrial unrest worsened in 1978. By February most of the telex lines used by the Met Service were out of action,[110] though luckily no D-Day landings were being planned. The Department accused its own technicians of sabotage and locked them out of key installations.[111] Airlines around the world lost access to their reservation systems when data circuits connecting them to the ASTRAL computer system operated by Aer Lingus went out of service.[112] A Dublin hospital relied on local residents to relay messages after its own phone lines went out of order,[113] while a Dublin businessman, unable to make a call to London, hired a private plane to Belfast, made his 40p phone call and flew back home.[114]

Strikes were not the only factor exacerbating the problems within what was already a crumbling and overloaded network. The 1970s also witnessed two oil crises. As had occurred during the fuel shortages of the Emergency 30 years before, those lucky enough to

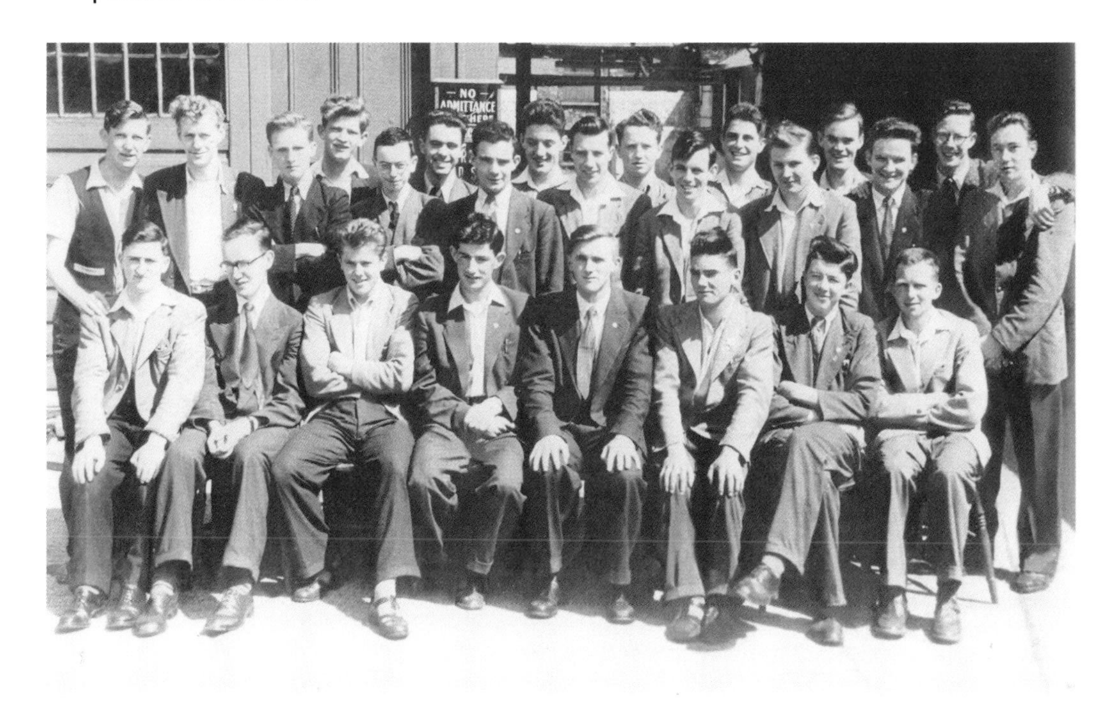

Good careers for young men

TECHNICIAN TRAINEES

are needed by the Post Office Engineering Branch

A large number of young men will be taken on for training as **Technicians on Telecommunications work** in the Autumn of 1971.

Weekly pay during training varies, with age, from £7-17-0 to £10-10-0. You will start at not less than £9-17-0 if required to live away from home.

On completion of training each successful candidate (if 20 years of age then) is normally appointed as Technician, Class II. There are prospects of promotion to Technician, Class I and higher posts, and some opportunities of qualifying as a professional engineer.

(Technicians Class II at present receive £20-£22 a week (approx.). Technicians Class I, £23-£26, and Inspectors £1,655-£2,005 a year).

Age-limits—16-18½ on 1st September, 1971.

Selection—By written examination, interview and oral test in Irish. Some places are reserved for boys from the Gaeltacht.

If you would like to hear more about this worthwhile career, the Civil Service Commission will be glad to send you further details and an application form. Write immediately (on a postcard) to the Civil Service Commission S(6), 45 Upper O'Connell Street, Dublin 1. The latest date for receipt of completed application forms is:

5.30 P.M. ON 10th DECEMBER, 1970.

Cuirfear failte roimh chomhfhreagras i nGaeilge.

Thousands of technical staff joined the Department through its trainee programme. Figure 8.12 (top): 'Youths-in-Training' from 1955–57, the first three years of operation of the programme *(Courtesy of Padraig Mitchell)*; Figure 8.13: 'Good careers for young men' – but not young women – in 1970 *(Courtesy of the Irish Examiner)* have a phone started to make more calls.[115] But this only added to the suffering as the extra traffic further taxed an overloaded network. In 1978 it was found that only 61.1 per cent of local calls and 37.7 per cent of STD calls were successful on the first attempt.[35]

Part of the reason for this chronic performance was the high level of line faults. In 1979 there was an average of 1.13 faults per line per year, a high rate by international standards. A large proportion of the faults was caused by water damage to underground cables, particularly the older cables that had a lead casing, within which each copper wire was encased in paper. If water penetrated the casing, the paper inside became wet and the wires inside shorted. As late as 1988 the *Farmers Journal* described how it was possible to tell the weather conditions in Dublin simply by lifting the phone:

> How you may ask? Very simple. If it rains in Dublin, then it is impossible to reach a Dublin number. So [the telephone] ... can also operate as an early warning on weather conditions in Dublin without even the cost of speaking to anybody. This level of technology must surely be in the realms of 'Star Wars'.[116]

In April 1978 the *Financial Times* ran a report about Ireland's woes, quoting the president of the Galway-based refrigerated transport company, Thermo King: 'Anyone who would move into Ireland now would have to be crazy.' The journalist added, 'I didn't speak to him directly. I attempted to contact him, but, predictably, I couldn't get through.'[117]

An all-out strike that began in February 1979 paralysed the network. With only a skeleton staff remaining to deal with emergency calls, subscribers connected to manual exchanges effectively lost their telephone service. Switching equipment and cables went without repair and maintenance, while the waiting list reached a new record high. The 127-day-long dispute marked the nadir of the crisis-ridden telephone service. It also finally precipitated a rethink of the control and management of the state's telecommunications infrastructure.

For 67 years the entire telecommunications network had been under direct government control. This gave politicians, most especially the Minister, a valuable form of political patronage. This attitude may have derived from Ireland's colonial past, so that for 'generations, Irish people saw that to get the benefits that public authorities bestow, the help of a man with connections and influence was necessary.'[118] There was an entire political class willing to help, or rather to be seen to help, Mrs Murphy get her telephone.

By the late 1970s, however, this level of control had morphed from a valuable form of patronage into a liability. The Department of Posts and Telegraphs was soaking up large amounts of capital but the service just seemed to worsen. Tucked in between abolishing car tax and rates and free telephone rental for Old Age Pensioners, Fianna Fáil's 1977 giveaway election manifesto contained a less noticed commitment to examine giving autonomy to telecommunications.[32] The incoming government, unsurprisingly, commissioned a review to examine the problems and structure of the telecommunications service. The report delivered a damning review of the service and structure. The Dargan Report, as it became known, did not mince its words:

> The state of the telecommunications service constitutes a crisis and a heavy dose of realism is now needed. Policies and attitudes need to be changed fundamentally and quickly. We have striven to bring out our report quickly so that no time may be lost.[35]

It attributed this crisis to the governance structure:

> The telecommunications service is a business. It has a rapidly changing technology and an increasing need for a marketing orientation. In those circumstances the governmental and civil service structure, which has to accommodate too many restrictions, is unsuitable for management of telecommunications.[35]

The report concluded that telecommunications should be set up as a semi-state body. Curiously, this report was sent to cabinet just as Margaret Thatcher's government came to power in the UK, and announced it was proceeding with its manifesto pledge to create separate entities for post and telecommunications. However, apart from that initial decision to split the postal and telecoms services, the policies of Ireland and the UK diverged for the next 20 years.

With the Minister taking flak for a strike-bound department lurching from crisis to crisis, the government leapt at the opportunity to rid itself of a painful liability. In November 1979 it created a new entity called An Bord Telecom as an interim step in creating a new semi-state company to take over the running of the country's troublesome telecommunications service. For the government, at least, the suffering would soon be over.

CHAPTER NINE

'AN INSTRUMENT SO POWERFUL'

Tuesday lunchtime, 2 June 1953, Dolan's pub on Dublin's Marlborough Street.

The pub was unusually packed for a weekday. Many of the patrons were peering awestruck at a snowy image on the brand-new television placed high up on the wall at the end of the bar.

The lunchtime drinkers were not engrossed in football or racing. This was a special day. The Queen of England, Elizabeth II, was being crowned, and there she was, magically rendered in a fuzzy, flickering image on the TV.

As the patrons debated the merits of the new contraption over pints of Guinness and glasses of Paddy, few noticed the arrival of a man in a long white dustcoat.

Just as the crown was held aloft over the head of the young queen, there was a commotion in the pub. The man in the long white dustcoat forced his way through the crowd to the end of the bar, pulled a hammer out of his coat pocket, reached up and smashed the TV screen with it.

Everything shattered (Figure 9.1). The blurry image of the queen vanished. Dolan's fell silent.

'Get out of my way!' the television-smasher shouted. 'Ireland is still free!'

Unmolested, he strode out of the pub and turned into Abbey Street. Within a few minutes a Garda arrested him and took him to Store Street station. When later charged with malicious damage, the TV smasher, Gearoid O'Broin, described his action as 'not malicious, but a protest against the denigrating influence of this type of thing'.[1]

As we saw in previous chapters, telecommunications and politics go hand in hand, from big national debates about the nationalisation of the telegraph service to petty squabbles about the location of a phone box in Waterford. Television was no exception, as demonstrated by Dolan's getting its TV smashed by the staunch nationalist in the dustcoat.[2]

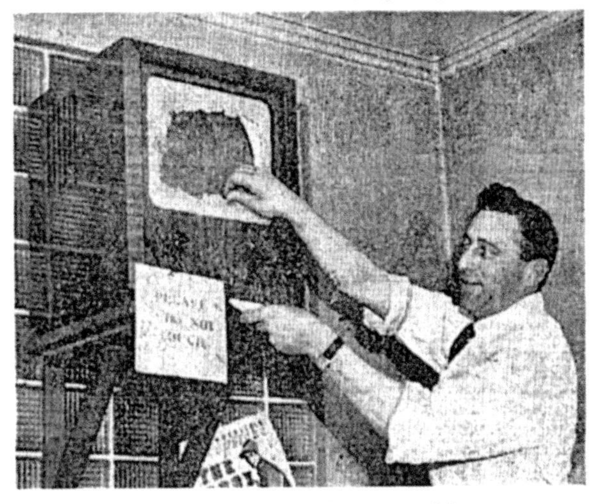

Television Set Was Smashed

Man Strolled In, Drew Hammer

A MAN was arrested in Abbey Street, two minutes after the television set in Dolan's public-house in Marlborough Street, Dublin, had been smashed yesterday.

At 1.10 the bar on the premises was crowded when a man, aged about 35, walked in, wearing a long white dustcoat. He strolled to the end of the bar and when in front of the set which was flickering, but not receiving, drew a hammer from under his coat and smashed the screen.

He turned around and calmly said: "clear the decks" and walked through the customers and out to the street.

Our picture shows Barman Tommy Farrelly picking pieces of the broken screen from the set.

Yesterday afternoon Gearoid O Broin, mechanic, 58 New Street, off Kevin Street, was charged in Dublin District Court with having caused malicious damage to the television set.

Asked if he had anything to say, O Broin said: "Nothing for the present. I don't want bail, and I will not allow anyone to sign bonds for me."

He was remanded in custody until Tuesday.

Figure 9.1: The smashed TV in Dolan's. *(With thanks to Irish Newspaper Archives)*

We also saw that physics has profound effects on telecommunications, both over wires and radio waves. Because of the physics of transmission along a wire, that single, fat coaxial cable linking Dublin with Cork, Limerick and Athlone and completed in the early 1950s could carry 600 phone calls, far more than the bare strands of copper cable strung along poles that it replaced. Furthermore, the physics of radio transmission respect no national boundaries as demonstrated by the worries about the use of the Radio Éireann transmitter at Athlone as a navigation aid for Luftwaffe war-time air-raids on British cities. On the other hand it was also the limitations caused by the physics of radio waves that rendered Queen Elizabeth's image 'fuzzy' and even indecipherable by the time it reached Dublin TV screens.

Over the decades from that incident in Dolan's pub, the dependence of television on this heady mixture of politics and physics led to death threats, surprise election results and dodgy donations.

Those pioneering viewers in 1953, such as those in Dolan's pub, were bending the rules of physics. They were tuned to the BBC television transmitter at Holme Moss in the north of England. As a general rule in physics, the higher the frequency, the shorter the range: this is the main reason why the RTÉ Radio 1 transmitter at Summerhill Co Meath on 252 kHz is audible 500 km away in London but a 5G mobile phone base station operating at 3.6 GHz (a frequency over 14,000 times higher) has a maximum range of a few kilometres at best. Holme Moss transmitted on 51.75 Mhz at the lower end of the VHF band, midway between those two extremes (Figure 4.7). Physics would suggest a maximum range of 180 km for a signal at such a frequency. It was no wonder that the image of Queen Elizabeth received in Dolan's or elsewhere in Dublin, 300 km away, was a blurry mess.

The reason why Irish viewers were prepared to defy physics was simple curiosity. The arrival of television in Ireland had been forecast since the early days of the State; indeed John Logie Baird, the Scottish 'father of television', had delivered a lecture on the subject to a packed Theatre Royal in Dublin as early as 1927.[3] By 1929 a headline in the *Irish Independent* breathlessly proclaimed 'Television To Be Introduced Soon Into The Saorstat'.[4] Though the prediction proved wildly optimistic, throughout the late 1920s and 1930s Irish newspapers gave extensive coverage to the development of early TV systems in Britain, France, Germany and the US. By the 1950s the expansion of the BBC's transmitter network persuaded some Irish people to pay up to £100 (€127) for a television for their home or pub.

In April 1952 the *Irish Times* reported that about 200 televisions had been installed in and around Dublin, though due to the physics of radio waves 'for very many it has been a case of pretending to enjoy a screen full of capering wavy lines'.[5] Undaunted, by June the same year, Flavin's public house in Sandyford (now the Sandyford House) advertised that they were showing racing from Ascot by television.[2] By March of the following year, a house in Dublin's 'select Terenure district' was on offer complete with all furnishings – including a television.[6] It was quite a bargain by today's standards; at a monthly rent of £23 it would have taken four months to pay for the television set alone.

The problems caused by the physics of radio waves diminished at lunchtime on 21 July 1955 when the BBC's transmitter at Divis outside Belfast was officially switched on. The range of the new transmitter included most of Northern Ireland – and a considerable part of the Republic. The first programme to be shown, the test match between England and South Africa, was unlikely to have had viewers in Ireland, north or south, glued to their sofas but perhaps more enjoyed the feature about breeding rabbits to make a fur coat on *About the Home* at 3 p.m.[7] Residents of border counties such as Cavan were quick to exploit the new service so that within weeks the *Anglo-Celt* was able to report that 'Television has come to Killeshandra. The first set has been installed at the residence of Dr C. J. Bourke by his radio dealer son, Geoffrey. Large numbers gather at the Bourke home each evening to watch and wonder.'[8] By May of 1956 it was reported that one in ten Dublin pubs possessed a television.[9] Dolan's was facing competition.

The prospect of better reception and the advent of rental and hire-purchase schemes for television receivers broadened the appeal of the medium. Not everyone was happy with the results, with the Dublin correspondent of the *Wicklow People* lamenting that:

> the most remarkable thing is the extent to which television aerials have sprung up in the city mainly in the Corporation housing areas. Every second house in some of the new housing colonies now has a television set and those contractors who specialise in erecting them have their staffs working over-time. In what used to be considered the better-off suburbs the aerials are few and far between because the people in those districts cannot afford them.[10]

One 'better-off' area that boasted a television aerial was the Mespil Estate, a private apartment development located beside a particularly verdant section of the Grand Canal on Dublin's southside. Available for rent at £156 to £180 per year, each apartment in newly-constructed Cherry House boasted not just individually controlled central heating, electric water heating and an intercom – but an even more unusual feature for Ireland in 1958 – a television aerial socket in each flat[11] supplied from a single communal aerial on the roof by means of a coaxial cable.

Though the residents moving into Cherry House in 1958 never considered it, that coaxial cable coming into their flat could, in time, offer a lot more than just the televisual delights of the BBC. That system of coax cables in Cherry House was the first cable television system in Ireland. It was also the beginning of a technology that, thanks to the physics of transmission along a wire, would eventually supply homes around the country with broadband and telephone services, as well as multiple television channels.

All these early televisions viewers from rural Killeshandra to urban Dublin were watching British television for the simple reason that there was no Irish station. The reason for this was largely political. As early as 1950 the Department of Post of Telegraphs had established a Television Committee to examine a TV service for the Republic. But the committee quickly ran into resistance from the powerful Department of Finance, sometimes over seemingly trivial requests. When the committee sought approval to buy a television and aerial to view BBC broadcasts, the Department of Finance was unimpressed with the idea of spending the vast sum of £180 on a television set:

> It would ... be ridiculous to think of a Television Service in a country which has manifested no interest in it and whose people would probably be opposed to the spending of considerable sums of public money on such a luxury, available, of course, to a very limited number because of geographical and financial reasons ... Television is a long way off here.[12]

Despite such views, the Department of Posts and Telegraphs kept a close eye on developments, requesting postmasters around the State to conduct periodic counts of TV aerials. The first such count, which took place in 1955, gave a national total of 2,200 televisions, half of them in Dublin. The postmasters also asked television owners to report on the quality of reception they received. Of the nine television owners in Cork city, six categorised their reception as 'good most of the time' with the best reception, unsurprisingly enjoyed by residents of Mayfield,[13] on the elevated eastern fringes of the city. Those intrepid viewers in Cork were picking up a signal from Wenvoe in South Wales, 362 km away. They really were bending the rules of physics, though undoubtedly what a television viewer considered 'good' in 1955 would probably not suffice for a modern viewer used to 4K definition. The difficulties involved in receiving British television in Cork city became a political issue for decades.

The opening of the Divis transmitter above Belfast in 1955 reignited the political debate about television in the Republic. The residents of the lost 'fourth field' enjoyed not only universal free healthcare and free secondary education but now also a television service.[12] South of the border, in March 1956 the Television Committee issued a report on the delivery of a home-grown television service. The report outlined why this was now of urgent importance: Irish viewers were becoming hooked on British TV. The Irish daily newspapers were publishing BBC television listings and the Republic had more television sets than Demark, Norway and Sweden[12]; though, unlike those three Scandinavian countries, the country did not have its own television service. Furthermore, echoing the concerns expressed in the 1920s about vulnerable Irish people having their heads turned by 'British music hall dope and propaganda' from BBC radio, the Irish authorities were concerned about the content of the broadcasts. The Television Committee explained why they considered the BBC programmes to be unsuitable for the Irish viewer: 'Some are brazen, some "frank" in sex matters, some merely inspired by the desire to exalt the British Royal Family and the British way of life.'[12]

The views of the Committee did not accord with all political opinions in the Republic. A resident of Sandycove on Dublin's southside enjoyed this emphasis on 'British achievements', writing to the Prime Minister of Northern Ireland, Lord

Brookeborough, in 1957 (Figure 9.2) to commend the 'fight you are putting up in the North against the disruptive element from the South', adding 'I would like to emphasise the terrific propaganda value of a good BBC signal being received in Eire.'[14]

THE MOORINGS, HARBOUR ROAD, SANDYCOVE, Co. DUBLIN.

November 29th 1957.

Lord Brookeborough,
Stormont,
BELFAST.

Dear Lord Brookeborough,

I would like you to know how very much I admire the fight you are putting up in the North against the disruptive element from the South.

I know that this must have occurred to you, but I would like to emphasize the terrific propaganda value of a good B.B.C. signal being received in Eire. The number of aerials going up in this district round me is really very heartening and many, many more people would follow suit if the signal was only a little bit stronger and better reception could be assured. Is Divis as strong as you can possibly make it?

Yours truly,

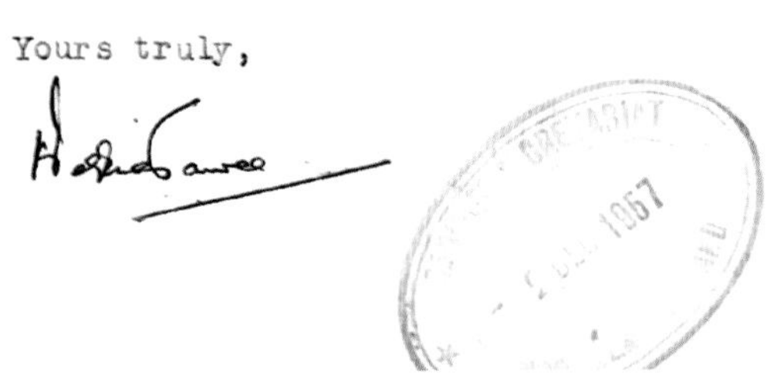

Figure 9.2: A letter from a viewer in South County Dublin in 1957 to Northern Irish Prime Minister Lord Brookeborough. *(Courtesy of the Deputy Keeper of the Records, PRONI, CAB/9/F/165/13/1)*

Finally, in 1958 the government decided that the Republic should have its own television station. Ignoring the findings of its own commission,[15] it determined that the television service would be owned and operated by the State, with revenue coming from advertising as well as licence fees. RTÉ, as it was eventually named, would be another semi-state company in the mould of the ESB and Aer Lingus.

Unlike many of the earlier private sector proposals, which were confined to Greater Dublin, the State's television service would be a national one. To ensure that this political decision would not be thwarted by physics, the Department's technical staff started work on planning a network of transmitters to bring signals to all 26 counties, creating a 3-dimensional map of Ireland with mountains built from papier-mâché in order to model the range of different possible transmission sites. A lightbulb was placed on each possible location and the amount of map it illuminated compared with others to determine the optimal network.[16] A site atop Kippure, 'the king of them all'[17] in the words of Brendan Behan, was selected to serve Dublin and the north-east.

Thanks to physics, signals from the new Irish television service could pass unimpeded across the border just as those from the UK passed into the Republic. Network planning thus involved bilateral discussions between the Department of Posts and Telegraphs and its UK counterpart. Perhaps as a result of the desire to avoid interference with BBC and ITV, all of the five main locations selected were well away from the border and thus did not cover the more populated parts of Northern Ireland.[18] The poor coverage drew protests from nationalist politicians, the GAA and private citizens in the North. In 1964 a Mr Liam Mulrenna wrote from Belfast to the Minister of Posts and Telegraphs about the perils of BBC and UTV:

> Sometimes we have to turn off our T.V. sets because the programmes are not suitable for the innocent minds of our children and indeed for grown-ups as well. You don't know how much we beg for to see your good clean programming.[19]

This scrupulous avoidance of Northern Ireland, which also left parts of the border counties of the Republic ill-served, would later become a political issue for two governments.

As well as planning transmitter locations, the Department's staff had to decide on the technical standards to be used, a decision that became also a political issue. While most European countries were using the 625-line standard, the UK stations used the 405-line standard, which had been adopted by the BBC in 1936. The 625-line standard provided higher definition and appeared to be the better long-term option. However, the tens of thousands of Irish households who were already using 405-line sets to receive BBC and UTV were dismayed at the possibility of shelling out for a new receiver for the domestic service, and made their views clear to politicians. Meanwhile some 'cultural nationalists' argued in favour of 625 lines precisely because it would make it more difficult to view UK broadcasts.[12] The solution adopted was a pragmatic compromise: the new station was equipped for 625 lines[20] but to serve existing viewers, 405-line equipment was provided at transmission sites in the east and north of the State including Kippure and Truskmore in Co Sligo.

The decision was proved correct as, within a few months of the opening of the Irish station, the UK decided to move to the 625-line standard. The 625-line system was widely adopted around the world and remained in use in Ireland until analogue transmission was switched off in 2012 after the rollout of Digital Terrestrial Television.

Telefís Éireann, as the state's television service was initially dubbed, commenced broadcasting on New Year's Eve in 1961. On the first night, an awkward-looking President De Valera gave a less than equivocal welcome to the new medium:

> I must admit that sometimes when I think of television and radio and their immense power I feel somewhat afraid. Like atomic energy, it can be used for incalculable good but it can also do irreparable harm. Never before was there in the hands of men an instrument so powerful to influence the thoughts and actions of the multitude.[21]

For the first month, only the Kippure transmitter serving Dublin and the north-east was ready but the network was quickly extended. Within 18 months, all five of the main transmitters were in full service, linked by a microwave network to the mast adjoining the new television studios at Montrose, the ancestral Dublin home of Guglielmo Marconi's mother.

De Valera's concerns about the power of television to influence 'thoughts and actions' proved accurate, with the arrival of the state's own TV service widely credited as a major catalyst of social change. This did not find favour with all. When the conservative Fine Gael TD Oliver Flanagan declared that 'there was no sex in Ireland before television', he was not bestowing a compliment. Ironically the location of the first public lecture about television in Dublin, the Theatre Royal, fell victim to the medium it had once promoted. It closed in 1962, a casualty of falling audiences as an evening at home in front of the 'box' replaced one in a box at the theatre.[22]

While documentaries and debates discussed previously taboo subjects, a more subtle influence on Irish viewers came from the wide range of imported material. No longer was Ireland an island of uniqueness but a consumer of American comedy shows and more esoteric fare supplied from the Eurovision system of the European Broadcasting Union (EBU). Initially Eurovision broadcasts were simply picked up from the BBC or ITV on a regular aerial and relayed over the national TV network, a method used to bring coverage of Ireland's first entry to the Eurovision Song Content in Naples in 1965. Later that year a microwave link was inaugurated between the studios at Montrose and Belfast[23], hooking Ireland fully into what was fast becoming a global TV network. The new two-way link also allowed Ireland to display its wares to the world, the first broadcast to exploit this being the Irish Derby in the Curragh which was televised live in June 1965 to Britain, France and the US via the Early Bird satellite.[24]

To Irish viewers, the Eurovision network became synonymous with the Song Contest following Ireland's success in 1970 and subsequent hosting of the contest from Dublin's Gaiety Theatre. With the Troubles at their height, special security measures were put in place to ensure that coverage to viewers in 29 countries would not be disrupted by terrorist action. Multiple links were provided over different cross-channel routes, and all sites in Northern Ireland were guarded by the British Army.[25]

Ireland hosted the contest several more times, with particular complications in 1993 when the event was staged in the small Co Cork town of Milstreet. Politics was again the reason for many of the challenges as, thanks to the fall of the Iron Curtain and breakup of Yugoslavia, the number of participants had swollen. Juries in 18 countries had been connected to the Gaiety in 1971; for the 1993 contest the Green Glens Arena had to be linked to 25 juries, with Ljubljana, Zagreb and Sarajevo joining the roll call alongside the more familiar London, Paris and Hilversum. This was the first time the Eurovision had ever been staged outside a city and a new fibre optic link was installed to transmit the contest to Dublin[26] and from there to 300m viewers across Europe and beyond.

While transmitting the 1971 Eurovision had involved the support of authorities on both sides of the border, such cooperation was unusual for that period, again because of political considerations. During the 1960s the government in Stormont was, like its southern counterpart, put under pressure to protect advertising revenue by preventing cable companies relaying 'foreign' stations. In this case the pressure came from UTV which, while privately owned, had many a sympathetic ear at Stormont. With authority over cable companies the responsibility of central government in London, Stormont persuaded a reluctant UK Post Office to remind cable operators of an obscure clause in their licences that forbade the relay of foreign commercial television programmes.[27] The fact that the government of the Republic had severely restricted cable systems from relaying UTV and BBC was used by Stormont as justification.

Such rules about cable TV became a political issue in February 1968. Residents moving into the Rossville Flats in Derry, many of whom had moved from houses where RTÉ was available via rooftop aerials, complained that they had lost the channel as they were not allowed to erect their own aerials and the piped TV system serving them was not allowed to relay RTÉ.[28] Maintaining the restrictions on 'foreign' stations was a political decision by a Stormont government hostile both to media from the South and to the wishes of its own Nationalist citizens.

Soon after, the Troubles kicked off in Derry. As violence engulfed Northern Ireland, cable television became a life-or-death matter. The town of Newry, situated in a valley, was a reception blackspot and was wired up for cable television in the 1960s by local television dealer William Wylie.[29] In 1977 the IRA threatened to kill Mr Wylie if he cut off subscribers for non-payment or stopped (illegally) relaying RTÉ.[30]

The absence of RTÉ across most of Northern Ireland became a critical issue for the government in Dublin after the outbreak of the Troubles. A confidential plan to improve coverage recommended an extra transmitter at Clermont Carn, in the Cooley mountains in Co Louth, just 1 km from the border. It was estimated that this would increase coverage from about 14 per cent of the North's population to about 66 per cent.[31] The RTÉ report also noted that this would require approval of the UK government as, if Ireland were to proceed unilaterally, it would be the first signatory to break the Stockholm Agreement governing broadcasting. In response to the proposal, the British Prime Minister Ted Heath informed the Taoiseach Jack Lynch in 1971 that:

> whenever there is a situation in which there is a deliberate policy of broadcasting across national frontiers a series of technical and legal problems arise ... In addition there are, of course, important political considerations. I have come to the conclusion that direct transmission by RTE into Northern Ireland might increase rather than lessen tension and that this is a risk which we ought not to run.[18]

Ten years later a transmitter at Clermont Carn was finally turned on, with an agreement that the signal radiated to the north would be restricted.[18] A new RTÉ transmitter was also installed at Holywell Hill on the Donegal–Derry border, its location cleverly selected so that most Derry city residents could pick up both UK and Irish signals with a single aerial.[32] Belfast residents required an extra, taller aerial but these quickly sprouted in the most surprising of locations. At age 17 I recall meeting a group of lads from Belfast's Loyalist Shankill area as part of a peace and reconciliation initiative and being surprised to learn that most of their families had installed aerials to watch RTÉ, their favourite viewing being the American sitcom *The Odd Couple*. It was probably not what Jack Lynch had intended, though perhaps the sitcom's tale of two bachelors with radically different personalities coming to terms with sharing the same space was an apt metaphor for the political situation.

It was not until 1984 that the UK government removed restrictions on the channels a cable television operator could carry. The change was not intended to facilitate access to RTÉ but rather was part of a political policy of liberalising markets encouraged by Margaret Thatcher. The expansion of cable in the UK was slow at first but in 1996 NTL-CableTel started supplying television and telephone services to Belfast and other towns across NI using coaxial cable with a fibre backbone.[33] The new cable network arrived just as Sky started to aggressively market its satellite service so that the network never spread as it had done south of the border. In 2017 only 6 per cent of Northern households received their TV via cable[34] compared with 17 per cent in the Republic[35] (Figure 9.3).

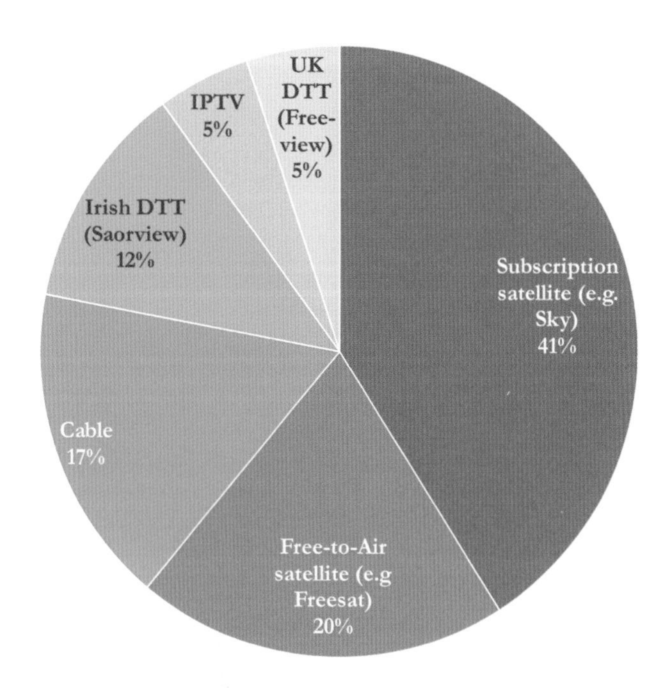

Figure 9.3: Television reception methods used by Irish households, 2017. Source: *Broadcasting Authority of Ireland, A report on market structure, dynamics and developments in Irish media* (Dublin, 2017)

To serve the majority in Northern Ireland without cable, in 2002 a new 120 m tall mast was installed at Clermont Carn specifically to boost coverage of the now four Irish stations across the north. Indeed signals for Clermont Carn travel so well into Northern Ireland that the BBC seriously considered installing a radio transmitter there in the early 1990s.[36] The latter plan did not proceed but the fact that it was contemplated was a measure of the improvement in Anglo-Irish relations in the 21 years since Ted Heath had blocked Jack Lynch's plan for an RTÉ mast there.

However, while the Irish government lobbied to allow citizens of Northern Ireland to view television from the Republic, it was reluctant to reciprocate. In 1961, to protect the state's nascent TV station, the government applied restrictions on the number of connections to a single aerial, effectively prohibiting the rollout of cable television. Those residents of Cherry House were in a lucky minority. The government restrictions effectively meant that each household had to have its own aerial (Figure 9.4) and the skyline of Dublin became crowded with tall aerials, eking out precious signals from Divis.

Figure 9.4: Ring Terrace in Dublin's Inchicore, about 1960. Like most parts of Dublin the skyline was crowded with TV aerials aimed at Northern Ireland or, later, Wales. *(Courtesy of Dublin City Library and Archives)*

The government's restrictions over cable systems did nothing to dampen demand for BBC and ITV from viewers in the Republic, however. Dubliners living in flats and houses where the required aerials could not be erected were vocal in venting their frustration. As a result in 1966 a special exemption was granted to allow all the flats and houses in Dublin Corporation's vast new housing scheme then under construction at Ballymun to be connected to a single aerial. However, this did not help tenants in other Dublin Corporation houses without chimneys and thus nowhere to mount aerials. Meanwhile private sector developers started to ignore the law by planning to cable houses as part of the construction works[37] (Figure 9.5). Popular pressure led to restrictions being eased in 1970 and lifted in 1974, replaced by a 15 per cent levy on cable subscriptions with the proceeds going to RTÉ. As a result, by the mid-1970s miles of coaxial cable connected thousands of homes.

FOR SALE
New Luxury 4-bedroomed Semi-detached Houses
(Contain full central heating)
(Five minutes walk town centre) No Auctioneer's Fees. No stamp duty. Remission of rates for ten years. Communal Aerial, giving perfect RTE, UTV and BBC.
Price £5,625 nett
Full details from :
G.H.Q. LTD., M.I.A.A.
Auctioneers and Valuers.
WINE STREET, SLIGO
Phone SLIGO 2957.

Figure 9.5: An advertisement for new houses from the Sligo Champion in 1970. The availability of 'perfect' UTV and BBC was clearly a key selling point. *(By permission of DNG Flanagan Ford)*

Thanks to the physics of radio waves, there was a distinct geographical pattern to this rollout of cable television. Cable companies found a receptive market in towns some distance from the border like Dublin, Sligo and Navan, where reception with a regular aerial was often indifferent. Where good reception of BBC and ITV was available from a modest aerial, there was little demand for cable and thus towns like Monaghan and Letterkenny were left out. Also omitted were villages and rural areas because it was simply uneconomic to roll out miles of coaxial cable to serve a handful of houses. However, the most vociferous group of those left behind lived in places beyond the coverage of the UK stations where even the tallest mast could not overcome geography and physics (Figure 9.6).

In response to the spread of cable networks across Dublin and other towns in the east and north, a group called the Multi-Channel Campaign was founded in Cork[38] to lobby for choice of viewing elsewhere. The chairperson of the group was quoted as saying that in 'the south of Ireland we are treated as second class citizens'.[39] The government was unmoved, replying to the Cork group to say 'what we are discussing here is "foreign" broadcasting and it is very arguable whether any special Government action to promote or encourage the dissemination of such programmes within the State could be defended.'[40] No-one pointed out the hypocrisy of a government that simultaneously agitated to extend RTÉ across the border, while attempting to prevent its own citizens from enjoying cross-border television.

In any case, the nationalistic language about not disseminating 'foreign' broadcasting was merely a disguise. There were more pressing reasons for the government's antipathy to the televisual demands of the south and west.

Figure 9.6: Multi-channel land, according to Cork's *Evening Echo* in 1970. *(Courtesy of The Echo)*

The first reason was again the physics of radio waves. In the early 1970s, with direct broadcast satellite television a futuristic fantasy, the only way to view television was by receiving it from a terrestrial transmitter. Many parts of the south and west of Ireland were well outside the range of UK transmitters. While a handful of doughty viewers in Cork might have been happy with their BBC picture in 1955, getting a reliable signal to Cork city would require an aerial many miles further east or north, from where the signal would have to be relayed over a coaxial cable or microwave radio link. The former would be difficult over a long distance – the Dublin–Cork coaxial telephone cable had taken three years to complete, while the latter would be opposed by RTÉ and the Department.

However, even if the authorities had relented and permitted microwave transmission, a second problem would have arisen. Television broadcasters had to pay royalties to their performers – this also applied to anyone, including in another country, retransmitting these broadcasts over the airwaves, even over a point-to-point microwave link.[41]

The third problem, which probably most occupied the government, was the potential loss to RTÉ's advertising revenue if more viewers were no longer captive to the broadcasts from Montrose. Protectionism was alive and well, at least when it came to state-owned industries.

A new government in 1973 seemed to herald a change of policy. The new Minister for Posts and Telegraphs, Conor Cruise O'Brien, held unconventional views in regard to Northern Ireland. He proposed an 'open broadcasting area' with complete mutual availability of television programmes across Ireland,[42] so that someone in Larne would have the same four channels as someone in Listowel. The proposal drew a mixed response. Playwright Hugh Leonard had a jaundiced opinion on how RTÉ would be regarded in Northern Ireland: 'There is nothing which unites people more than a common misery, and a sustained bombardment of RTE programmes is an act of inhumanity which should have Prods and Popeheads alike yammering for mercy in a matter of months.'[43]

As a first step it was proposed that the Republic would build a transmitter network to rebroadcast BBC1 Northern Ireland but the issue of royalty payments for rebroadcasting proved thorny. Cruise O'Brien cited examples from Italy where, by the mid-1970s, television from Yugoslavia and other neighbouring countries was being retransmitted.[44] However, the BBC was petrified that trade unions would seek a rate increase far higher than the 2.5 per cent increase in potential viewers that would be delivered by rebroadcasting their service across the Republic.[41] With industrial relations already fractious, the British broadcaster did not want to poke this particular union bear. Meanwhile trade unions within RTÉ viewed rebroadcasting BBC as a missed opportunity for jobs in Ireland. After an opinion poll commissioned in 1975 by the Department of Posts and Telegraphs and RTÉ showed a majority of viewers in favour of a second RTÉ channel rather than retransmitting BBC1, Cruise O'Brien bowed to public opinion. RTÉ 2 came on air on 2 November 1978 with a night of entertainment from the Cork Opera House. There was now a choice of viewing practically everywhere in Ireland.

The opening of the new station did not stop the campaign for multi-channel viewing, however. In March 1981 the Minister for Posts and Telegraphs, Albert Reynolds, announced that the Department would allow cable operators to supply Cork, Limerick and Galway,[45] the timing undoubtedly influenced by an upcoming general election. Within a year, cabling of Cork city commenced. Technology had advanced in the ten or so years since Dublin and other centres on the east had been cabled, allowing Cork to lead with features such as a popular local television channel. A less popular innovation was the decision to scramble signals so that a decoder was required by viewers, though this Cork novelty eventually became a requirement across the country. While royalty payments to the BBC and ITV would apply if the signals for Cork viewers were relayed by microwave, international convention was that royalties did not apply if television programmes were relayed solely by cable.[41] Thus a 66-km-long underground coaxial cable was laid to Cork city from an aerial site on the Knockmealdown Mountains, the second longest such cable in the world.[46] In the end this still was not long enough and, to provide a reliable signal, the cable had to be extended ever further to Seefin mountain in the more distant Comeraghs.

Such was the desire for British television that residents in the small town of Tallow in Co Waterford blocked diggers employed to bring the coaxial cable with the precious television signals to Cork city. 'We are entitled to the very same consideration as the people of Cork city', Frank O'Brien of Tallow told an unimpressed *Cork Examiner*.[47] The residents' strategy worked, with Tallow and the nearby towns of Lismore and Cappoquin being cabled in advance of many much larger centres. By the mid-80s, most major towns away from the border were being supplied with multi-channel television by cable. Ownership of the systems consolidated with those in Dublin, Galway and Waterford being merged in 1986 into a single entity called Cablelink.[48] Control of this cable network, one of the largest in Europe, was to shape the future of telecommunications in Ireland.

In the midst of this expansion of cable television, a rival technology was emerging, free of coaxial cables – and sanction from the Department. The growing availability of electronic components such as amplifiers for TV signals allowed enterprising individuals and community groups to erect a 'deflector'. Typically located on a hill, this picked up the UK channels, amplified them and then retransmitted them locally over the air. All that was needed by the viewer to pick up the retransmissions was a simple rooftop aerial: no cable, no satellite dish, no huge mast. By 1982 it was estimated that 50 such systems were in use from Donegal down to Galway.[49]

The name was somewhat disingenuous, for a 'deflector' was a self-made television transmitter and, like any transmitter, required a licence from the Department of Posts and Telegraphs. The advice from the Department's lawyer and the Attorney General was that the Minister had the power to licence such deflectors[50] but neither Fianna Fáil's

John Wilson nor his successor, Fine Gael's Jim Mitchell, saw fit to do so. Many backbench politicians were, by contrast, happy to champion the use of deflectors, viewing it as a rural-vs-urban, west-vs-east argument. Galway West TD John Donnellan told the Dáil:

> I have five children who love to watch this or that programme thanks to the deflector system. Much as we may despise the British for the number of things they do, it has to be said that they transmit some very interesting programmes one might like to view occasionally. Therefore it is necessary that we in the west have the same access to television programmes as those people living in the Dublin area.[51]

A 1988 bill to close down pirate radio stations originally also included powers to close down television deflectors but these were removed at the last minute to prevent a revolt by rural TDs. Meanwhile, Ministers did not licence deflectors for fear of disturbing interest groups, in this case the cable operators[52] who had paid the Department for their licences. Successive governments played a political dance. Deflectors were tolerated in areas where there was no licenced alternative – essentially villages and rural areas that were uneconomic to wire for cable TV. But when a deflector system began to serve parts of Galway city, just as its licenced cable system was being rolled out, it was quickly shut down.[53]

Rather than licencing the deflectors, the Department proposed a new technology called Multipoint Microwave Distribution Service (MMDS). This technology had been developed in North America and was sometimes called 'wireless cable'. It was based on the same principle as traditional broadcast TV and deflector systems, using a transmitter on high ground beaming out a signal to the surrounding area but at frequencies of around 2.5 GHz to 2.7 GHz, higher than those used for conventional television transmission. The use of higher frequencies allowed for a larger number of channels and reduced the possibility of interference with other transmissions, but also required a larger number of transmitter sites to provide nationwide coverage.

Launching the MMDS scheme as one of his last acts as Minister, Jim Mitchell said it would provide multi-channel viewing to every home in the country and that existing deflector operators would be allowed to apply for an MMDS licence.[52] Neither claim proved true.

MMDS was cursed from its inception. For a start, its name was somewhat unfortunate since, for most people the term 'microwave' conjured up images of cooking, which is probably why newer communications technologies using the same frequencies such as WiFi and Bluetooth have avoided the term.

The more intractable issue for MMDS was, however, competition from the by-now established deflector systems in many parts of the country. MMDS required the viewer to install a specific aerial and decoder as well paying a subscription fee. Partly because MMDS was not widely used elsewhere in Europe, the equipment was relatively expensive, while under an arrangement similar to that which had applied to cable companies since the late 1980s, MMDS operators had to pay royalties to the BBC and ITV.

Nevertheless the system was enthusiastically promoted by Jim Mitchell's successor, Fianna Fáil's Ray Burke. His Department ploughed ahead with a franchise scheme, dividing up the country into 29 areas, with licences awarded in 1989. Curiously, 19 were awarded to companies associated with Independent Newspapers, then controlled by Tony O'Reilly. Despite the assurances of the previous minister, not one of the franchises was awarded to deflector operators. Ray Burke also announced that the new MMDS network was to be used to distribute a new national television channel[54], thereby avoiding the cost of constructing a new transmitter network. Burke's enthusiasm for a new station, independent of RTÉ, was born of his desire to clip the wings of the state broadcaster which he reputedly regarded as hostile to Fianna Fáil.[55] Yet again, politics was influencing technical decisions. In the end, however, it would be a further ten years before TV3 would grace Irish screens – pipped to the post by state-owned Teilifís na Gaeilge, now TG4.

A reason for Burke's enthusiasm for MMDS became clear a few years later when it emerged that a sister company of Independent Newspapers had paid Ray Burke £30,000 in order to help secure those 19 licences.[56] It was not the last time that a government minister would be embroiled in controversy over awarding of a telecommunications licence.

Independent Newspapers must have contemplated looking for their donation to be returned as viewers continued to favour the illegal deflector systems over the more expensive MMDS. In a letter to the media group, Burke assured them that 'immediately MMDS service is available in any of your franchise regions, my Department will apply the full rigours of the law to legal operations affecting the franchise region'.[56] In reality, faced with widespread opposition from both opposition and government TDs, neither Burke nor his successors did much to curb the deflectors and it was normally left to the MMDS operators themselves to exercise their licence rights in the courts.

Through the 1990s, the deflector system remained the more popular option: in 1997 150,000 households were obtaining UK television from 100 illegal deflector systems in 12 counties while the legal MMDS system had only 95,000 subscribers.[57] In May of the following year, following a marathon High Court case that ruled in favour of the rights of deflector operators,[58] the Department belatedly started a licencing scheme for deflectors. As with the decision to cable Cork city 16 years earlier, the precise timing was undoubtedly due to politics: an election had been called and the issue of multi-channel television was a huge electoral issue in the west and south.

The issue of deflectors was particularly contentious in West Donegal where, in 1996, MMDS equipment at Ardara had been destroyed in an arson attack.[59] In the 1998 General Election, local man Tom Gildea ran as an independent candidate for Donegal South West with a primary objective of retaining the deflector system. The pre-election edition of the *Donegal News* contained an advertisement for Tom Gildea and, on the opposite page, another from the Department inviting deflector operators to apply for licences.

The outgoing government hoped that the long-awaited legalisation of deflectors would win over voters but the policy shift did not please some influential vested interests. Just before the election, in a break with tradition the *Irish Independent* published a front page editorial under the headline 'It's payback time', urging readers to reject the incumbent government and vote for Fianna Fáil. Independent Newspapers had a large stake in the MMDS system thanks to their £30,000 donation to Fianna Fáil's Ray Burke in 1989 and the unprecedented editorial was allegedly due to the prospect of the rival deflector systems being legalised.[60]

Donegal voters, evidently sceptical of the government's proposals, elected Tom Gildea to the Dáil. The politics of bringing foreign media to Irish screens was no longer confined to smashing a television in a Dublin pub; it was now affecting who governed the country.

The election of Tom Gildea is probably the only case in the world where the main objective of a member of a country's parliament was to ensure the availability of television from another country. Without a hint of irony, one of Gildea's other polices was opposing the construction of telecommunications masts.[61] Sadly his election literature did not explain why he considered a television transmitter built by volunteers to be a good thing while a telephone transmitter operating at similar power and frequency built by professionals was to be opposed. To be fair, Gildea's lack of consistency regarding the physics of radio waves was, and remains, widely shared among elected representatives.

The new government continued with plans to licence deflectors, administration now devolved from the Department to the newly formed Office of the Director of Telecommunications Regulation (ODTR). As part of their licence conditions, deflector operators now had to pay royalties to the BBC and ITV[62] like cable and MMDS companies.

The issuing of MMDS licences was not Ray Burke's only legacy to telecommunications. Late in 1988 RTÉ decided to sell off most of its share of Cablelink,[63] the largest cable operator in the state. The sale attracted considerable interest, with US-based PacTel offering the largest bid and a promise of £30m (€38m) worth of investment.[64] Minister Burke then intervened, instructing the RTÉ Authority to sell to Telecom Éireann, despite its poorer offer, saying that he wanted to protect Telecom Éireann from possible competition.[65] Burke was flying in the face of a European resolution to open up the telecommunications market and, eventually, Telecom was forced by European rules to divest itself of Cablelink. Unfortunately for the consumer, that sale came too late, for reasons that will became evident later.

Cable and MMDS slugged it out across the west and south of the country but increasingly both faced a common enemy from the sky. Satellite television, like its terrestrial predecessor, crept into Ireland slowly from abroad. By the early 1980s it was possible, for anyone prepared to install a receiving dish several metres wide, to watch television signals transmitted from a satellite 36,000 km away in the sky, though the difficulties involved and the lack of content in English limited its appeal. To further add to the difficulties, in predictable protectionist fashion, the Department of Communications warned in newspaper adverts that those in possession of satellite receiving equipment rendered themselves liable to prosecution. Equally predictably, no such prosecutions took place.

Satellite dishes thus remained limited to niche markets. A Limerick hotel installed a dish so that Aeroflot crews could catch up on Soviet TV before heading from Shannon to Havana or Moscow, while some enterprising Aran islanders paid £1,500 to install a 1.8 m diameter dish in 1986 on Inis Mór. This cosmopolitan choice of television provoked surprise from American tourists: 'Maybe they might have come over here to sample the wholesome, simple life of the Islands, but when they sat down to have a pint one night, they had no objections to the selection of channels available', an islander told the *Connacht Tribune*.[66]

A few months later, Cork viewers joined this cosmopolitan club led by the Aran islands when satellite stations, including the nascent Sky, were added to the cable lineup.[67] Other cable television operators around the rest of the country followed suit and by 1989, thanks to the popularity of cable in Ireland, Rupert Murdoch's Sky channel had more potential viewers in Ireland than in its target market of the UK.[68]

For a period in the 1980s, it looked as if the tables would be turned and Irish-made television would be beamed into homes in the UK and further afield after the government issued a licence to launch Ireland's own Direct Broadcast Satellite (DBS).

Such satellites operated at a higher power than earlier models so that viewers across Ireland and the UK would be able to pick up the broadcasts using a receiving dish of just 60-90 cm. A company called E-Sat Television, its name derived from Éireann Satellite, arranged to take one of the five channels[69] but thanks to a row between the other shareholders, the Irish satellite literally never got off the ground. That was not the end of E-Sat, however. Shorn of a hyphen but retaining its ambitious 28-year-old boss, Denis O'Brien, Esat went on to revolutionise telecommunications in Ireland.

Without its own satellite, Ireland effectively became part of the market for UK satellite broadcasters. DBS eventually arrived early in 1989 when the Astra satellite was launched into orbit, allowing viewers throughout Western Europe to pick up stations with a small receiving dish. While it still lacked the crowd-pulling BBC1 and ITV, as the Sky platform on Astra expanded to include dedicated sports and film channels, it started to become a viable supplier of choice for many Irish households. By 1997 an estimated 80,000 homes had subscribed to one of Sky Television's satellite packages.[57] Its attractiveness increased further in October 1998, just as deflector systems were receiving their licences, when Sky launched its digital service.

Much of rural Ireland now not only had multiple channels, but multiple means of viewing them. For a price, Sky offered a plethora of choice in digital widescreen quality, including, from 2001, the main BBC and ITV stations. For the budget-conscious, deflectors offered the four most popular UK stations, though often with indifferent quality. Stuck in the squeezed middle was MMDS, offering less choice than Sky but at a substantially higher cost than the deflector systems.

By 2003 satellite reception of the main BBC channels was available for free as well as through Sky's subscription service. Soon after ITV and Channel 4 also became available and almost every home in Ireland could enjoy crystal-clear reception of the main UK networks without paying a subscription fee. It was the death knell for both the cursed MMDS system and the rival deflectors. By 2016, with only 24,000 remaining customers,[17] MMDS services ceased. The last of the rival deflector systems had closed four years earlier as part of the closedown of analogue television.

That slow closedown of analogue television started in the late 1990s with the launch of Digital Terrestrial Television (DTT) in Northern Ireland and elsewhere in the UK. Like its analogue ancestor, DTT uses a network of transmitter sites, with signals picked up on an aerial. South of the border, the DTT service marketed as Saorview started on 29 October 2010.[70] And so, on 24 October 2012, after almost 50 years of service[71], the 625-line analogue transmitter networks both North and South were turned off. Not only could DTT accommodate more channels at higher definition, it also required less bandwidth so that part of the frequency spectrum formerly used for television was freed up for eventual use by wireless data services.

By the late 1990s those fat coaxial cables snaking their way around the housing estates of Irish towns and cities seemed like a relic of an earlier age. Cable was being hit by a triple whammy. In much of the country, cable operators had prioritised expanding their controversial MMDS systems over cable. Many of the new Celtic Tiger-era housing estates were served by MMDS and never cabled, while the infrastructure in existing cabled areas received little investment. In the towns of Tallow, Lismore and Cappoquin, the cable networks installed after the townspeople had blockaded the coaxial cable en route to Cork city were abandoned. Like MMDS, cable faced competition from ever-increasing choice of channels on satellite. The third, and arguably biggest, problem was a bewildering sequence of mergers and buyouts which left the country in 2000 with two cable providers, neither of which had sufficient capital for investment.

In time, that network of fat coaxial cables found a new raison d'être. Belatedly, from about 2007, the cable networks in towns and cities across Ireland were upgraded to carry digital television, telephony and broadband. The enhanced network was based on a model with a fibre optic backbone but still using coaxial cables for the last leg into homes and businesses. This hybrid fibre-coaxial (HFC) network was similar to the fibre-to-the-cabinet (FTTC) model being used to carry broadband over telephone lines. However, thanks to physics, coaxial cables offer far greater bandwidth than the copper pair used in traditional telephone networks. As a result, by 2015, most cable customers were able to enjoy 240 Mb of broadband, far faster than that available over DSL technology using thinner copper telephone lines. As we will learn in later chapters, however, those thinner copper telephone lines are being replaced by fibre-to-the-premises (FTTP). This may, eventually, supplant the venerable coaxial cable and lead to homes having a single fibre-optic connection used for all communications. Unless they also upgrade to a full-fibre network, this leaves cable operators stuck in another technological cul-de-sac; perhaps a reason why the country's dominant cable operator was reportedly put up for sale in 2021[72].

In the interim, the limited geographical extent of the cable network has given rise to a digital divide. It is partly the familiar geographical divide of urban versus rural, with cable available only in cities and larger towns, not villages or isolated houses. But there is another geographical divide, based on those sometimes faltering television signals from Northern Ireland and Wales. Even big towns close to the border like Monaghan or Letterkenny were never cabled because their residents could receive UK television with a simple aerial, and thus those towns could not benefit from broadband supplied by cable. What was for many years a blessing became a curse.

The battle to receive television from the other side of the border was effectively won in the mid-2010s. From 2005, TG4 and the two RTÉ stations were broadcast from transmitters within Northern Ireland while by 2015 almost all cable customers, North and South, were being served by Liberty Global under the Virgin brand. For many years subscribers to Sky's satellite service on either side of the border have received almost identical content, with viewers from Larne to Listowel glued to the same English Premier League matches. The Open Broadcast Area proposed in the 1970s is effectively in place, though probably not as Conor Cruise O'Brien imagined it.

The screen on the wall retained its potency, becoming a ubiquitous feature in bars and homes. As part of a global shift in the way in which television programmes were consumed from the 2010s, viewers across Ireland increasingly used streaming services such as Netflix and Disney Plus instead of watching programmes from broadcast TV stations. The distinction between a television and a computer became blurred. Even more so, the traditional division between the telephone and cable television systems evaporated, with both networks now carrying data ranging from telephone calls to television content – and everything in between.

The trend towards streaming services accelerated in 2020, when people became confined to home for long periods thanks to the global Covid-19 pandemic, giving Netflix an estimated market share of two-thirds of the Irish population.[73] One of the most popular Netflix series of the period was *The Crown*, a historical drama series about the reign of Queen Elizabeth II.

Unlike in 1953, however, there were no press reports of television screens being smashed due to the appearance of the British monarch.

CHAPTER TEN

WHERE'S MY BLOODY PHONE?!

In the pretty Co Clare village of Mountshannon, on the morning of 28 May 1987, postmistress Florence Bugler opened up the village post office and sat down at its manual switchboard (Figure 10.1) just as she had done almost every day for the past 13 years. The village post office was small, but it was a busy job operating the switchboard as well as dealing with customers at the counter. In fact, for her first six years on the job Florence worked an 84-hour week[1] from 8:00 in the morning till 10:00 at night six days a week, with respite only on Sundays when she was allowed to close the switchboard for most of the day. But she did not mind the long hours too much: she and her family had a long history of providing a service to the people of this part of east Clare. She was the great-granddaughter of the village's first postmaster who had been appointed in 1888 and when the village's telephone exchange opened in 1937 it was her great-aunt who operated it until her death in 1972.[2] Working a switchboard was second nature to Florence who had been helping her great-aunt from a young age.

However, today Florence would be finishing early. This was no ordinary day for her, or for Mountshannon. The post office's 88-line magneto switchboard, little different to the one opened by Agnes Duggan in Dublin's Crown Alley in 1900, would soon be defunct. The last manual exchange in Ireland, it was about to be replaced by a digital exchange as part of a massive programme to revolutionise telecommunications in the country.

At noon, the switchboard bell rang with a call – a call that Florence and in fact everyone else in the village was expecting. Normally sleepy Mountshannon was alive with reporters and cameras with a loudspeaker system set up specially to broadcast this call across the village. Even the confirmation ceremonies in the local church were rescheduled to allow the children and their parents to witness the event.[2] Everyone poured out onto the streets to listen and watch, with a crowd forming outside the post office.

Figure 10.1: Florence Bugler at the magneto switchboard in Mountshannon shortly before its closure in 1987. *(Noel Gavin/Press 22)*

Florence, with a twinge of nervousness, saw the line flap drop and plugged a cord into the indicated socket. 'Mountshannnon exchange. Number please?' she asked, even though she already knew the answer.

'Get me An Taoiseach', replied the caller, who was none other than the chair of Telecom Éireann, Michael Smurfit, making this call standing on a podium directly in front of the post office.

Florence plugged the other end of the cord into a line to Limerick, and dialled the number. The call was answered by an even more familiar voice, that of Taoiseach Charlie Haughey.

'You're through now caller, go ahead,' Florence said, as she had done so many times before.

Everyone listening outside the post office heard through the loudspeaker Smurfit begin by warning the Taoiseach that their call was being broadcast across the village and 'not to say anything you shouldn't say' – possibly an indication that previous phone conversations between them had been less than cordial. But this call was to celebrate a happy event and not an unpleasant word was exchanged.

Haughey congratulated Smurfit and all involved, reminding them that the replacement of the last manual exchange in Ireland was 'the end of an era in one way and the start of a new era in technology in another', before he finished the call saying, 'God bless you. Bye, bye',[3] his words echoing across the village.

Florence unplugged the cord and hung up her headset for what was the very last time. She walked out the door of the post office and stepped onto the podium. Under the gaze of hundreds of well-wishers, reporters and a television crew, she was handed a Waterford crystal lamp by Michael Smurfit himself.

Smurfit then addressed the assembled crowd. The Irish phone network, he announced, was 'technologically one of the most modern in the world'.[1] The crowd applauded. It was a claim that no-one would have dared utter even three years earlier. But it was true.

The Irish telecommunications network had been in a state of crisis. In 1980 94,000 people were waiting[4], some of them for years, just to get a phone line installed. Ninety years after Almon Strowger had invented the automatic exchange, almost half of the telephone exchanges in the State were still manual (Figure 10.2). And for those lucky enough to have a phone on an automatic exchange, the chance of getting through first time on a trunk call was about one in three – if it rained they were lucky to have any service at all.

The crisis in the State's telecoms network, its full horror exposed in the 1979 Dargan Report, led to two critical government decisions. The first of these was the approval of a £650m (€825m) five-year Accelerated Development Programme (ADP).[5] This was an enormous amount for the early 1980s, equivalent to 2 per cent of GNP[5] during a period largely marked by austerity. Indeed, the £235m (€298m) spent in the single financial year 1980–1 exceeded the total amount spent on the telecoms network over the entire 58-year period since 1922.[6]

The other key decision was to take the country's troublesome telecommunications service out of the clutches of the civil service and move it to a new semi-state body solely responsible for telecoms. As a first step, in November 1979 the government created a new interim body to plan this transition. An Bord Telecom, as the temporary entity was called, was rather toothless, with the telecoms budget and staff still controlled by the Department of Posts and Telegraphs. But the appointment of Michael Smurfit as chairperson lent it a higher profile than suggested by its actual powers

Smurfit was – and still is – seen as one of Ireland's most successful businessmen: in 1979 his company had 100 paper and packaging plants dotted across the world. His appointment was an attempt to bring business acumen to a service that severely lacked it and Smurfit was happy to be the public face of telecommunications, including making the last call from Mountshannon's manual switchboard with the aid of operator Florence Bugler. That in itself was a radical departure for a service traditionally viewed as being run by anonymous functionaries.

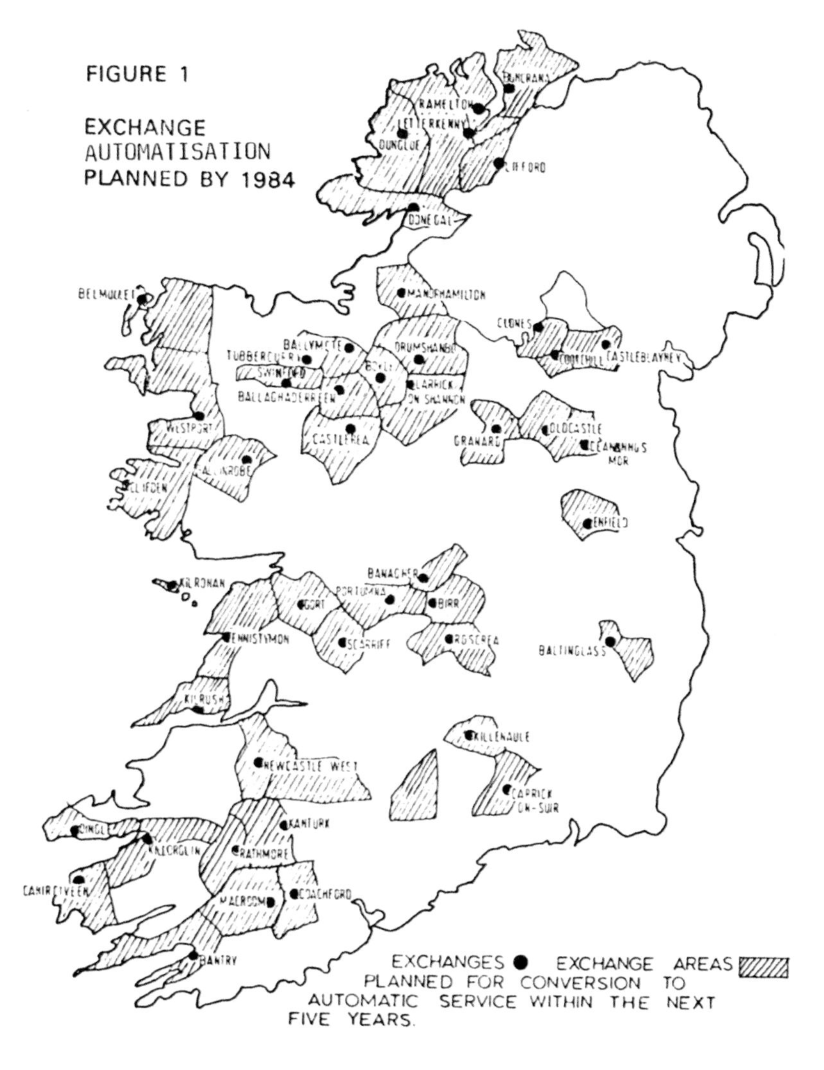

Figure 10.2: Map of manual exchange areas in December 1980. Despite the huge level of investment, the planned date of 1984 was not met and it took until 1987 for manual switching to be eliminated.

Soon after becoming chair of An Bord Telecom, Smurfit went on a 'mystery shopper' exercise. In his 2014 autobiography he describes how he walked into the P&T public office on Marlborough Street in Dublin, completed a form to apply for a telephone line and handed it to the clerk behind the counter, who did not look at it, or appear to recognise the man standing in front of him. When Smurfit asked him what he did with completed forms, the clerk replied unambiguously, 'It's none of your f*****g business.'

Rather than taking offence, Smurfit was sympathetic, blaming such attitudes on the choking bureaucracy that staff had inherited from the civil service. Smurfit described the staff as demotivated and undervalued, 'ashamed to tell anyone where they worked, because they knew what they would get: "You work for the telephone company? Jesus Christ! Where's my phone?"'.[7]

Smurfit was not exaggerating. Isolde Goggin recalls that her father, a telephone technician, instructed his family never to tell anyone that he worked for P&T lest they be plagued with demands to be pushed up the waiting list. She herself won a Student Engineer scholarship from P&T in 1977, studying in Trinity before joining an old-fashioned organisation where engineers were served cups of tea by more junior civil servants.[8] Her arrival was heralded in the staff newspaper, *PagusT,* by the headline 'At last a woman engineer'.[9]

Smurfit must have winced at a story in the *Irish Times* in 1983, when a businessman told the paper about how he had bribed a P&T technician to fix his phone line. He was £50 (€63) down but his single phone line was restored to life after being dead for three weeks.[10] Perhaps it was not surprising that faults took so long to be repaired for, in the same article, a mother described her son's 'working' day since joining P&T. After a leisurely breakfast each morning:

> the crew driver ... and the foreman then study form in the newspaper and then it's dinner time. By the time the money is put on the gee-gees and the results discussed it is time to go home often as early as 3.30 p.m.[10]

The poor attitudes and morale amongst some of the staff were not surprising. Industrial relations had been toxic within the Department of Posts and Telegraphs and the relationship with customers was antagonistic. Sheer frustration led to one Dublin travel agent placing an advertisement in the *Irish Times* in 1984 (Figure 10.3). Addressed to the CEO of the state-owned telco, it offered a free holiday in Crete in exchange for a guarantee that their phones would work for at least one week.[11]

Not only were staff involved in conflicts with subscribers who were waiting years for a phone line or months for a fault to be repaired, they were also engaged in constant disputes about money, in the form of a myriad of complaints about telephone bills. Subscribers did not trust their bills and, by extension, the entire Department. It was no wonder that staff were afraid to tell people where they worked.

The system was completely broken. Fixing it would be an enormous task, requiring not only the huge investment promised by the £650m ADP but also a complete change in the mindsets of staff, politicians and the long-suffering customer. But it had to be done.

WAVIN PIPES LIMITED
BALBRIGGAN

We regret that our telephone is temporarily out of order.

For urgent orders please telephone

(01) 4 3 7 7 9 3

WE APOLOGISE FOR ANY INCONVENIENCE CAUSED.

TO MR. BYRNES
OF BORD TELECOM
If you can guarantee that our phones will work for at least one week, you or anyone you care to nominate will be given a holiday in Crete this weekend.
GREEK HORIZONS
55 Capel Street, Dublin 1.
Tel. (01) 740234

THE CENTRAL WASTE PAPER CO. LTD.
Station Road, Portmarnock, Co. Dublin, can now be contacted on the following lines:
460052, 460961, 462040, 460742, 460538.

DESPERATE !
Have you got a spare telephone line in the 71 (Crown Alley) exchange area ?
If so, we need it !
Contact Box 80707.

Figure 10.3: During the early 1980s, newspapers made considerable revenue from businesses forced to notify their prospective customers of their telephone woes, sometimes with inventive promises.

Since the foundation of the State, successive governments neglected telecommunications such that the country was put in the 'position of dray-horses as against race-horses' as warned about by minister J. J. Walsh in 1923. By the 1970s it was obvious to everyone that the condition of the country's telecoms network was a huge barrier to economic development. Ireland was competing in a race with other countries for foreign investment but it was way down the field, overtaken by sleek thoroughbreds with modern infrastructure. Finally, 56 years later, the state had taken decisive action to allow it to compete with 'up-to-date countries' in this race for economic development.

The following 20 years from 1979 to 1999 were the period of greatest change to the telecommunications network in the country's history. If we picture the period as a 20-year race between the telecoms networks of different countries, Ireland ran amazingly well for 14 years, coming from way behind to catch up and even outrun the other countries. Unfortunately, however, Ireland became distracted, lost much of its lead and finished somewhere in the middle of the field. The exception was in the mobile phone race, which is a story unto itself.

The early part of the race was simply spent catching up with the other horses. There was an awful lot of ground to make up. The waiting list needed to be cleared, ancient equipment replaced and the network completely automated. There were also the thorny problems of mending staff relations and building up trust with customers. All of this required making important and sometimes difficult decisions. But, remarkably in view of the previous 56 years, Ireland actually caught up and, for a time, outpaced the field.

The performance of Ireland during the first five years of this telecoms race from 1980 to 1985 was massively boosted by the aforementioned Accelerated Development Programme. Probably the most important improvement brought about by this investment was the virtual elimination of the waiting list for phones, though this was definitely an uphill race. The number of people waiting for a line remained stubbornly high, largely due to the high demand for new phone lines. This demand continued in spite of P&T resorting to the old trick of jacking up the payment required before installation. In 1982 an applicant located more than three miles from the exchange faced a bill of £980 (€1,244) merely to have a line installed.[4] While the waiting list proved particularly slow to clear in Dublin due to problems securing suitable locations for new exchanges, it was no longer necessary to relieve a telephone engineer's toothache in order to get a phone installed.

A far-reaching technical decision was taken at the start of the ADP, one that was to give Ireland a huge advantage in the race: the new telecommunications infrastructure would be digital. Digital technology means that information – such as a human voice – is encoded in binary format as a series of zeroes and ones. As a result, multiple subscriber calls can be transmitted over the same transmission medium, allowing for much greater capacity. In his autobiography, Michael Smurfit describes persuading An Bord Telecom to change tack and replace orders for further crossbar equipment with ones for digital.[7] This was somewhat fanciful as the decision to go digital predated Smurfit's arrival by a year[12] but, regardless of who was responsible, the selection of digital technology quickly proved to be the correct one.

Within a few years of the decision in Ireland, countries all over the world were going digital too, sometimes ditching relatively new crossbar equipment. For example, the step-by-step exchange in Sligo town was replaced with a digital exchange in 1983.[13] Three months later, 70 km across the border, the step-by-step exchange in Enniskillen Co Fermanagh was also replaced – with an electro-mechanical crossbar exchange[14], which was itself replaced within seven years.[15] The days of southerners looking longingly across the border at the better planned and more modern telephone network up north were over. The decision to go digital allowed Ireland to move right up the field by effectively leapfrogging a generation in telephone technology.

With the selection of digital technology made, the next move was to decide on who was to supply the equipment. In another wise move, the Department decided not to put all its eggs in one basket but to standardise on two exchange types: Alcatel's E10 and Ericsson's AXE.[16] The choice of Ericsson was not unexpected; the Swedish telecoms giant had been a supplier to the P&T since the Economic War in the 1930s[17] and had signed a bulk supply agreement in 1972.[18] The selection of French-owned Alcatel was a departure but one with good reason.

In France, throughout the 1950s and 60s, the telecommunications system had a similar reputation to that in Ireland. By the 1970s these deficiencies had become a liability for the French government, which embarked on a massive telecoms investment programme.[16] With traditional Gallic statist chauvinism, the programme used almost exclusively French equipment suppliers including Alcatel, which had pioneered digital exchanges based on time-division multiplexing (TDM),[16] a concept first used on the telegraph network (Figure 2.12). The plan proved a remarkable success, with the number of telephone lines in France doubling in the four years from 1975 to 1979. The success of this French programme, and the technology it used, was studied back in Dublin. A soft loan from the French treasury and an offer of technical assistance sealed the Alcatel deal.[19]

Ireland's first digital exchange went live on 4 December 1981 in Athlone.[20] Appropriately for the location, it was an Ericsson AXE, partly built at the company's plant in the town. The first Alcatel E10 went live in Kells Co Meath five days later.[20] The importance of the Alcatel deal to France was demonstrated when the later official opening was attended by the French post and telecoms Minister Louis Mexandeau, together with 35 French officials and journalists, as well as the Irish minister.[21] To streamline installation, mini-exchanges called Remote Subscriber Units were fitted in containers complete with batteries and wiring, ready for installation at smaller locations. As a result, a slew of digital exchanges opened over the following years and manual switching was finally eliminated.

Equally important were the wholesale additions of new lines to link the new exchanges to homes and businesses. From 1982 fibre-optic cables were deployed to link exchanges together. Fibre-optic cables are comprised of a strand of glass just slightly thicker than a human hair, along which light is transmitted (Figure 10.4), allowing transmission of more bandwidth over longer distances than copper cables. By 1993, 70 per cent of the transmission capacity comprised of fibre-optics.[22] Meanwhile existing microwave and coaxial links were upgraded to digital transmission, greatly increasing their capacity. This was not the traditional P&T pattern of make-do-and-mend, of adding a few lines here and an extra switchboard there. Ireland was no longer just catching up but was racing ahead.

The selection of Alcatel and Ericsson as suppliers was also due to their willingness to establish facilities in Ireland. The long-established Ericsson plant in Athlone grew to over 500 employees[23] and was involved both in the delivery of the digital AXE exchanges and more humble devices like telephone handsets (Figure 10.5). The first

E10 exchanges, such as that opened in Kells in front of a delegation from France, were supplied to P&T by a joint venture between French-based Alcatel and Irish firm Telectron. However, within days of the opening ceremony in Kells, the partnership fell apart after US telecoms giant AT&T took a large share in Telectron.[16] Two years later Telectron was closed down, the loss of 500 jobs at its main site dealing a body blow to Tallaght.[24] Alcatel continued alone, developing its own plant in Bandon Co Cork, which operated for 20 years.

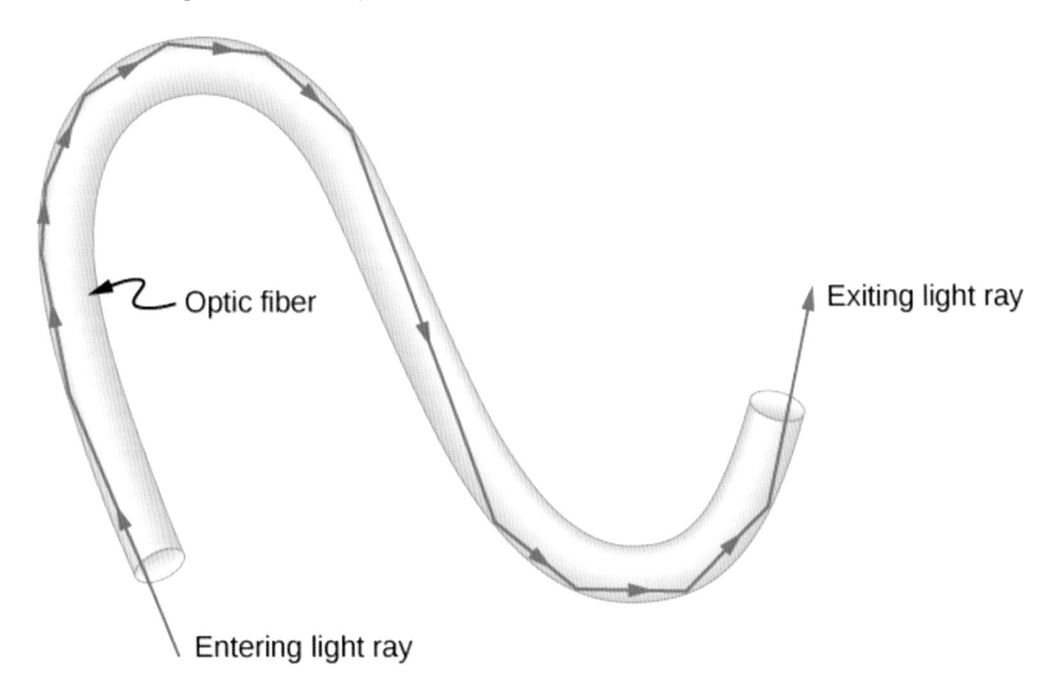

Figure 10.4: A fibre-optic cable allows transmission of large volumes of data, including phone conversations, by light, typically generated by a laser. The light rays are reflected within the fibre, even following it around corners. *(Steven Mellema via CC BY 4.0)*

Canadian firm Northern Electric, which had established a factory in Galway in 1973[25], also benefitted from the investment program. In addition to making components for PABXs, it produced telephone handsets for the state-owned telco (Figure 10.6). The Galway location successfully survived multiple changes of name and ownership, first to Northern Telecom, then Nortel, changing focus in the 1990s from manufacturing to services. Under the control of Avaya, it became a centre for development of software for business communications.

These two telephone models were manufactured in Ireland and installed in homes and businesses across the country from the early 1980s. Figure 10.5 (left): Originally called F78 and marketed in Ireland as 'Shannon', this handset was designed by Danish designer Henning Andreasen and is now considered a design classic. (*Norsk Teknisk Museum, CC BY-SA 4.0*); Figure 10.6: This handset was made by Northern Telecom in Galway under the name 'Harmony' also used by its Canadian parent. *(Courtesy of John Mulrane)*

In a direct sense, the decision to use suppliers with a local manufacturing base was not a long-term success: Telectron shut down and the other players ceased manufacturing in Ireland, as part of a global move of production towards lower cost economies. However, companies such as Ericsson and Nortel/Avaya were able to move up the value chain from manufacturing into services and in doing so play a part in establishing the country as a hub for Information and Communications Technology (ICT), helping in Ireland's economic race.

Another huge problem that the generous capital funding tackled was the number of line faults and their management. The long-standing problem of water damage to cables was addressed through a programme of cable pressurisation.[26] Air was pumped into the underground cables at the exchange with the aim of minimising water seepage into cables and junction boxes. This was definitely a case of catch-up, as the UK had started such a programme in 1960.[27] But a new fault logging and testing service, developed in conjunction with computer vendor Nixdorf, moved Ireland up the field. Introduced in 1986,[28] the system checked for duplicate records before passing the report to the repair centre and could scan lines for faults at night-time, detecting an issue before the customer had even noticed.[29]

The public telephone service finally began to catch up from 1981 when P&T started replacing the venerabl A/B coinboxes with a more advanced payphone (Figure 10.7) from Danish manufacturer GNT Automatic.[30] This permitted long-distance calls to be dialled direct, improving service to users and removing a huge burden from operators.

Figure 10.7: Minister for Posts & Telegraphs, Albert Reynolds, launching the first new 'Payphone' in Tallaght on 3 March 1981. The new model was from Danish company GNT Automatic, a partner of Ericsson. *(By permission of irishphotoarchive.ie/Lensmen Collection)*

While more durable than its predecessor, the model was still susceptible to determined vandals and dependent on being emptied of coins regularly. To try to address some of these problems, in 1988 a trial of phones that accepted prepaid cards was belatedly commenced.[31] The trial proved a success: with fewer moving parts and no money inside to attract vandals, they were more reliable than their coin-operated predecessors. Again, a French solution was selected using chip cards (Figure 10.8) and from 1990 many coin-operted public phones were replaced with cardphones.

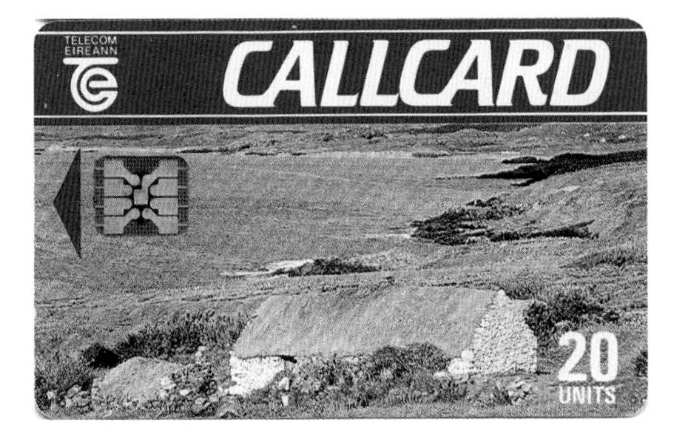

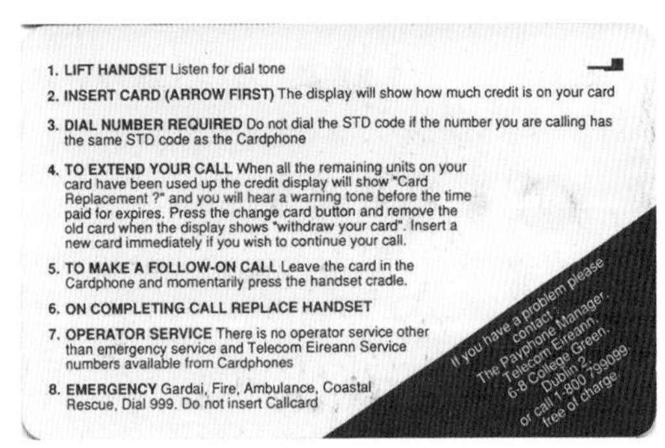

Figure 10.8: During the 1990s, 'callcards' found their way into the purses and wallets of many Irish people. This design, featuring a traditional Irish cottage, was used for many years. The last callcards were produced in 2003. *(Courtesy of irishcallcards.net)*

The range of destinations that could be dialled direct from the new payphones, and private phones, also expanded. While direct dialling between Dublin and London and Dublin and Belfast had been possible since 1971 and extended to some other British routes during the 1970s, calls to most parts of the UK still had to be connected manually. At the end of 1982 direct dialling was introduced to all of Northern Ireland[32] and a few months later this was extended to all parts of Britain.

These two developments removed a huge strain from operators, allowing for more catching up by introducing services that had been available elsewhere for years. In December 1984 Telecom Éireann implemented a Freephone service,[4] initially on a manual basis. Callers dialled 10 and asked the operator for 'Freephone' and the name of the business using the service, for example 'Freephone Social Welfare'. The operator would look up the corresponding phone number from a list and place the call with charges reversed. It was not exactly a novel idea: a similar service had been in use in the UK for 24 years.[27] An automated 'toll free' service using the prefix 800 had existed in North America since 1967[33] and the idea that a prefix that included the digits 800 indicated a free call slowly spread across the globe. In April 1990 Telecom Éireann caught up and introduced an automated Freephone service using the prefix 1800, alongside 1850 which charged callers one unit regardless of the call duration.[34] Telecom had been faster off the mark with premium rate numbers where some of the revenue was shared with the call recipient. These were introduced in 1989 using the prefix 03000, before being moved in 1991 to the number range 15xx, with the latter 2 digits determining the charge.

As Ireland moved up from the back of the field in the telecoms race, it finally achieved independence from the UK for international telecommunications in the form of a satellite ground station. Its location at Elfordsdown near Midleton Co Cork was selected partly because of the particularly mild climate in the area, as snow collecting in the satellite dish would degrade the signal.[35] Planned since 1980, the station cost £8m (€10m) and was equipped by Japanese firms Mitsui and NEC. When it opened in 1984, it provided 192 circuits to the US.[36] Calls were transmitted from its 300-tonne, 32 m diameter dish up to the Intelsat satellite, 36,000 km above the earth[37] and then down to an earth station at Etam in West Virginia. By 1987 there were two other satellite ground stations in use for telecoms purposes; one at Galway for use by Digital,[38] the other at Terenure in Dublin, and satellites were carrying over 50 per cent of the country's non-UK international telephone traffic.[39]

The heyday of satellites for two-way communication was brief, however. In 1988 the first transatlantic fibre-optic cable went in service between the US, UK and France. In the same year, a fibre-optic cable was laid between Portmarnock in north Dublin and Porth Dafarch near Holyhead. The 118-km-long unrepeatered BT-TE1 cable consisted of 12 fibres and had a capacity of 140 Mb, enabling it to provide 4,000 voice circuits.[40] This was 167 times greater than the previous two cables laid between Ireland and Wales in 1948 – but similar to the bandwidth of a typical domestic broadband connection for a single house in the late 2010s.

A year later, on 24 May 1989, a branch of the new PTAT-1 fibre-optic cable was brought ashore at Garretstown strand in Ballinspittle, Co Cork,[41] connecting Ireland with North America for the first time since the closure of the last transatlantic telegraph cables from Valentia in 1966. Fibre-optic cables offer two advantages over satellite: greater bandwidth and lower latency. The 17,000-telephone-circuit-capacity of PTAT-1 dwarfed the capacity possible through satellites. Furthermore, calls transmitted over the 7,520 km cable did not suffer the annoying delays imposed by a 72,000-km-long return journey up to a satellite and down again. Soon after, Elfordsdown and its satellite links were demoted to use as back-up to the fibre-optic cables.[42] As more cables were laid, satellite back-up was no longer required and the station was closed, its dishes becoming something of a white elephant among the rolling hills of east Cork. It was not the end of the road, however. In 2011 the redundant station was taken over by the National Space Centre, which uses the site for space research and commercial purposes.[43]

The short life of the Elfordsdown satellite station indicated both the pace of technological evolution and the level of investment taking place. Both the PTAT-1 and BT-TE1 cables were out of use by 2007,[44] having served less than 20 years, replaced by newer cables with much greater capacity. It was quite a contrast to the P&T tradition of patching up and making do.

But to enable Ireland's telecommunications service to catch up with other countries required more than just vast capital investment. It was also essential to reform the governance of the telecommunications service. The Dargan Report had proposed that the telecommunications and postal services should be taken out of the hands of the civil service and moved to two new semi-state bodies, similar to the ESB. An Bord Telecom had been created as an initial step to plan this transition. However, the actual transfer of budget and staff from the clutches of the civil service took several years, becoming bogged down in the political uncertainty of the early 1980s.

Ironically, part of this political uncertainty was due to a controversy that had the telephone service at its heart. In December 1982 the *Irish Times* broke the news that the telephones of two journalists, Bruce Arnold and Geraldine Kennedy, had been tapped officially with warrants signed by former Minister for Justice Seán Doherty. An inquiry found that neither journalist had been connected with criminal or subversive activities. Instead the taps had been instigated by Seán Doherty, who was concerned about leaks to the media from cabinet.[45] In the resulting case the High Court ruled that the telephone tapping was an invasion of the journalists' constitutional right to privacy and an abuse of power by the State.

The Postal and Telecommunications Services Act of 1983 legislated, albeit imperfectly, for who could, and who could not, authorise tapping of a telephone. More significantly, the act finally created two separate state-sponsored enterprises for telecommunications and for postal services (Figure 10.9). The rump of the Department of Posts and Telegraphs survived, renamed as the Department of Communications to supervise general policy on telecommunications issues.[16] As part of the legislation, workers would retain the same terms and conditions as they had enjoyed as civil servants. The size of the Irish civil service was halved[22] as employees were transferred to two separate new semi-state entities for postal and telecommunications services.

Telecom Éireann, as the new telecommunications entity was called, came into existence on 1 January 1984. With 18,000 staff, it was the largest employer in the State[46] and its establishment was a national event with its CEO Tom Byrnes addressing the staff over a special RTÉ television broadcast at 11 a.m. on the day of vesting. As well as its staff, Telecom inherited a large debt burden from P&T,[22] though the Exchequer retained liability for pensions for the years during which Telecom employees had worked in P&T.[47]

One of the reasons for moving responsibility for telecoms to a commercial semi-state company was to make it easier to raise the funds required for capital investment. To assist this, another French idea was copied. This was the creation of a subsidiary,

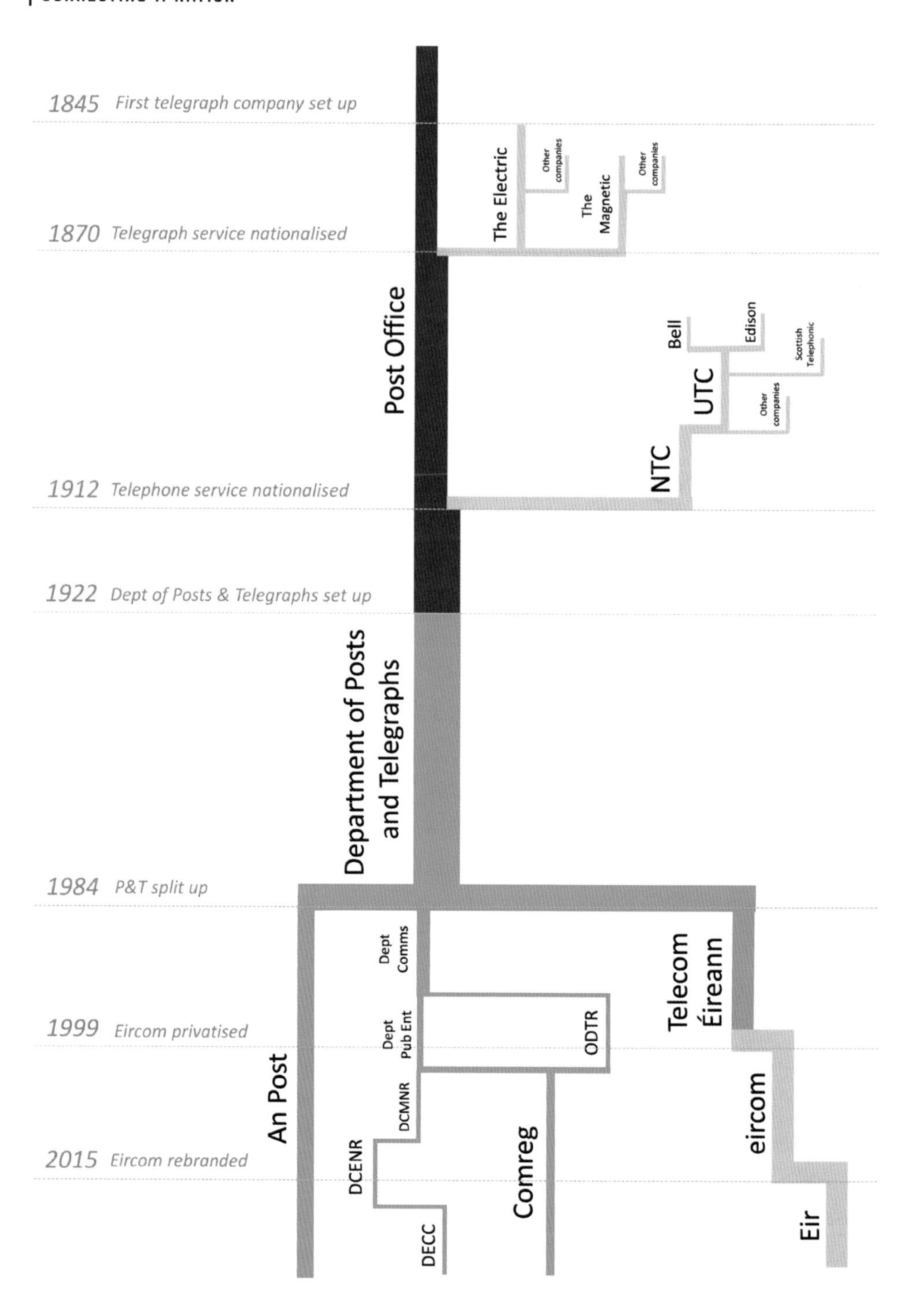

Figure 10.9: A simplified family tree of the Department of Posts and Telegraphs, starting with its ancestors in the postal service before independence and finishing with its successors including Eir.

Irish Telecommunications Investments (ITI)[16] which raised money privately to buy assets that it then leased to its parent.[48] Financial help also came from the European Community through the STAR development program.

By the late 1980s, thanks to continuous investment, Ireland was pulling ahead of its competitors, eliminating outdated step-by-step exchanges in advance of many other countries. At Ship Street in Dublin, the step-by-step equipment at the country's first automatic exchange was replaced on 20 April 1989,[49] 62 years after the exchange had been opened by W. T. Cosgrave. Nine days later, step-by-step switching was eliminated from the network when Clontarf exchange went digital. Ten years later, the local network became fully digital when Quaker Road exchange in Cork city was converted from crossbar to digital on 30 March 1999.[50] The replacement of older equipment and vast expansion of transmission capacity brought immense benefits to the quality of service. In 1978 only 61 per cent of local calls and 38 per cent of STD calls were successful on the first attempt[51]; by 1993 the equivalent figures were 99 per cent and 98 per cent, respectively.[4]

As well as improving service, the wholesale replacement of outdated equipment allowed for savings in current expenditure. The number of telephone operators required was slashed in a completely automatic network. The old step-by-step exchanges were tended to by a team of technicians who regularly oiled the moving parts and replaced worn-out components (Figure 10.10). By contrast the digital switches had no moving parts and needed no routine maintenance. There was a joke in Telecom Éireann that the only staff required at a digital exchange was a man and a dog: the dog was employed to bark at the man if he touched the equipment and the man was employed to feed the dog.[8]

For years Ireland had been behind in the race to offer new telecoms services. The Speaking Clock reached Dublin 37 years after it had arrived in Paris, while International Direct Dialling arrived 19 years after Brussels. Now, thanks to the decision to go digital and the massive investment programme, Irish users were able to enjoy enhanced services, such as call forwarding and three-way calling, ahead of many other countries. Marketed as 'PhonePlus', these services were introduced in the Cork area in 1987,[52] just two weeks after Florence Bugler had hung up her headset for the last time at Mountshanonn's magneto exchange. The services were then rolled out to all digital exchanges in the country, with Caller ID becoming available in 1998.

Cork was also the location for a pilot of voicemail at two of the city's exchanges in 1994; the system from Octel was the first implementation of voicemail on a public telephone network in Europe.[53] This led to additional revenue, both as a result of calls answered by the voicemail system that otherwise would have received no answer or an engaged tone, and calls returned to those leaving messages. As a result voicemail became a standard part of the telephone service.

Figure 10.10: A P&T technician tending to the group selectors at Crown Alley 2000-type Strowger exchange in during the 1950s. Electro-mechanical exchanges such as this required constant attention.

Like the choice of digital technology, some other, smaller, decisions allowed Ireland to skip a generation and avoid technological cul-de-sacs. Just like telephone users in Mountshannon who were catapulted from winding a handle to pushing buttons, from July 1981 operators answering Directory Enquiries calls in the Drogheda region moved from consulting paper directories to a computerised system,[54] cutting the average waiting time from 2 minutes to 20 seconds.[55] At the same time, BT was completing moving its Directory Enquiries service – to microfiche.[56] It was a further case of Ireland racing ahead of the others on the field.

Capital investment solved another major problem for both staff and customers – the thorny problem of disputed bills. P&T had been slow to update the processes around billing its customers. For decades, details of calls were written out in longhand by operators and subsequently copied on to statements prepared manually on accounting machines,[57] a tedious and labour-intensive process. In 1970 the Department struck a deal with Aer Lingus to prepare bills on the airline's computer system. Operators now marked call details on pre-printed 'mark sense' cards using soft pencils. The information on the cards was processed by the airline's system which then produced the customer bills.[57] It was not until 1973 that P&T installed an IBM 370 computer and took billing back in-house.[58] The 'mark sense' cards were later replaced by more sophisticated versions using optical character recognition.

Computerisation did not, however, solve the fundamental problem from a customer's perspective. Their quarterly bills baldly stated how many call units had been registered on the meter at the exchange, leaving them mystified as to how this figure had been achieved. The bills were, in fact, generally correct but customers often did not understand, or want to understand, the highly varied cost of calls. On a call to Australia, for example, the meter advanced every 1.38 seconds[59] so that a 10-minute call was the equivalent of 435 local calls. It was easy for a family member to surreptitiously make expensive long-distance calls and then let the bill-payer dispute the amount due with Telecom. Following the creation of the office of Ombudsman in 1984, that office was kept busy with complaints about phone bills, with the Ombudsman remarking sagely that 'in cases where there is a telephone bill problem in a family with teenage children, the chances are the problem lies at home and not with Telecom.'[60]

Trust in the bills issued was so poor that in 1984 it was reported that ten per cent of phone bills in Ireland were queried, more than ten times the rate in Sweden.[61] As a result in the early 1980s an enormous amount of Dáil debating time was consumed by questions to the Minister along the lines of why the phone bill issued to Mrs Murphy from Ballygobackwards was twice her usual bill. The public office of the accounts department on Dublin's North Frederick Street was perpetually filled with customers clutching bills that they could not fathom.

Eventually, towards the middle of the race in 1988–90, detailed billing was gradually introduced following completion of a huge project that involved the transfer of customer information from cards, some of which dated back to the 1920s.[62] Billing disputes eased off markedly but worries about running up a huge telephone bill remained for many people, as we will discover later in the story.

While most of the technological advances in the period were the result of the large investment programme, some came from outside, with little input from P&T or Telecom Éireann. The most important of these was the facsimile machine, or fax. The idea of being able to transmit a picture had existed almost as long as the telephone or radio. Various systems were introduced from the late nineteenth century but all remained expensive, slow and cumbersome, limited to use by big newspapers and press agencies.[63]

The Irish Press became the first fax user in Ireland when it installed an Associated Press 'wirephoto reception machine' in September 1947. The impetus was the All-Ireland football final between Kerry and Cavan which was played that year in New York. Photos of Sunday's game were ready for the Mondy morning edition,[64] which must have boosted sales to no end in Co Cavan, home of the winning side (Figure 10.11). The Irish Press fax machine was later joined by those in other newspaper offices but fax remained a specialist product until the 1980s.

CAVAN 2-11, KERRY 2-7 IN GREAT FINAL

As Mayor O'Dwyer throws in the ball to start the Polo Grounds Final, Teddy O'Sullivan (Kerry) tries to grip, while J. Stafford (Cavan) watches developments before rushing to his position.

A.P. Photo (by radio and wire)

IRISH PRESS Pictures Sent By Wire

THE IRISH PRESS made Irish newspaper history last night when pictures of the Gaelic Football Final in New York were being processed in its offices in Dublin before the people were home from the grounds. This is the first time that photographs have come over the wire into a Dublin office.

The decision to cover the match pictorially was taken as soon as the G.A.A. had decided to hold the final in New York and arrangements were begun.

As there was a five-hour time-lag to beat—New York time being five hours later than ours—flying the pictures across the Atlantic was automatically ruled out and wire-photo service arranged.

Associated Press, world-wide news agency supplying THE IRISH PRESS, was asked to co-operate.

EQUIPMENT FLOWN FROM NEW YORK

It was discovered that the agency's London headquarters had no wire-photo machine available for shipment to Dublin, that the only way to obtain one was from the New York A.P. Office.

The equipment was thereupon flown from New York to Dublin and installed in the IRISH PRESS offices.

With the transmitter in New York and the receiver in Dublin, the next problem was to provide a circuit over which to transmit the pictures. Associated Press, mindful of the fact that no wire-photos had ever been sent over a circuit to Dublin, checked with the British General Post Office, while the Irish General Post Office was consulted on this side.

The complete circuit was arranged, and Mr. J. H. Williams, A.P. representative, came to IRISH PRESS, Dublin, to carry out the tests.

They proved successful.

The organisation for the day of the match was then put into motion. This is how it worked—yesterday—to produce on our paper of to-day the pictures of Kerry and Cavan in their historic final:—

The pictures, taken by A.P. photographers, were rushed by motor-cycle to their offices at the Rockefeller Plaza. Here they were immediately developed and printed, despatched to the radio-transmitting station and radioed to cable and wireless headquarters in London. Despatched immediately to the A.P. Office, London, they were transmitted on a special telephone circuit in the G.P.O., London, and the G.P.O., Dublin, into the Art Department of the IRISH PRESS.

The Art Department staff were standing by to re-develop and print the photos. In a matter of minutes the Process Department took over and the blocks were made, paged and gone to press.

Had the match been played at Croke Park, Dublin, at the same hour (8.30 p.m. Irish time) as it was in the Polo Grounds, New York, the pictures could scarcely have been processed sooner.

The operators at each point of the transmission were working to a time schedule that left hardly any margin for delay, and there was the chance of a technical hitch to be guarded against, since this was the first time the New York-Dublin wire-photo link was completed.

There were technical hitches—yet the pictures came through in time.

The success of the transmission was due to the efficiency and co-operation of the five separate organisations that took part.

HOW IT WORKED

THIS is a brief description of the process involved in the wire-photo system: A light, reflected on the original photos, was reflected to a photo-electric cell, which converted the light into electrical energy.

This was transmitted over the transatlantic wireless system and re-transmitted over the wire-photo equipment on the telephone circuit to Dublin.

In Dublin the electrical energy was re-converted into a beam of light, which was reflected on to photographic paper, thus reproducing the original photographs.

How Cavan Heard The News

(IRISH PRESS Staff Reporter.)

CAVAN, Sunday Night.

Cavan is en fete to-night with bonfires, dances and cheering crowds signalling the announcement of All-Ireland victory.

Since about 7 the town was like the Deserted Village, young and old being clustered round every available radio set.

On the broadcaster's announcement that Cavan had won, every door was thrown open and the people hurried into the streets. Bonfires were lighted and children and their parents joined in singing local ballads and cheering.

The first message to go out from Cavan was that of Most Rev. Dr. Lyons, Bishop of Kilmore, who wired Mayor O'Dwyer of New York: "Bishop of Cavan heartily congratulates the Cavan team."

Mr. Jack Williams, of Associated Press, superintending the reception of a picture of the G.A.A. Final at New York, by the wire-photo instrument in THE IRISH PRESS Office last night.

Figure 10.11: Excerpts from the front page of *The Irish Press* of 15 September 1947. That year's All-Ireland Final, played in New York, made heavy use of telecoms, with the commentary as well as facsimile photographs transmitted by transatlantic radio-telephone from the US to England and thence by cable to Ireland. *(With thanks to Irish Newspaper Archives)*

The widespread adoption of fax from the 1980s was driven by three factors. The first was the telephone network itself. Throughout Ireland, and across much of the world, it was now easier to get a telephone line, charges were generally falling and the network was almost totally automated. The second factor was the adoption in 1980 of a new global standard for fax transmissions. Called Group 3, it digitised the scanned image, allowing for image-compression methods to be employed, and then transmitted the image file by modem at data rates between 2,400 and 9,600 bps. The third factor was the Japanese language, which lent itself to fax much more than telex. Mass production in Japan led to a fall in prices and the fax machine became a consumer product sold in office equipment and electrical stores around the world.

The concept of buying a fax machine in a retail store upended the old view of telecommunications technology. Traditionally, subscribers were provided with equipment by the Department as part of their line rental and for decades the only choice offered to residential users was whether the phone was black or cream in colour. You took what the Department gave you and felt grateful to have a phone at all. For many people, the fax machine was the first time they were able to select a piece of telecommunications equipment themselves instead of having to accept whatever the Department foisted on them. Fax created a new line of consumer telecoms products, a market that mushroomed in the mobile era.

Despite its different heritage, fax shared many of the advantages of telex but was simpler and often cheaper to operate. One of the advantages in common was that a fax could be received when no-one was present, a feature that could be exploited in surprising ways. In mid-1992 an English-based group called Feminist Forum mounted a 'fax-a-thon' where they sent details of English abortion services to fax machines in workplaces across Ireland.[65] This evaded a 1989 Irish court ruling that providing information on the availability of abortions in other countries violated the Constitution.

A further catalyst in the adoption of fax was the belated introduction of Direct Dial In (DDI), allowing a PABX extension to be directly accessible by an outside caller, without the intervention of a switchboard operator. The first installation was at RTÉ early in 1990, when its 1,200 extensions became reachable by simply prefixing the extension number with the digits 64.[66] DDI had been around in Germany for decades, so this was definitely a case of catch-up.

The increased number of phone lines and introduction of fax and DDI caused pressure on the availability of new phone numbers. To alleviate this, Dublin numbers were converted from six to seven digits in phases between 1990 and 1994, with similar changes in many other areas around the country. Further numbering changes occurred in 1998 when the national numbering plan was revised. This involved the merging of several areas, especially in the north-west, and increasing the length of local numbers. As part of this programme of changes, the prefix for landlines in Northern Ireland was changed from 080 to 048. The use of the code 048 was deliberate, as it implies that the six counties in Northern Ireland are close to their southern Ulster counterparts of Monaghan (prefix 047) and Cavan (049). The choice demonstrated that telephone numbering schemes are not always completely determined by engineering requirements.

The sea-change in attitudes towards investment in telecommunications that had occurred since 1979 had largely been driven by a stark realisation that the dismal condition of the country's telecommunications network was thwarting industrial development. In the late 1970s the quality of the telecoms network was 'hurting the IDA effort daily'. Once Ireland began to catch up with its competitors in the telecoms race, the IDA quickly began to exploit the potential of this new infrastructure to attract inward investment so that 'the new promised land of digital phones on demand was added to the IDA gospel around the globe.'[67]

As Ireland started to overtake many of its competitors on the field, the IDA was able to use the advancements in infrastructure to lure more sophisticated business sectors to the country. Six days after receiving that call on his mobile from Michael Smurfit in Mountshannon, Charlie Haughey launched the Custom House Docks redevelopment.[68] The project was aimed at attracting financial services to Dublin and in his speech Haughey described how it would capitalise on the latest telecommunications technology.[69]

As the telecoms race gathered speed, the huge investment programme corrected a long-standing regional imbalance. For the first time ever, rural Ireland enjoyed telecoms services that were at least as good as those enjoyed in the bigger towns and cities so that the IDA was abe to use telecommunications as a tool to promote regional development. One example of this was the office in Castleisland Co Kerry, created in 1988 by insurer New York Life. Claim forms from their customers in the US were flown to Shannon and taken to Castleisland for processing with the captured data being sent back to the US over a leased data circuit.[70]

The IDA also began to exploit the vast increase in international capacity provided by the new undersea fibre-optic cables to the UK and US to promote Ireland as an attractive location for call centres.[71] By 1996, 26 call centres had been established including American Airlines, Hertz and Sheraton Hotels, employing over 3,000 people. Forty per cent of the multilingual staff employed came from abroad, helping to reverse

Ireland's traditional tide of emigration[67] and earning Dublin the title of 'call-centre capital'.[72] By 2002 the sector employed up to 19,000 people.[71]

In a development that would have seemed unbelievable even ten years before, Ireland began to export its expertise in telecoms. From 1988 hundreds of Telecom Éireann staff were seconded to work in Britain on contracts for BT and its rival, Mercury. Soon after, the collapse of the Iron Curtain created new opportunities to sell Irish expertise. Newly democratic Eastern European governments knew that economic development depended on modernising their decrepit telecoms networks and sought Irish expertise to help achieve this. Telecom Éireann's treasury wing, ITI, set up a joint venture with the Hungarian state-owned telco to raise finance from the capital markets using the same model successfully adopted in Ireland.[73] Dr János Láng, the Hungarian CEO of the joint venture, told the *Irish Times* that they wanted to emulate Ireland's success: 'You have your digital network and services of a high quality and any applicant can get a line within a few days.'[74]

There were some dissenting voices in the midst of the adulation. A 1987 report prepared for accountancy firm Ernst and Whinney was optimistic about the prospects for the financial services centre at Custom House Docks but considered Telecom Éireann's service to be the project's weak link, with transatlantic calls costing twice as much from Ireland as from the UK.[75] The report's author argued that there was no technical reason why firms in the Custom House Docks area could not set up their own telecommunications links, bypassing Telecom's network, adding 'if the Pope was coming to the dock site, they'd have a satellite dish down there in two days.'[76]

That report from Ernst and Whinney was not the only one to point out that telephone users in Ireland paid more than elsewhere. The issue had been bubbling away for many years but as the telecoms service improved the issue of charges became the main focus of complaint for customers. A 1981 comparison by the *Irish Times* of telephone charges across nine European countries found that Ireland was in the top three in terms of price under every heading.[77] A three-minute call from Denmark to Ireland cost 88p (€1.12), but from Ireland to Denmark cost £2.25 (€2.86).[77] Matters were little better in 1988 when a survey commissioned by the then European Community found Ireland to be the most expensive country to have a phone installed, while long-distance phone charges were almost five times greater than in the Netherlands.[78] Telecom was caught in a dilemma: its charges had to cover repayments for the loans it needed to deploy new technology but by charging so much the technology was not fully utilised.

Part of the reason for the high charges for long-distance and international calls was that, unlike many other countries such as the UK and Germany, local calls were untimed. A fee was charged once the call was answered but after that the caller could yammer for as long as he or she wanted at no extra cost. These low rates were cross-subsidised by the much higher charges for long-distance and international calls which were a hangover from the days when trunk circuits were expensive to provide and limited in number. By the end of the 1980s, however, the advent of fibre-optics had rendered the distinction irrelevant. It did not really cost Telecom any more to connect a call to the other end of the country than to the phone next door. It was just that customers expected to pay less for the local call.

The high charges for trunk calls did not help Ireland at the telecoms race. Over the period 1991 to 1993, some steps were taken to reform tariffs through a reduction of long-distance and international call charges, expansion of the area classified as a local call and starting to charge for local calls based on duration. The changes were of particular help to business customers who tended to make more long-distance calls. They also benefitted the more sparsely populated Western counties with, for example, the number of phones that could be reached at the local call rate from Belmullet increasing from 1,200 to 10,500.[79]

The changes exposed a tension between the need to provide competitive international rates to attract multinational companies and the expectations of consumers for cheap local calls. In 1993 further planned reductions in the cost of long-distance calls and increases to local calls created outrage in the Dáil chamber, in the letters columns of the newspapers and even on the streets.[80] The level of protest was greatest in Dublin and the other large centres where telephone users were used to making most of their calls at cheap local rates. Nonetheless the changes were implemented as planned.

Such 'rebalancing' of tariffs did not help address a more fundamental problem – Telecom Éireann's staff costs. A 1989 study estimated that the state company's operating costs were the highest of any EU state.[81] In 1993 Telecom's 13,000 staff looked after 1.1m lines[4], a ratio of 85 lines per employee – the equivalent figure for BT was 152 lines per employee.[82] These higher costs were not just due to Ireland's low population density – the costs per line in the sparsely-populated Canadian provinces of Newfoundland and Manitoba were 40 per cent lower than Telecom Éireann's.[83]

The Irish government was still tacitly using the telecommunications service as an employment creation mechanism, with, as usual, the customer picking up the tab. It was an understandable reaction with unemployment at over 15 per cent. Lowering Telecom Éireann's cost base would lead to lower calls charges which, in turn, would lead to increased foreign investment and thus increase net employment. But they were probable future benefits, whereas dealing with Telecom's staff numbers would cause definite immediate pain. It was a case of what sociologists call a 'commons dilemma' where the short-term interests of a particular group are at odds with long-term group interests and the common good.

Some other countries, facing similar problems, had introduced competition with the aim of driving down costs. This was not a straightforward task as the provision of telephone lines is clearly a natural monopoly. As we saw earlier the chaos caused at Belfast's fire station in 1880 when there were two separate, unconnected networks showed the potential drawbacks of unbridled competition. However, by the early 1980s some governments began to consider how to allow competition for telecoms services without ending up like Belfast in 1880. A key moment occurred in 1982, when the US Federal Trade Commission reached a settlement with AT&T.[22] As a result, the giant US telecoms company was broken up into a number of regional companies that were responsible for lines to subscribers' premises in their area and local calls, while AT&T itself was confined to providing long-distance service.[22] Subscribers could also choose between a range of other companies for their long-distance calls, quickly leading to lower rates. Similar moves happened in the UK where competition for long-distance traffic arrived with the launch of Mercury in 1985. Customers could dial a prefix that routed their call from the local BT exchange onto Mercury's own network and thence to its destination.

In most of the rest of world, telecommunications remained a government monopoly, even if, as in Ireland from 1984, the relationship was an arm's length one. However, from the late 1980s the European Commission commenced a journey to liberalise the telecommunications sector within the European Community. It was felt that more competition in telecoms would encourage development in the ICT sector where Europe was seen to be slipping further behind the US and Japan. Liberalisation was also enthusiastically championed by a UK government that saw opportunities abroad for British firms that were already experienced in dealing with competitive markets at home.[84] A first step was a 1990 European directive to open up the market for value-added services such as data and fax.[22]

On foot of this directive, a company called Esat applied to the Department of Communications in 1991 for a licence to operate as a telecommunications company. This was the same company that had been involved in a bid for one of the TV channels on the abortive Irish satellite. Their initial application was not even acknowledged by the Department[85]: the protectionist instinct had asserted itself again. But eventually the government enacted the directive and granted Esat a licence. In 1994 Esat Telecom started to provide long-distance and international services to corporate customers who were connected to its own network over lines leased from Telecom Éireann. By the end of the year Esat had 94 corporate customers.[85]

As with the early days of both the telegraph and telephone, this liberalisation, limited as it was, attracted several new entrants. Within three years, Esat had been joined by other companies chasing the corporate market such as Cable & Wireless, BT, TCL (later Worldcom) and Stentor, which in an ironic twist was led by Patrick Cruise O'Brien, son of the former Minister for Posts and Telegraphs. Stentor ran up huge losses, ditched most of its management and board and eventually disappeared into a maze of mergers.

The licencing of Esat and others to provide limited services to corporate customers did not represent any sea-change in the mind of the government. Pressure from Telecom and the vocal Communication Workers Union led it to seek a derogation for Ireland from further liberalisation of telecoms. Instead of facing competition from 1998 as in other EU counties, the incumbent operator, Telecom Éireann, would be protected for ten years until 2003 in order for it to 'prepare for the rigours of full competition'.[86] Not everyone in the government that had sought the derogation was thrilled with the news. Mary O'Rourke, then a junior minister, recalls thinking to herself 'Why is this marvellous?'[87]

In any case, the derogation did not protect Telecom from all competition. Not only were companies such as Esat starting to pitch for corporate customers, by the end of 1993 a competitor emerged touting for international calls from residential customers. This was Swiftcall, founded by actor-turned-businessman from Clones Co Monaghan Tom McCabe and his wife Bridget.[88] Users dialled a regular number which connected them to Swiftcall's system, then entered their account number and the number they wanted. It was a bit of a rigmarole, but even after the controversial 1993 tariff changes, Telecom was still charging an eye-watering £1.84 per minute (about €3.75 in 2021 prices) for calls to countries like South Africa while Swiftcall charged 71p a minute. Many people were happy to press all the buttons required for such savings.

With networks requiring continued heavy investment and EU policies encouraging competition, state-owned telecoms were no longer seen as strategic assets – and cash cows – by governments. The idea of privatising the telecommunications network had been pioneered by the UK which sold off 51 per cent of BT in 1984 largely to small investors in the largest ever public share offering to that point. The idea of privatising Telecom Éireann was mooted by Michael Smurfit four years later, after the telecommunications service had reported an annual profit for the first time in 17 years, saying that it was not a question of 'if, but when' Telecom would be privatised.

The idea got little traction at that point, with privatisation being regarded by many as a wild excess of Thatcherite government policy. Union leader David Begg reacted angrily to the concept: 'I can tell you that he will be waiting for hell to freeze over before workers in Telecom will agree to their contribution to making a profitable public enterprise being squandered to make profits for multi-millionaires.'[89]

Slowly at first, however, other countries started to adopt the policies pioneered in the UK. In 1990 the New Zealand government followed the British lead and during the mid-1990s there was a flurry of sell-offs of state-owned telecoms companies across the world. Some were floated on the stock markets while others were partly bought by other telcos, creating a web of strategic alliances. While Telecom Éireann revenues were increasing thanks to a growing economy, its overall financial position remained weak due to heavy borrowing undertaken in previous years to fund its vast programmes of infrastructural improvements.

In 1993–4, it still had a debt mountain of £939m (€1.19bn), resulting in £93m (€118m) of interest charges – but was also expected to pay a £28m (€36m) dividend to the government. This dividend payment for a single year exceeded the total value of EU Structural Funds the company was due to receive over the next five years.[90] It looked like the government was back to its old tricks of treating telecommunications as a cash cow, hobbling the country's performance in the telecoms race.

In the early 1990s all attention seemed to be on the financial position of Telecom Éireann with endless discussions about potential partners buying a minority stake of the company from the state. UK-based Cable & Wireless spent many months in discussions but eventually a 20 per cent share of Telecom Éireann was sold to Dutch–Swedish consortium KPN-Telia in 1996 for £183m (€232m).[91]

The focus on finding a suitble partner distracted attention from Ireland's performance at the telecoms race. Capital investment had plummeted since the dizzy years of the ADP (Figure 10.12): the 1989–90 figure of £106m[92] was, adjusting for inflation, the lowest since 1974. Ireland had so convinced herself that her lead was unassailable that she was overlooking the fact that the race was now being won largely on the basis of data services and her competitors were racing ahead of her. An example of this was ISDN. Standing for Integrated Services Digital Network, ISDN is a set of communication standards for simultaneous digital transmission of voice, fax, data and other network services. For example, a small business could be provided with two voice circuits and ten DDI numbers – all carried over one physical line, comprising of a copper pair, little different to those laid by the NTC in the 1890s. BT started a pilot of ISDN in 1985[93], and by 1992 ISDN was available in a majority of countries in the then

European Community.[94] However, it took Telecom Éireann until 1994 to launch the service, an early customer being Gairmscoil Éinne on Inis Mór where German pupils received lessons by video conference from their teacher on the mainland.[95] While the reignof ISDN as the shiny new thing in telecoms was brief, Telecom's lack of commitment to the technology was not helping Ireland in the telecoms race.

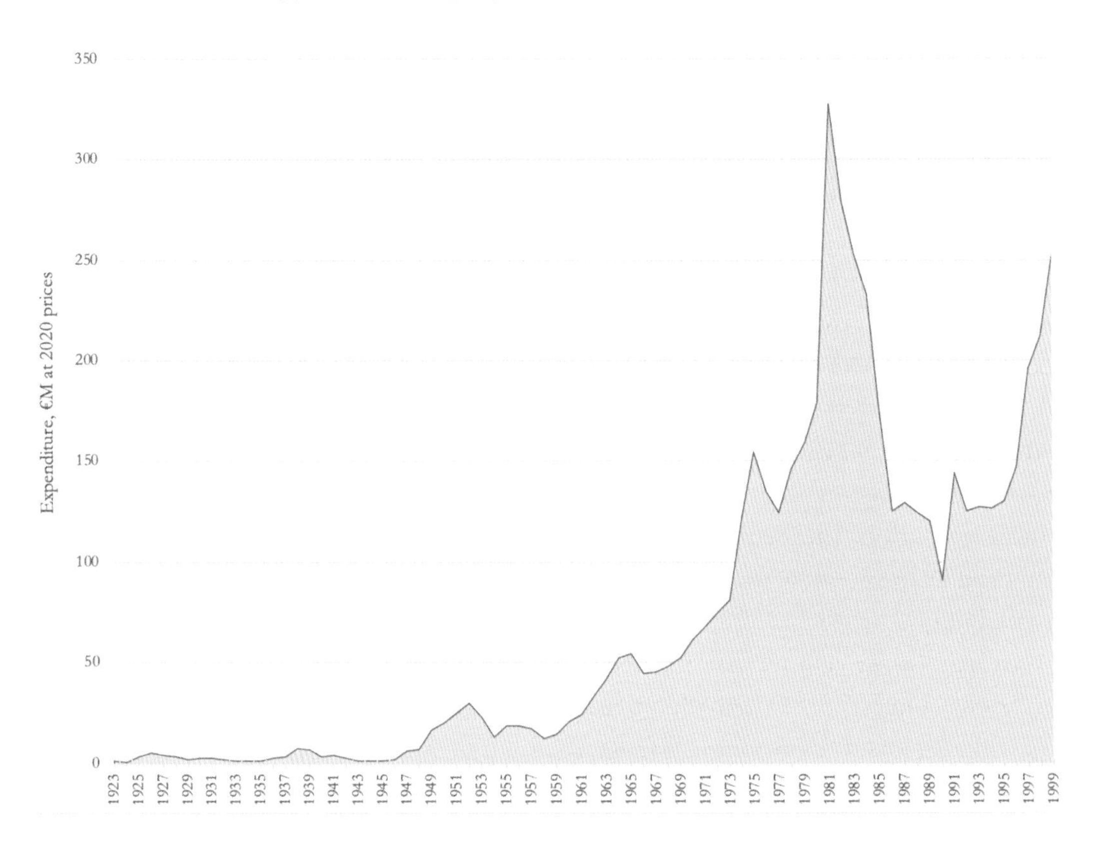

Figure 10.12: Annual capital expenditure on telephones. The massive increase in 1980, and decline in the late 80s and early 90s are all clearly evident. Sources: Flynn, Roddy, *The Development of Universal Telephone Service in Ireland 1880 - 1993* (1998) DCU, Doctor of Arts thesis; Telecom Éireann Annual Report, various dates.

The mid-1990s saw the dawn of the internet age and a rapid expansion in data communications. But this seemed to be ignored by successive governments, who were consumed by endless debates about ownership and competition, while happy to rake in dividends from the state-owned telco. In other countries ISDN was a popular way to access the internet as, amongst other benefits, it allowed the customer to surf the web and take voice calls at the same time. Thanks to its late introduction here, however, few people in Ireland had that option.

From an economic development perspective, the problems faced by larger customers who wanted high-speed data links were more serious still. Early in 1998 a report from Forfás, the state's advisory board for enterprise, trade, science, technology and innovation, delivered a sobering report on the state of the country's telecoms infrastructure. The report concluded that Ireland, despite having one of the most modern phone systems in the world in the late 1980s, was now lagging behind its competitors, creating a potential disincentive to investors. It criticised the paucity of high-speed leased lines for data and the lack of commercial Asynchronous Transfer Mode (ATM) services. Finally, it called for the government to abandon the derogation it had obtained from the European Commission and to liberalise basic voice telephony as soon as possible to encourage potential investors in telecoms infrastructure in Ireland.[96]

It looked like Ireland had been competing in the race still dreaming of Michael Smurfit's words from 1987 that her telecoms network was 'technologically one of the most modern in the world', while being overtaken by other countries. The world wanted cheap high-speed data connections but Ireland was still talking about push-button phones and call waiting. A few weeks after publication of the Forfás report, its warning became a reality when newspapers reported that Ireland had lost out on a new Microsoft data centre because of the country's broadband shortcomings.[97] Microsoft had been in the country for 13 years at this stage, employing 1,000 staff, so its decision created major ripples.

In May 1998 a cabinet meeting decided to end the derogation for Telecom Éireann and allow competitors to provide voice services. The decision was taken just four days after the Microsoft decision was reported, though Minister for Public Enterprise at the time, Mary O'Rourke, recalls that the timing was coincidental: 'We realised that competition was coming. We thought we better get moving or we're not going to be at the races.'[87]

Competition for residential customers was to arrive on 1 December 1998. At the launch party for its domestic product, Esat Telecom flew musician and TV host Jools Holland over to entertain guests.[98] As an indication of how government attitudes had changed towards liberalisation, the first phone call from a residential line over a privately owned network in 86 years was made that night by the minister, Mary O'Rourke. Within a few months, Ocean, a joint venture between the ESB and BT, and cable TV operator Irish Multichannel, were also touting for business. Esat's charging structure applied a flat rate across the country, an innovation for the time. However, customers still had to deal with two bills: one to the provider for calls and another to Telecom for line rental.[99] This had a major deterrent effect – three months after Esat was launched, a survey found that only 0.7 per cent of users had moved from Telecom Éireann to one of its rivals.[100]

With the market now partially liberalised, the issue of continued state ownership came back into focus. By 1997 a global slew of part or full privatisations of formerly state-owned telcos had proved wildly successful with both governments and investors making a killing. The stock market was booming: the ISEQ gained almost 50 per cent in 1997 alone, with technology companies, such as Irish-based security company Baltimore Technologies, riding the crest of the wave. In the preceding five years, Telecom's debt mountain had been slashed from £1bn (€1.27bn) to £148m (€188m).[101] With the Celtic Tiger in full sprint, the time seemed perfect to jump on the privatisation bandwagon.

Just before Christmas in 1998, the cabinet approved the sale of the state's share in Telecom Éireann. Union opposition had been bought off by a commitment to allocate 14.9 per cent of Telecom to an Employee Share Ownership Trust (ESOT). The KPN-Telia consortium acquired a further 15 per cent of the company, increasing their stake from 20 to 35 per cent. The government's remaining share was thus 50.1 per cent. The only question was the price of the share offering.

That decision was ultimately in the hands of Finance Minister Charlie McCreevy, who wanted to extract as much as he could for the exchequer. In her 2013 autobiography, Mary O'Rourke describes a tense meeting to thrash out the offer price at which McCreevy told the meeting that his Secretary General 'would resign if such and such a price was not achieved'.[102] The Secretary General of the Department of Finance got his way, with the launch price set at £3.07 (€3.90).

Despite being at the upper end of expectations, the price did not deter public interest, whipped up by media hyperbole. Shane Ross, later to become Minister for Transport, told his *Sunday Independent* readers: 'The downside is almost zero ... this is a one-way bet ... Hungry financial institutions will still be buying long after the flotation ... The price of Telecom shares, bar a global market disaster, will be virtually underwritten.'[103] Ivan Yates, then a TD for Wexford, told listeners during an interview on RTÉ's *Morning Ireland,* 'this is not a risky investment; this is a sure bet ... money for old rope.'[102] The exhortations worked: some 574,000 small investors bought into the stock, some taking out loans from banks only too willing to lend on this 'one-way bet'.

On 8 July 1999, shares in Telecom Éireann were launched at a tricolour-bedecked New York Stock Exchange, with the Minister ringing the famous opening bell. Outside, dancers from the Galway-based street theatre company Macnas, dressed up as insects and accompanied by drummers, surprised Wall Street stockbrokers on their way to work.[104] Eighty-seven years of public ownership of the telecoms network was over. Within two weeks, the share price had risen from the initial €3.90 to €4.91.[105] The exchequer was €4.21bn richer, 574,000 citizens were quids in and the unions had been bought over. Ireland, having lost her lead in the telecoms race, seemed poised to overtake her competitors again.

CHAPTER ELEVEN

A MOBILE NATION

The damn Vikings were late and they were going to miss the deadline.

Sarah Carey, Brian Noble and the rest of the bid team from the Esat Digifone consortium stood fretting on the steps of the Royal College of Physicians on Dublin's Kildare Street[1] taking shelter from the unusually warm sunshine. For months they'd been locked away in a basement office on Grand Canal Street, slaving away on a proposal to win the licence for Ireland's second mobile phone network. The deadline for proposals was today, midday – and they had done it, they were ready to deliver their proposal, all 12 copies of it, to the Department of Communications in the building opposite where they were standing.

But their proposal documents were nowhere to be seen. The Vikings were supposed to be delivering them, but they were late, probably lost – not on the high seas but in Dublin's fearsome traffic.

Esat Digifone was born of a hasty marriage between Irish upstart Esat Telecom and the Norwegian state-owned telephone operator Telenor. The consortium's sole purpose was to win this second mobile phone licence. A Viking theme for this moment, intended to showcase the Telenor involvement, was a theatrical move dreamt up by Esat's ebullient chairman Denis O'Brien.[2] But now, with just minutes to go before the deadline at noon, no one was looking too ebullient.

There was a lot riding on this licence. Up until this moment Ireland's mobile phone market was a monopoly. The only operator was Eircell, a subsidiary of state-owned Telecom Éireann, which itself had a monopoly over residential lines. But the tide was turning against monopolies in telecoms, both mobile and landline. The EU had mandated that member states had to liberalise the telecoms market; while Ireland had successfully sought a derogation from competition in the landline sector, a competitor to Eircell had to be allowed into the mobile market. The Irish economy was booming, with the moniker 'Celtic Tiger' coined just six months before. Whoever got the licence for the second mobile network would surely make a killing.

On Kildare Street the deadline ticked closer. A taxi pulled up outside the Department, depositing some men in grey suits and 12 cardboard boxes of documents.[2] It was one of the rivals, delivering their bid on time and with no theatrics.

Finally, just before noon, the anxious party standing on the steps saw a fleet of Esat-branded vans turning the corner at Nassau Street. Behind them a 40-foot container lorry, its barely-dry paintwork carrying the colourful though unfamiliar logo of Esat Digifone, awkwardly rounded the corner and pulled up outside the Royal College of Physicians. The truck's back doors swung open and, to the astonishment of the tourists and civil servants strolling along Kildare Street, a girl with a violin emerged from inside, engulfed in mist swirling out of the truck. Some of the growing crowd of onlookers recognised the mysterious tune as 'Nocturne' and the violinist as Fionnuala Sherry, the combination that had won that year's Eurovision for Norway.[3]

Two by two, like a surreal Noah's Ark, they emerged, 12 besuited men paired with 12 bare-chested Vikings, each pair lugging between them a sleek aluminium and glass container. The 'suits' were from Esat side of the Esat Digifone consortium while the 'Vikings' represented the Telenor side. Inside each of the fancy boxes was a copy of the damn licence proposal. Led by Esat's Denis O'Brien and Telenor's Per Simonsen (Figure 11.1), the suits and the Vikings hauled the containers across the street into the Department of Communications and deposited them on the floor of the lobby.[4] The whole spectacle was witnessed by the bemused porters of the Department, who no doubt longed for the simpler days of P&T.

The Esat Digifone team on the steps opposite, though relegated to spectators after all their efforts on the bid, rejoiced. Now they just had to wait and see if their proposal was successful.

This day, 4 August 1995 was a significant moment in the story of telecommunications for a few reasons. Firstly, offering the second mobile phone licence was the first major step taken to liberalise Ireland's telecommunications network. This step, however, was not without trauma. Esat Digifone, and not one of the five other bidders, did win the licence, but this in turn led to years of recrimination, legal battles and political intrigue. But most important of all, the arrival of the Vikings heralded the conversion of Ireland from a somewhat reluctant consumer of telecoms to being an enthusiastic adopter.

For all the money lavished on the Irish telecommunications network between 1979 and 1995, the telephone remained for many an exotic device. Even though the quality of the fixed-line network in Ireland had caught up with, or ever surpassed, its peer countries, arguably the fixed-line network was still associated in the public mind with waiting lists, unpredictable bills and indifferent service. The network might have been almost entirely digital, but only 76 per cent of Irish households had a phone line in 1995,[5] the lowest rate amongst her peers (Figure 11.2). For many people, there was little reason to think that Ireland would adopt mobile technology with any greater enthusiasm.

Figure 11.1: Denis O'Brien and Per Simonsen lead a party of Vikings delivering the Esat Digifone bid on 4 August 1995. *(RTÉ)*

In fact, it turned out that Irish people loved – and continue to love – modern telecommunications. It was just that the telephone service had been so poor and so expensive for so long that many of them had become used to doing without it. Once a telecoms service perceived as being useful and affordable came along, Irish people took to it with abandon. Ironically, the low rate of adoption of fixed lines provided a ready pool of customers for mobile phones. As a result, within five years of the Vikings' arrival at the Department of Communications, mobile phones had moved from the plaything of businessmen to a democratic communications tool, in the pocket of virtually every citizen in the country. Ireland, so often the telecommunications laggard, was near the top of the league table for ownership of mobile phones. It has remained near the top of such league tables ever since.

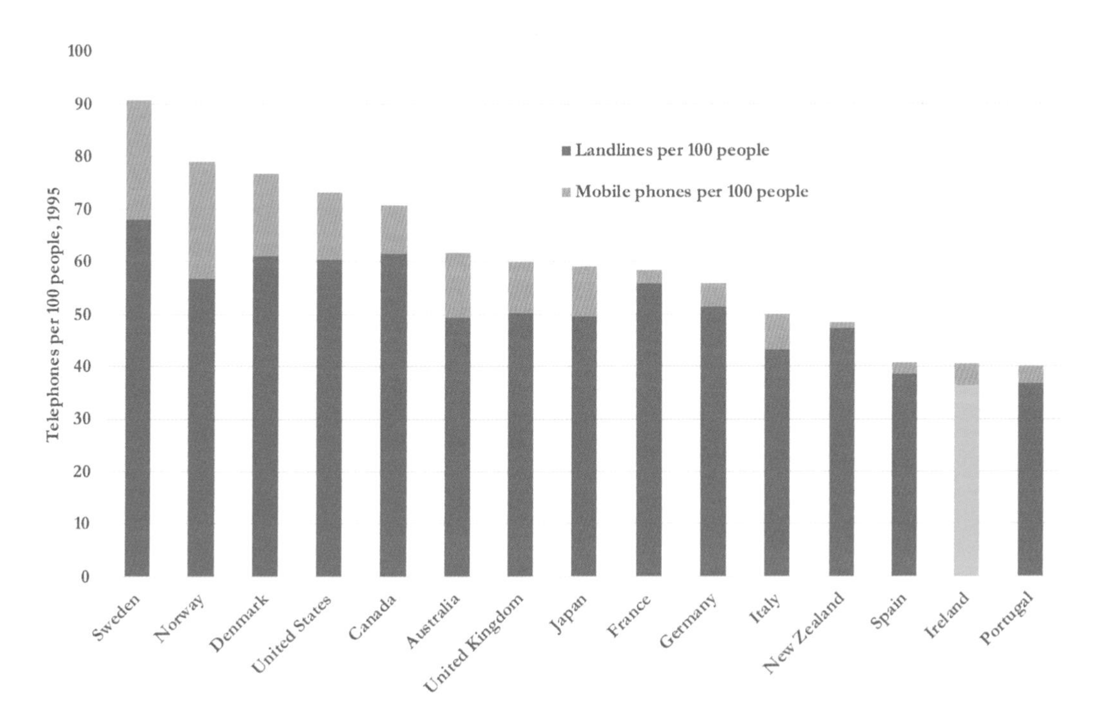

Figure 11.2: Telephone adoption rates in 1995. The low rate of landline penetration in Ireland is evident. Source: ITU, Yearbook of Statistics: Telecommunication Services (Geneva: 2004)

In order to appreciate Ireland's uptake of mobile technology it's necessary to look first at how mobile telephony came into being and developed. In fact, mobile telephony began earlier than one might think. As early as 1926 the German National Railway introduced the *Zugtelefon* onboard trains plying the busy Berlin–Hamburg line,[6] where transmitters and receivers at both ends of the line were connected to a wire that acted as an aerial strung along the existing telephone and telegraph lines alongside the length of the railway line. As for the long association of mobile phones and cars, this started just after the war when AT&T started a car telephone service in Saint Louis, Missouri. This was extended to another 24 US cities, while state-owned telecoms providers in several European countries set up similar networks. All of these systems, whether on a train or in a car, suffered the same constraints. Each base station covered a large geographic area, typically 40 km, requiring each phone to have a large, power-hungry transmitter. Capacity was very limited as each conversation required its own frequency. In New York in 1965, with only 12 channels to serve 2,000 users, the average waiting time to make a call from a car was 20 minutes.[7]

A solution had been identified as far back as 1947, when a Bell Labs engineer with the fortuitous name of Douglas H. Ring proposed the cellular concept. Instead of using a single high-power transmitter to reach users within a radius of 40 km, Ring proposed using 'cells' as small as 1 km, with lower-power transmitters and an assignment of channels to each cell (Figure 11.3). This technique would allow the use of

U.S. Patent Mar. 13, 1979 Sheet 2 of 8 4,144,411

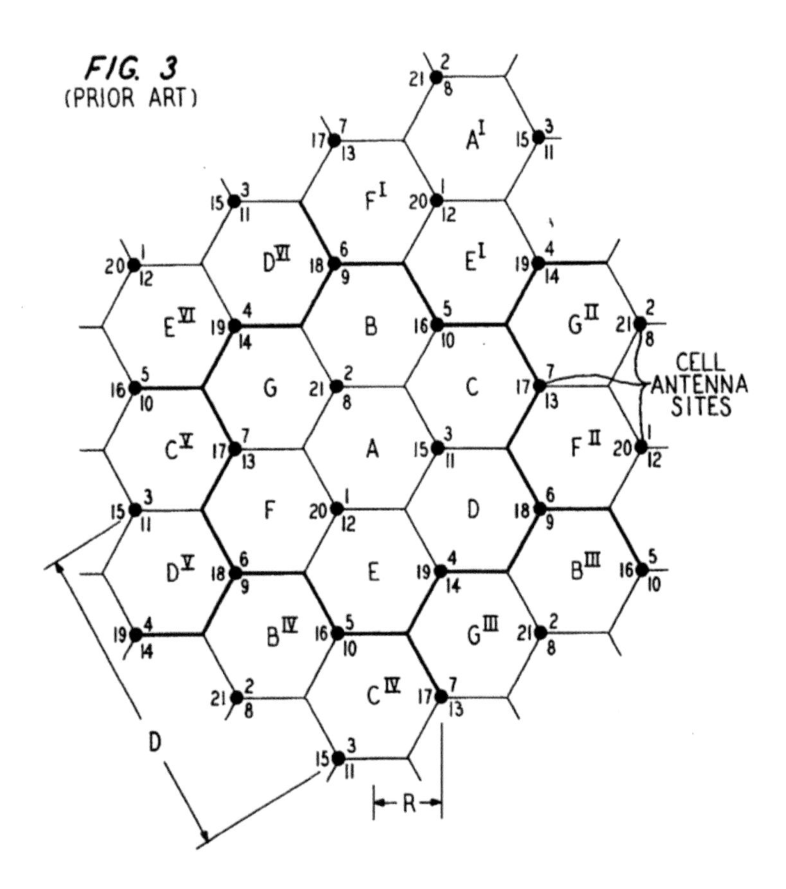

Figure 11.3: The principles of cellular telephony. Typically, each hexagonal cell is served by three base stations and each base station serves three cells. The three numbers around the base stations are channel numbers which, as can be seen, are reused across the network. *(Richard Frenkiel, US Patent 4144411.)*

lower-powered, and thus smaller, mobile phones. It would also permit frequencies to be reused, increasing capacity significantly.[8] However, the small coverage area of each cell imposed a requirement that each call be 'handed off' from one cell to another as the user moved. This was beyond the technical capabilities available in the late 1940s and the concept remained on the shelf. The idea was not forgotten, however.

A breakthrough arrived in the late 1960s when computer technology started to be applied to telecommunications. Computer technology could do something new: it could 'hand off' a call seamlessly from one cell to another thus enabling cellular telephony to become a reality, firstly in Tokyo in 1979.[9] Two years later, the NMT (Nordic Mobile Telephone) network was launched in Sweden and Norway. NMT was based on standards agreed between several telecoms operators and equipment vendors, including a facility for roaming, so that, for example, a Telenor customer could use their phone in Sweden.[10] These early developments gave the Far East and the Nordics a head start in mobile communications which they retained for decades.

Despite having invented the concept of cellular telephony, it took until 1983 for a commercial service to launch in the US, using a system called AMPS. But the delay paid off, because 1983 also saw the introduction of a device that would transform the concept of telecommunications. This was the Motorola DynaTAC 8000x, the world's first hand-held mobile phone. The 1 kg brick-shaped phone became a symbol of the era, used by the amoral financial broker Gordon Gecko in the 1987 film *Wall Street* to bark orders and issue threats.[11] Martin Cooper, an engineer with Motorola on the team that developed the first hand-held mobile (Figure 11.4), correctly foresaw its far-reaching consequences:

> We believed people didn't want to talk to cars and that people wanted to talk to other people ... [You] could assign a number not to a place, not to a desk, not to a home, but to a person.[12]

In Ireland, however, the tens of thousands waiting for a landline to be installed would probably not have been impressed had the Department decided to use its limited resources to put phones in the cars of rich businessmen or politicians. A number of private companies such as Radio Alert set up 'car phone' services, but these were effectively closed networks that could relay messages but could not connect their customers to the public phone network.[13] In the early 80s, however, with a massive investment programme in telecommunications underway, the Department of Posts and Telegraphs began to tentatively plan for mobile telephony. The first question was which of the available cellular standards to choose. Initially the Department of Posts and Telegraphs favoured the NMT system because of its close relationship with Ericsson, who were involved in the NMT project. However, a trial conducted by Ericsson in Dublin in the middle of 1983 indicated a problem of interference with existing two-way radio systems, such as the Garda network, which was discovered to be also utilising the 450 MHz band (Figure 4.7) used by NMT.[14] As a result the Department plumped for the TACS system which was being developed in the UK and was derived from the American AMPS standard. Ericsson was informed of the change of direction via a two-line telex message.[15]

Figure 11.4: Martin Cooper pictured in 2014 holding a prototype of the Motorola Dynatac phone he invented. The photo was taken in Dublin outside a building called Marconi House, then the studios of radio station Today FM. *(WENN Rights Ltd / Alamy Stock Photo)*

Research concluded that the market for mobile phones would be limited to a small number of business customers.[16] As a result, when the Eircell network was officially launched in Dublin on 11 December 1985,[17] it had an initial capacity for just 1,000 users, with pricing reflecting the view that this was a service for the select few, not the masses. Just to get started, the user had to buy a handset for £1,395 to £2,100 (*c.* €1,770 to €2,700), then pay a once-off network access charge of £50 (*c.* €63) and an installation charge of £100 (*c.* €127). That was all before a quarterly rental charge of £105 (*c.* €130) and then calls at 32p (*c.* 41c) per minute. At these prices it was little wonder that the *Irish Independent* described the service as being intended 'for the businessman and not the Sunday motorist'.[17]

Ironically for a mobile telephone network based on a cellular concept, the initial Eircell system consisted of a solitary base station on Three Rock mountain, covering a radius of up to 50 km from Dublin. Within a year, more base stations were added to cover North Leinster and by June 1987 Limerick and Cork were reached. In a pattern similar to the expansion of the landline network by the Free State in the 1920s, network planning was sometimes influenced by Telecom Éireann's district structures, so that single base stations were scattered thinly in a desire to avoid any region feeling neglected.[15] This scattergun approach could create problems, with an unimpressed *Evening Echo* criticising the initially patchy service in Cork under the headline 'Dial U for Useless.'[18]

Despite such hiccups, pretty quickly it became clear that the market for mobile phones was a lot larger than that initial forecast of 1,000 customers. Within five years Eircell had 11,300 customers[19] with its network covering 65 per cent of the landmass and 75 per cent of the population.[20] Prices also started to fall, with the cheapest phone a mere £900[19] (*c.* €1,140). It turned out that the typical mobile phone customer was a salesperson or building contractor rather than a Gordon Gecko-type financial broker. When Eircell was launched in 1985 there was still a waiting list of *c.* 35,000 for a landline connection.[21] By 1992 this had been eliminated, not just because additional landlines had been installed, but also because many business users such as small builders and site contractors moved to using mobile phones.[14] In the same year card-operated mobile phones were introduced on Iarnród Éireann trains.[22] Unlike the *Zugtelefon* in Germany 66 years before, these train-phones could be used by passengers in second class as well as those in first. Mobile phones were becoming more democratic.

The mobile phone did not just substitute for landline phones but also largely replaced other forms of telecommunications. One of these was the pager. First used in closed settings, such as Dublin's Richmond Hospital from 1961,[23] a service to users across the capital was launched by several private companies from 1972.[24] To contact someone with a pager, you phoned the number of the paging company and gave the name of the person and your message to an operator. The operator instructed the paging system to contact the relevant pager device which would 'bleep' to alert the user. Typically, the user would then have to find a phone, call the switchboard and collect the message. Despite this somewhat cumbersome process, pagers were relatively popular in Ireland in the 1970s and 80s within a niche market of business people, with many being used by those waiting for a landline to be installed.

State-owned Telecom Éireann set up a joint venture with Motorola in 1988 called Eirpage[21] in an attempt to expand the market beyond this base. They were no doubt influenced by the continued popularity of pagers in countries such as the US where eight per cent of the population had a 'bleeper' in the early 1990s.[25] However, Eirpage arrived just as the waiting list for landline phones was disappearing and, more importantly, mobile phones were becoming ubiquitous. Despite almost nationwide coverage thanks to use of frequencies of 153–4 MHz, similar to broadcast FM radio, take-up was modest. An attempt to break into the consumer space in 1995, with a product called Minicall, also failed despite an expensive advertising campaign featuring French-born TV star Antoine de Caunes. Pagers became confined to specific sectors such as hospitals, but with a 2010 survey in Waterford finding that 98 per cent of hospital doctors preferred using mobile phones to pagers for work communications,[26] pagers seem destined to become another technological cul-de-sac, bypassed by the onward march of the mobile phone.

Some big technology companies have also found themselves bypassed, such as Motorola, which made the pagers for Eirpage and Minicall at the facility it opened in Swords Co Dublin in 1991. The factory had a relatively short life due to the global shift of manufacturing to lower wage economies in the Far East and manufacturing ceased in Swords in 2003.[27, 28] Motorola's software centre at Blackrock in Cork, set up in 1981, had a longer and happier history, employing up to 500 people and becoming the worldwide centre of excellence for development of Operations and Maintenance software used to monitor the performance of mobile telecoms networks.[29] Unfortunately after Motorola lost market share to competitors, the company was dismembered, with the Cork office closing in 2007.[30] By contrast, one of Motorola's rivals, Swedish giant Ericsson, better managed the structural changes that affected the industry, successfully reinventing its Athlone factory as a software development centre.

On Ericsson's home turf in Stockholm, a meeting held back in December 1982 changed the course of human communications. The meeting was the inaugural gathering of the *Groupe Speciale Mobile* (GSM) committee comprising of 12 European telecommunications administrations[31], which had been given the task of creating a future pan-European mobile communication standard. The committee agreed the standards that would fix many of the problems with the analogue cellular systems then in use. With GSM, voice was digitised and compressed before transmission, so that each call required much less bandwidth than on analogue systems. GSM also utilised Time Division Multiple Access (TDMA). This concept, which had been pioneered on the transatlantic telegraph cables, allowed multiple users to share and use the same transmission channel by dividing signals into different time slots (Figure 2.12). These two technologies allowed GSM much greater potential capacity than its analogue predecessors such as NMT or AMPS.

In 1987 telecom administrations in Ireland and 11 other European countries signed a Memorandum of Understanding committing them to introducing GSM.[32] The name was also changed to 'Global System for Mobile communications' as non-European countries started to become involved. The new name proved prophetic, with GSM becoming the dominant standard across the globe. Thanks to the economies of scale this dominance provided, the adoption of GSM led to the democratisation of mobile communications.

The advent of GSM left analogue cellular phones in another technological cul-de-sac by offering far more secure communications that its predecessors. Analogue phones were susceptible to being cloned, whereby the phone's details were copied onto another phone, with calls from the cloned phone being charged to the original. A cloning fraud perpetrated by Dublin-based gangs in 1993 cost Eircell a total of £292,000[33] (*c.* €371,000). To prevent this, GSM included the concept of the SIM (Subscriber Identity Module) card. The chip on a SIM card, rather than the device, securely held data about the customer's mobile subscription details; there was nothing on a GSM phone itself to identify the customer's mobile account. A further security feature was the use of encrypted transmission, unlike most analogue systems which transmitted calls unencrypted over FM, allowing anyone with a scanner to listen in.

GSM was launched in Ireland on 24 June 1993 using the 087 prefix. Despite its inherent advantages, GSM was not an immediate success in the country, largely due to coverage issues. The initial network comprised only of cells serving Dublin, Cork, Limerick and Galway.[34] Furthermore, while the frequencies and power used were similar to the existing 088 TACS network, the effective range of each base station was less due to an inherent characteristic of digital radio. With analogue communications, even when there is low signal strength, it is usually possible to comprehend what is being said, whereas a digital system typically supplies either perfect audio quality or nothing at all.[7] Furthermore, as they required more complex electronics, GSM handsets were initially larger[15] and more expensive than their analogue equivalents. In 1995 only 2,000 of Eircell's 80,000 customers were using its GSM 087 system.[35]

That soon changed. Indeed, the period from 1996–7 was a tipping point for mobile telephony in Ireland. Eircell was established as an independent entity, still owned by Telecom Éireann but with its own management and budget. Its new CEO, Stephen Brewer, having previously worked with UK operator Cellnet, knew how to prepare Eircell for eventual competition[36] and set about aggressively recruiting customers. The operator introduced new price plans with inclusive minutes, a pattern that has persisted to the current day. Eircell also introduced another concept that became almost universal when it began to subsidise handsets so as to attract new customers. On top of this, the gross cost of GSM handsets fell as production volumes scaled up thanks to the large potential market across Europe and beyond. The net result was that a Motorola TAC 7500 on offer in October 1996 was smaller, lighter and had far more features than the £2,100 'brick' offered ten years earlier – but it cost the customer nothing.[37] Eircell also inked deals with partners so that, for example, AA members could avail of a free phone as part of their membership.

Ironically this massive growth in customer numbers occurred just as it became more difficult to expand the network. In 1994 Telecom Éireann had been given a six-month window during which it could erect masts up to 15 m high without obtaining planning permission[38], an exemption it had sought based on similar rules in the UK.[15] Telecom, and its Eircell unit in particular, made the most of the exemption period to expand the network as obtaining planning permission was a fraught process. Once that window expired, however, network expansion became much more difficult.

Almost every new form of telecommunications infrastructure has a visible manifestation, often leading to controversy. In particular, proposals to erect mobile phone antennae frequently attracted hostility with politicians falling over each other to display their ignorance of physics. In Macroom, nestled in the valley of the Lee in Co Cork, the Urban District Council wrote to Telecom Éireann in 1995 to complain about poor mobile phone coverage in the town. Council chairperson Denis Kelleher claimed that 'you may as well be talking into a packet of cigarettes as talking into a mobile phone in Macroom.'[39] A year later, councillors were up in arms again, this time to complain about the construction by Eircell of a mobile mast to serve the town.[40] It was a pattern repeated across the country, one that continues to the present day.

These were big changes to the mobile telecommunications market, but a bigger one was yet to come. This was the advent of competition. On 2 March 1995, Michael Lowry – described by the *Irish Times* as having 'risen almost without trace to take on the portfolio of Transport, Energy and Communications'[41] – announced a competition to decide on the allocation of a licence for a second mobile operator. Lowry said that he expected mobile phone ownership to increase to around 250,000 by the year 2000 and stressed that the competition was not an auction, explaining that 'There is no cheque-book policy',[35] an assurance which indirectly later came back to haunt him. The successful bidder would have to pay the government £15m (*c.* €19m) – a snip when compared to later licences.

The competition for the second mobile licence attracted huge interest with six consortia lodging bids, involving a panoply of international telecoms companies, local partners from the private and semi-state sectors[42] and, of course, Esat and those 'Vikings' from Telenor. The media tipped the Persona consortium, thanks to the presence of mobile-phone pioneer Motorola as partner. The Esat Digifone bid, led by Denis O'Brien, with Telenor added to the ticket quite late, was regarded as a long shot. Nor was much attention paid to the bid from Cellstar, 60 per cent owned by US cable and mobile phone giant Comcast, with minority stakes held by RTÉ, Bord na Móna and a small Irish company called Ganley Communications International.[43] A Danish consultancy firm, Andersen Management International, was employed to decide who would be awarded the licence.[44]

In a surprise move on 25 October 1995, several weeks ahead of the anticipated decision date, Michael Lowry convened a press conference and made the unexpected announcement that Esat Digifone had won the competition for the second mobile phone licence.[45] The decision, and the process that led to it, caused a furore so great that in 1997 it became the subject of a Tribunal of Inquiry which was to take 14 years and cost €100m.[46] During the hearings it emerged that Digifone's bid had in fact come first in the scoring.[47] However, the Moriarty Tribunal's final report was damning, concluding that the Minister, Michael Lowry, had 'secured the winning' of the 1995 mobile licence for the consortium led by Denis O'Brien.[47] Amongst other misdemeanours, the Tribunal determined that Lowry had passed on information to O'Brien to help him secure the licence, had shut down departmental discussion of the bids, had bypassed consideration by his Cabinet and that after the licence competition he had received payments totalling *c.* €635,000 from O'Brien.[48] The Tribunal also found that Telenor had made a US$50,000 donation to Fine Gael on behalf of Esat Digifone.[48] Both O'Brien and Lowry disputed the Tribunal's findings.[49]

Meanwhile, independent of the Moriarty Tribunal, the decision to award the licence to Digifone was also challenged by some of those involved in unsuccessful bids. One case was taken by Sigma, part of the hotly-tipped Persona consortium[50], which had come in second place.[47] The other was taken by Comcast and Declan Ganley, parties in the Cellstar bid[51] which had received the lowest score.[47] The two separate legal cases have ground their way slowly through the courts for, so far, 25 years.[50, 52]

The legal proceedings and surrounding controversy did not affect Esat Digifone's launch plans, with the consortium formally awarded the licence by the Department on 16 May 1996. Digifone faced a different problem – building the network of masts required to reach the stipulated 80 per cent of the population. While its bid had included draft leases with landowners across the country, getting approval for masts through the planning process proved difficult.

A draft deal in November 1996 with the Department of Justice to use radio masts at up to 700 Garda stations around the State for mobile phone base stations initially floundered due to uncertainty around planning and rental payments.[53] A solution came the following February, when environment minister Brendan Howlin amended planning regulations to allow up to 12 mobile telephone antennae to be attached to an existing radio mast, or the replacement of an existing mast, without requiring planning permission.[54] It was an attempt to level the playing field with Eircell, which had been able to exploit a temporary exemption from planning permission a few years before.

With this agreement in place a date was finally set for the launch. The state's second mobile phone company would go live on 21 March 1997. At a huge party in Dublin's Point Depot the night before, Denis O'Brien took centre stage. Michael Lowry, though now deposed as minister following a scandal about tax evasion, was there too (Figure 11.5). In a speech to his 1,000 guests, O'Brien made specific mention of Lowry, describing him as the minister 'who did so much to encourage competition in the telecommunications sector'.[55]

The next morning is not one Margaret Furnell, who was a team leader dealing with new activations at Digifone's customer care centre in Limerick, will quickly forget. Customer applications flooded in from mobile phone dealers around the country:

> There were six fax machines set up to receive applications. From the moment the stores opened, the machines were churning away, with the printouts spilling onto the floor. We couldn't keep up with them. Then, to our relief, at about 11:00, all the machines went quiet. We thought we had a chance to catch up, until someone noticed that all the fax machines had run out of paper![56]

Part of the reason for the initial flood of customers to Digifone was, paradoxically, due to the success of the sales campaign mounted by Eircell in the months preceding competition. The incumbent's customer base grew far faster than its network capacity, resulting in a flood of complaints to RTÉ's *Liveline* show about congestion, reception blackspots and poor customer service.[57] Digifone also introduced 24-hour customer care[57] – a revolution in a country where some places had only gained 24-hour phone service just ten years before.

Figure 11.5: A delighted-looking Denis O'Brien (right) meets a contemplative-looking Michael Lowry at the launch of Esat Digifone in the Point Depot on 20 March 1997. *(Eamonn Farrell / RollingNews.ie)*

The new network had attracted 105,000 customers by Christmas 1997. Not all of those early customers immediately grasped the concept of a mobile phone complete with battery and charger, with one phoning to complain 'mobile my a**e, it won't reach more than three feet from the wall'.[56] This rapid growth brought its own set of problems. The billing system, persistently rumoured to have been developed from a bakery management system, struggled to cope and bills were issued months late. Ironically, Digifone's own network soon started to suffer similar problems to Eircell's, with the MSCs (exchanges) becoming overloaded during busy calling periods so frequently that Digifone's technical staff used to place bets on which of the exchanges would go down first. The problem was particularly acute on Friday evenings as office workers arranged after-work pints. Mobile telephony was no longer solely the preserve of the financial broker or the building contractor.

If Digifone thought that the state-owned former monopoly would not know how to react to competition, it was wrong. Eircell had a trick up its sleeve, one that would transform the market at least as much as the arrival of competition. Within seven months of the Digifone launch, and perfectly timed for the Christmas market, Eircell introduced its 'Ready-to-Go' prepaid product.[58] Prepaid packages eliminated the need for a credit check and made the cost of a mobile connection more predictable and controllable than a bill-pay mobile or a fixed line. Arguably the 'anticipal pain' of waiting for a – possibly huge – phone bill[59] was a bigger factor in Ireland, where there was still a lingering memory of large, unexplained telephone bills from the 1970s and 80s. Parents could gift them to teenage children, transferring responsibility for managing costs to the teenager and eliminating the traditional internecine rows over the landline phone bill. Teenagers also enjoyed the independence of having their own number.[59] The vision of Martin Cooper, the pioneering Motorola engineer, had arrived – a phone number was now associated not with a place or a home, but with a person.

Eircell was an early adopter of the prepaid concept, launching it soon after earlier successful introductions by networks in Portugal[59] and Italy. Ready-to-Go was launched first on the by-then ageing analogue network: a deliberate decision as GSM had finally taken off with the result that the older 088 network had some spare capacity at last.[15] The analogue network also still covered more of the country[15], an important factor for a consumer product that was being sold through retail chains as well as dedicated mobile phone dealers. The initial Ready-to-Go package of a Motorola A130 handset with a top-up scratchcard (Figure 11.6) for £20 (*c.* €25) of call credit[14] ended up under Christmas trees throughout the country.

Within just two months of its launch, Eircell's prepaid mobile service had been taken up by 2 per cent of Ireland's population.[59] Soon after, the *Irish Times* wrote that the 'streets are alive with the sound of mobile phones and now they are as likely to belong to laid-back teenagers as they are to high-powered men in suits.'[60] This process accelerated further after Digifone launched its own prepaid offering, called Speakeasy, in 1999. By early 2000, there were 2,461,000 mobile phones in use[61], ten times greater than the total predicted by Michael Lowry just five years before. Two-thirds of these were on prepaid plans. Even more significantly there were more mobiles in use than fixed lines (Figure 11.7).[14] Ireland was one of the first countries where this happened.

Figure 11.6: With the advent of prepaid services such as Eircell's 'Ready-to-Go', consumers became used to buying scratchcards to top up their account balances. *(Courtesy of irishcallcards.net)*

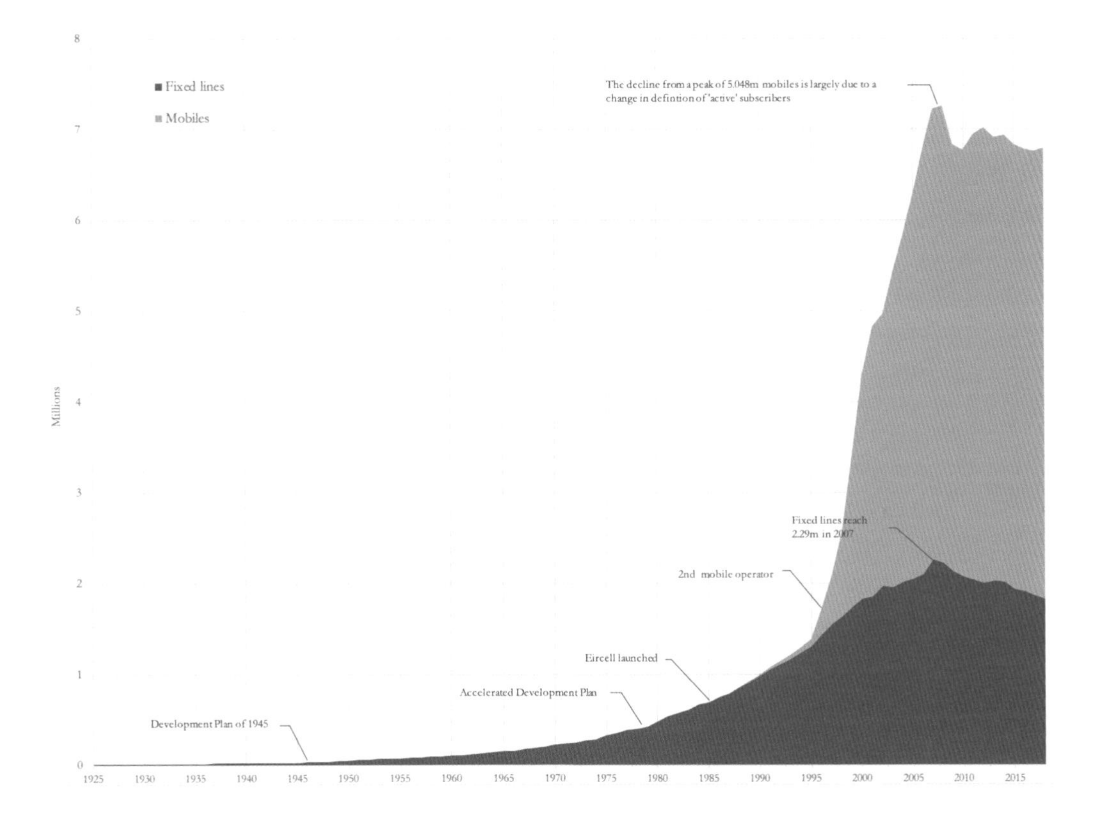

Figure 11.7: Growth in fixed and mobile telephones since the foundation of the state. The explosion from 1995 to 2007 was largely due to the massive growth in mobiles. Sources: Flynn, Roddy, The Development of Universal Telephone Service in Ireland 1880–1993 (1998) DCU, Doctor of Arts thesis; International Telecommunication Union, Yearbook of Statistics: Telecommunication Services (Geneva: 2004); Comreg, Quarterly Key Data Report (Dublin: Various).

The introduction of prepaid mobile services, by making it easy and affordable to get a phone, brought telecommunications to the masses. From about 2006, however, prepaid started to lose some of its popularity, probably a result of the rising adoption of more expensive smartphones. While the up-front cost of phones was generally subsidised by the networks for their bill pay users, the same was not the case for prepaid. However, prepaid plans continued to be popular in many sectors of society, such as those on uncertain incomes, providing communications and internet access to people who would otherwise be left incommunicado.

Within nine months of the Esat Digifone network going live, a competition for a third mobile phone licence was announced.[62] This time round the competition was arranged by the newly established Office of the Director of Telecommunications Regulation (ODTR). As with the controversial competition for the second licence, a 'beauty contest' assessment was used and, again, the process became a gravy train for lawyers. In June 1998 the regulator awarded the licence to Meteor, owned by US-based Western Wireless and Ireland-based RF Communications. The only other bidder was Orange, which at that point was a mobile operator in the UK. Orange won a High Court case quashing the award of the licence but this was overturned unanimously by the Supreme Court, leaving Orange with a £2m (*c.* €2.5m) legal bill.

Thirty-nine long months after the competition, on 22 February 2001, Meteor was launched[63] into a market dominated by prepaid plans. One of the terms of the third licence was that in urban areas, where the 900 Mhz band was becoming full, it would use the 1800 MHz band[62] (Figure 4.7). With its shorter range, more cell sites would be required. Partly as a result, Meteor's network at launch reached only 55 per cent of the population (Figure 11.8) and it took many years to convince potential customers that its coverage was adequate. The legal delay also cost Meteor dearly in terms of potential customers. In those three years between winning the licence and the launch, the proportion of the population with mobile phones had grown from 26 per cent to 77 per cent. Meteor's potential market had shrunk by 1.9 million before it had even carried its first call.

The obvious solution was to target the majority of the population who already owned a mobile phone. But that faced a problem. Each network had its own prefix, which was often used as shorthand by customers in place of the often-changing network names. People would say 'I'm on 087, there's no 086 coverage where I live'. If a customer moved between operators, they could ask to keep the last seven digits of their number but the prefix would change, for example from 087 to 085. It was just too clumsy and few customers budged. All this changed on 25 July 2003, when Full Mobile Number Portability (FMNP) was introduced. Anyone with a mobile phone could move network and keep their full number. The process, which involved changes to the systems within each operator as well as a new central database managed by Ward Solutions, was considered to be the largest private sector logistical project ever implemented in Ireland.[64]

Figure 11.8: Meteor's coverage at launch in 2001 was relatively limited, hampering sales.

Despite this change to allow mobile phone customers to easily move between networks, a set of humorous television adverts and a 2004 national roaming deal with O2, formerly Digifone, to expand coverage, Meteor struggled to gain market share, with a mere 6 per cent share of the market after three years in operation.[65] But as Ireland's love affair with mobile deepened, Meteor's market share grew.

During those early years for Meteor, major changes occurred in the ownership of all the networks. Late in 2000, newly privatised Eircom agreed to sell Eircell to Vodafone for €4.5bn. Vodafone had been the first privately owned mobile telephone company in Europe when it commenced operating in the UK in 1985, and by 2000 it had become a huge global player. Many argued that the sell-off was short-sighted as it was estimated that 70 per cent of Eircom's market value was derived from its Eircell subsidiary which had almost 1.9m subscribers.[66] Furthermore, as part of the deal, Eircom was precluded from entering the mobile telephony business for three years. For Eircom, however, it provided an opportunity to placate shareholders nursing significant losses from the fall in the company's share price in the months following its flotation. This was because the proposed deal offered the 400,000 investors in Eircom 0.4739 Vodafone shares for each Eircom share held. In May 2001, the shareholders approved the sale[66] and the Eircell brand was consigned to history.

Something else consigned to history that summer was the original analogue 088 system. At its peak in 1996 the 088 analogue network had almost 125,000 subscribers but only 12,000 phones were still using the system by the time it was shut down by its new owners on 31 July 2001.[67] Some of the last remaining users were fishermen, as the analogue network could be used much further out to sea than GSM which was technically constrained to a range of 35 km from a base station.[68] In its 16-year history, the 088 TACS network had covered almost the entire country and played its role in democratising mobile communications.

Big changes were afoot at the country's second-largest operator too. Early in the new millennium, BT agreed to purchase Esat Telecom for €2.4bn, the sale personally netting Denis O'Brien in excess of €300m. It was widely believed that most of the value placed on Esat was due to its 49.5 per cent stake in Digifone, rather than its own fixed line business.[69] BT then bought out Dermot Desmond's small stake in Esat Digifone and, the following year, Telenor's 49.5 per cent share.[70] The Vikings were gone, though with 77 per cent of the Irish population now possessing a mobile phone[5], they had certainly left their mark. Saddled with a huge debt mountain, BT quickly moved to hive off its mobile businesses across Europe, branding the networks as O2. After a few years of trading independently, O2 was acquired by Spanish incumbent Telefónica in 2006.[71]

Meanwhile Meteor was also changing hands. The first change came in 2004 when Western Wireless International bought out the other shareholders. Then in the following year, with its non-compete clause with Vodafone now expired, Eircom re-entered the mobile market by buying Meteor.[72]

These changes came about because Ireland was an attractive market for global telecommunications companies, with plenty of business for everyone. In 2001 Ireland had the twelfth highest rate of mobile phone ownership in the world, just behind Norway.[61] In her *Irish Independent* column in 2003, Mary Kenny observed that 'New Yorkers do not move around with a mobile phone clamped to their ear as is now the common practice in Dublin.'[73] This was undoubtedly true, largely because the rate of mobile phone ownership in the US (Figure 11.9) was less than two-thirds the rate in Ireland.[61] For decades, Ireland had lagged behind the rest of the Western world in the deployment of telecommunications technologies: as we have seen everything from the Speaking Clock to International Subscriber Dialling seemed to arrive in Ireland decades later than elsewhere. But with mobile it was different. New products and technologies were typically rolled out in Ireland in line with other rich countries, and were quickly lapped up by an appreciative audience.

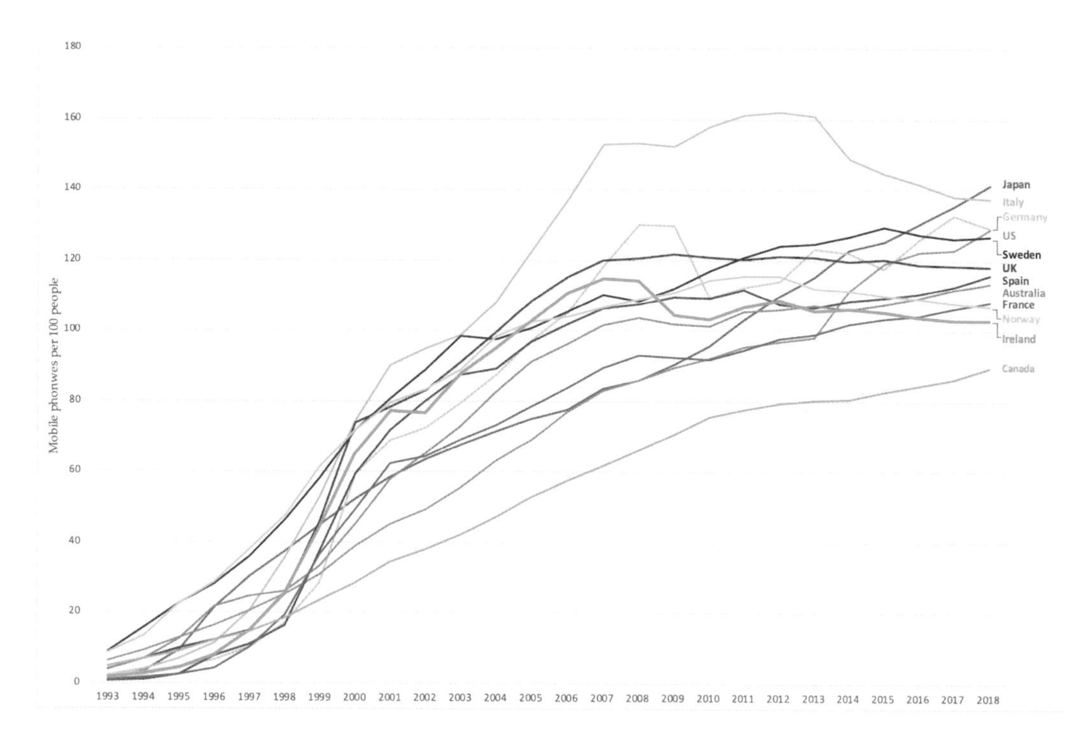

Figure 11.9: Mobile phone diffusion in selected countries. Sources: ITU, Yearbook of Statistics: Telecommunication Services (Geneva, 2004); ITU, Mobile-cellular subscriptions (Geneva, 2019).

In Ireland this effect was amplified by the booming economy of the period, with the mobile phone becoming a totem for the Celtic Tiger era. In a letter to the *Irish Times* on the first days of the new millennium, a Corkwoman bemoaned the soulless 'New Ireland' as she saw it: 'If Ireland were to be personified now, it would no longer be [as] a misty-eyed, red-haired maiden, but as a smug, mobile-phone-toting youth in a suit, driving a soft-top Merc.'[74] Of course the democratisation of telecoms meant that the 'mobile-phone-toting youth' could equally be found on a bus as well as in a soft-top Merc. A few days further into the new millennium, Ross O'Carroll-Kelly, the fictional South Dublin rugby playing hero of the eponymous newspaper column, was thrilled to receive an award from Denis O'Brien for the category of 'Most Irritating Mobile Phone Ring on the 46A'.[75]

The Celtic Tiger brought customers in Mercs and on buses – but also occasional problems. Particularly in the Digifone/O2 network, cell sites were connected mostly by microwave links. The frenzy of construction activity in the capital led to the skyline being dotted with cranes. It became a common occurrence for a base station to go out of service because a crane had been erected which interrupted the microwave beam connecting it to the rest of the network. When this happened, transmission links had to be rerouted or sometimes cells left off the air until the offending crane was taken down. It was not a problem that had troubled the first microwave link between Athlone and Galway in 1961.

Ireland's love affair with the mobile phone deepened further in the twenty-first century, for the new millennium saw the mobile phone become a device for accessing data as well as for talking on. For many people, the year 2000 was heralded by the arrival of a text message from a friend or acquaintance. SMS (Short Message Service) was the unexpected 'killer application' buried in the depths of the GSM protocol from 1991, alongside now-forgotten services like fax and Videotex. The quest to attribute its origins to a single moment or person have spawned a number of stories, such as that it was invented over dinner in a Copenhagen pizzeria or single-handedly by a modest Finnish engineer.[76]

The real story appears to be more prosaic. SMS was not 'invented' by any one individual but was defined by a group of people working together to agree the GSM protocol through an endless series of workshops across Europe during the late 1980s. SMS uses the signalling capability of the GSM system, allowing a text message to be transmitted independently of a voice call but restricting the amount of data to 160 characters.[77] SMS was conceived of as a way to notify a subscriber that they had a voicemail message, or for a millionaire to keep a watch on the changing value of their stocks.[43] Its creators had no idea that it would become one of the world's most widely used communications methods, with Ireland playing a key part in its early development.

Adding the ability for customers to send and receive SMS was relatively simple for the telephone networks; the main extra element required was called an SMSC (Short Message Service Centre). Joe Cunningham, CTO of a small Dublin-based software company called Aldiscon, realised that none of the established players such as Ericsson had developed an SMSC and set about creating one. The company launched the world's first commercial SMSC platform with Telia in Sweden in 1993[78], the start of a global domination of a niche market with huge potential. By the time of its sale to Logica in 1997, Aldiscon supplied over 100 mobile carriers worldwide.[43] Nevan Bermingham fondly remembers his time as a young engineer travelling between mobile phone operators across the globe:

> Logica-Aldiscon found a niche with text messaging that led to it expanding so fast sometimes the operations staff couldn't keep up with the SMSC installations on operator sites. But it was a young company that fostered loyalty among its young workforce, and that meant long work days but even longer nights celebrating their success.[79]

Early in the new millennium, however, telecoms operators began to struggle financially after the bursting of the dot-com bubble. The bigger players caught up and started to offer their own SMSCs,[79] Logica wound down its Irish operation and this brief episode of Ireland in the telecommunications hall of fame drew to a close. However, Aldiscon left a legacy in the form of the Short Message Peer-to-Peer (SMPP) protocol still used for exchanging SMS messages.[80] Millions of automated SMSs sent every day by businesses all over the world to keep in touch with their customers use technology developed in Dublin in the 1990s.

Ironically, it took a little while for Irish mobile to benefit from the SMS technology being pioneered in Dublin as Eircell did not launch SMS until early in 1997.[81] Prepaid customers of both networks had to wait until December 1999[82] – just in time for those millennium greetings. The introduction of SMS on prepaid was heralded by an Esat Digifone television advertisement about the advantages of using text to communicate. The ad's only line of dialogue 'it's me – the guy from the bar' became a well-known saying, with the ad being classified as a legend in Irish advertising history, 15 years after its first transmission.[83] The campaign clearly worked. In January 2004 it was reported that Irish mobile users sent more text messages than those in any other country, and that on New Year's Eve, two million text messages had been sent in a period of just one hour.[84]

SMS use peaked in 2012, when Irish mobile customers sent a total of 3.1bn messages. From 2011 smartphone apps such as WhatsApp and Apple's own messaging service led to a drop in traditional SMS traffic (Figure 11.10). Such apps, however, were built on the success of SMS in popularising the concept of text messaging.

During the noughties, Irish mobile phone customers were not just sending more texts than their international peers, they were also talking more. In 2004 O2 revealed that its Irish customers talked for an average of 188 minutes each month compared with

109 minutes in Germany and 107 in the UK.[85] Just as the United Telephone Company had discovered in 1883, the stereotypes about the Irish being talkative seemed to be true. All this talking helped the two main networks enjoy handsome profits.

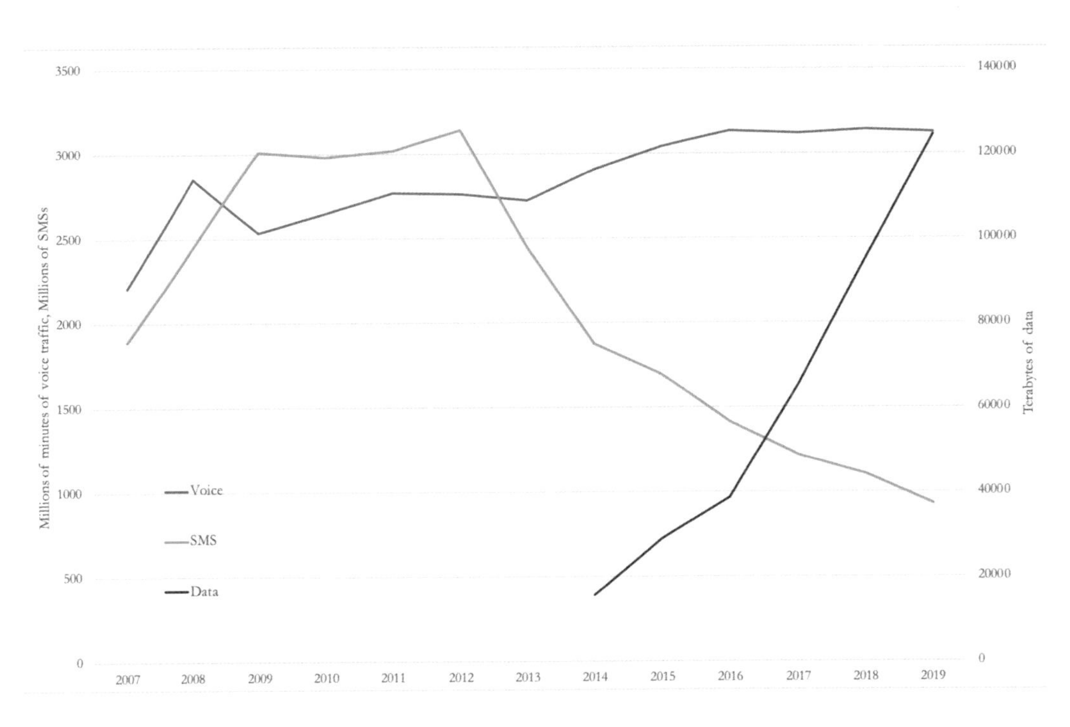

Figure 11.10: Mobile phone usage in Ireland. The substitution of SMS by over-the-top services such as WhatsApp from about 2011 is clear. Source: Comreg, Quarterly Key Data Report (Dublin, Various)

These profits were helped by the absence of competition from MVNOs (Mobile Virtual Network Operators). The concept of an MVNO was that it bought capacity at wholesale rates from a network and sold it to the MVNO's own customers. In many countries competition from such operators had driven down charges but Ireland's early history with the MVNO concept was not a success. In the late 1990s a company called Imagine, run by Sean Bolger, bought airtime in bulk from Eircell and resold it to its own 20,000 customers.[86] Like the licence applicants, the parties ended up in court and the arrangement ceased in 2001, just after Meteor had opened for business. For the next few years the Irish networks resisted attempts by the regulator to force them to host MVNOs. It certainly worked to their financial advantage: in 2005, Vodafone declared a pre-tax profit of €340m on a turnover of €1.21bn while O2 declared €214m on a turnover of €858m.[87] These were margins that most other businesses could only dream of.

With plentiful cash flow, and buoyed by the success of SMS, the mobile networks spent millions enabling another mobile data service, which came to the market late in 1999. This was WAP (for Wireless Application Protocol) and was marketed as the 'internet in your pocket', though it proved to be nothing of the sort.[88] Amongst its many limitations was its use of circuit switched data access, so the phone had to dial a number every time the user started to browse the 'internet'. It was a solution, of sorts, looking for a problem. Unfortunately the habit of over-hyping technologies rather than consumer products continued to dog the mobile telecoms industry well beyond WAP.

Considerably more successful was the introduction of the BlackBerry smartphone in 2002. This device used a new data standard, called GPRS (for General Packet Radio Service). Often called '2.5G', GPRS was based on packet transmission, which divided data files, such as emails, into smaller units known as packets that could be transmitted more efficiently. (Packet transmission will turn up again in Chapter Twelve). Unlike WAP, GPRS was 'always on' and thus the device using it was permanently connected to the internet. The BlackBerry became wildly popular amongst a small but valuable market of highly mobile business people, particularly in the US, earning it the nickname 'CrackBerry'. GPRS, and an enhancement called EDGE, were effectively overlaid on GSM, using the same frequencies. The speeds possible were similar to dial-up internet – fine for sending email from your BlackBerry but tediously slow for internet browsing. The industry had been working on a set of technologies to provide much faster data, technologies that became known as 3G.

Having seemingly learnt little from the over-hyped launch of WAP technology, the telecoms industry did it again with 3G, only this time there were other snouts in the trough. As 3G required the use of a new frequency band, governments around the world eyed up potentially huge windfalls to be made by auctioning off licences for this band to telecoms operators. The UK Treasury struck particularly lucky by extracting £22.4bn (*c.* €25bn) from its spectrum auction,[89] the equivalent of over €550 per adult.

Ireland, however, prevaricated about how best to allocate licences. Minister for Finance, Charlie McCreevy, adopting similar logic to the setting of the Eircom share price, wanted a straightforward auction to maximise the take for the Exchequer. The regulator, Etain Doyle, wanted another 'beauty contest' so that licences would be allocated based on the quality of the applications. After a one-year stand-off, the beauty contest went ahead[90] with a deadline for applications of 27 March 2002. Each licence would cost €114m, to be paid over 15 years. On top of that cost, the mobile phone companies would have to construct a new 3G network, which, because of the frequencies used, would require more cell sites than before. It was to prove an expensive exercise for everyone.

While different arms of the government were debating the best way to gather the golden eggs of 3G, the goose dropped dead. Mobile phone operators, having spent heavily on licences in other countries, and stricken by the collapse of the dot-com bubble, stayed away from the Irish market in droves. Only three applications were received. Two were from existing networks, Vodafone and O2, with the other from newcomer Hong Kong-based Hutchison Whampoa.[91] Meteor did not even bother applying. The fourth licence for 3G was allocated to Smart Telecom in 2005 but then, following a predictable court battle with the regulator, it was eventually acquired by Meteor.

Vodafone launched its 3G service in 2004[92] to an underwhelming response. Eventually in 2006 O2 followed suit, introducing 3G handsets with i-mode, a 'walled-garden' cut-down internet. While a huge hit in its native Japan, it proved as popular in Europe as WAP had done. Hutchison Whampoa launched its network in Ireland under the Three brand on 26 July 2005[93] with a completely 3G network – used by its customers largely for voice and text. Three was Ireland's fourth mobile network and, like Meteor before, it struggled to gain market share despite keenly priced products, clocking up losses of €507m over its first eight years.[94]

For several years 3G seemed to be another expensive solution chasing a non-existent problem. Then along came two game-changing ideas about mobile phones from northern California. The now-iconic Apple iPhone was introduced in 2007 by Steve Jobs who told the audience 'today Apple is going to reinvent the phone'.[95] Bizarrely, the first model used the already outdated EDGE and GPRS data technology, but the concept of a touch screen interface, occupying almost one entire side of the phone, was a game-changer. The first iPhone model was soon followed by a 3G version, which reached Irish shores on 11 July 2008.[96] The second idea from California was the Android operating system, largely developed by Google. Phones running Android went on sale in Ireland in time for Christmas 2009[97] so that by 2010, a range of smartphones to suit all pockets was available.

Thanks to the smartphone, the amount of data carried by the mobile phone networks grew at an exponential rate, increasing 13-fold in the six years between 2011 and 2017. After several years carrying little traffic, those expensive 3G networks started to become congested, as people read the news, checked their social media feed and streamed videos on their phones.

Along came 4G LTE (for 'Long Term Evolution') data technology to cater for this increased demand. 4G also benefitted from another technological advance: the replacement of analogue television with DTT which freed up part of the UHF band for use for mobile data, in what was dubbed the Digital Dividend. In 2012 the regulator offered 4G licences across three frequency bands. Unlike previous licencing rounds, there was no beauty contest, just a straightforward auction. Each of the four existing operators made a bid, netting the State a cool €855m in up-front fees and instalments.[98] The upfront charges were particularly welcome, coming as they did when the country was in the midst of the post-financial crisis bailout. 4G networks started to go live towards the end of 2013,[99] offering typical download speeds of around 26 Mbps,[100] faster than those available over fixed lines in many places.

The financial crisis of 2007–8 and subsequent bailout caused many consumers to start looking at their personal finances more closely, including their mobile phone bills. It was the perfect time for some extra competition in the market. Right on cue arrived the first successful MVNO, Tesco Mobile Ireland, which started as a joint venture between the supermarket giant and Telefónica, owners of O2, on 30 October 2007.[101] An Post's Postfone MVNO followed in 2010,[102] with Lycamobile in 2012, the former using Vodafone's network and the latter O2's. They targeted the value-conscious segment with Lyca particularly focused on the immigrant market. In addition to the competition provided by MVNOs, some of the networks created sub-brands, such as 48 which was launched by Telefónica in 2012 and aimed at the youth market.

The global financial crisis also strained many telecommunications companies, leading to sell-offs and consolidations across the world. Faced with a large debt burden, in 2013 Telefónica agreed to sell O2 Ireland and 48 to Hutchison Whampoa, owners of Three, for €850m.[103] Because the merger would reduce the number of networks from four to three, the deal required European Commission approval. Amongst the conditions of the approval was Three providing capacity for two new MVNOs.[104] The acquisition was fully completed on 15 July 2014 and the process of merging the two networks then began in earnest. One of the MVNOs hosted as a condition of the deal, iD Mobile, had an unfortunately brief life. Despite a flexible pricing model and a wide distribution network through the stores owned by its parent, Carphone Warehouse, it closed in 2018 after less than three years in operation. The other new MVNO that emerged was Virgin Media, allowing it to expand beyond its cable network to offer a 'quad-play' of broadband internet, television, landline phone and mobile phone.

The advent of the MVNOs accelerated the downward trend in charges. The early adopters of mobile phones in 1985 paid the equivalent of €44.45 per month for the privilege, before making a single call, but prices steadily fell as mobile phone ownership became more democratic. Competition reached new peaks in 2019 with the launch by Eir of its GoMo sub-brand, successfully targeting the value-conscious segment with a simple offer of €12.99 a month. At 2021 prices, that's seven times cheaper than the

£105 per quarter paid by the first mobile users in 1985, but included all national calls, texts and data – the latter two being of course unknown in 1985. The net effect was dramatic, with a 2018 survey finding that the cost of mobile data in Ireland was the eighth lowest of 28 European countries and one-sixth of the average cost in the US.[105] It was quite a change from the 1980s and 1990s when Ireland was one of the most expensive countries for telecoms services.

There were massive changes too in the charges for using a mobile device abroad. Roaming was an integral part of the GSM specifications. This was due to its heritage as a European project, with the GSM Memorandum of Understanding signed by 13 European countries in 1987 just two months after the Single European Act had come into effect. But before a single GSM network had actually been built, Ireland had implemented roaming with the UK. The primary reason for this was technical, with Ireland using the same TACS standard as the UK, and Eircell sharing the same frequencies as Vodafone there. With TACS, calls were always set up using the base station with the strongest signal, even if a viable signal was available from another base station. It was thus possible that an Eircell customer, even when on the southern side of the border, could receive a stronger signal from a Vodafone base station than from an Eircell one. Without some form of cooperation this would block their service.[14] As Eircell and Vodafone expanded their networks northwards and southwards respectively, this probability increased.

To avoid this issue, an arrangement was made in 1992[106] between the two networks, so that a customer of Eircell could receive calls anywhere in Northern Ireland or Britain served by Vodafone's Glasgow exchange, with a reciprocal arrangement for Vodafone UK customers. Describing the system 21 years later, Tom Allen, Eircell's then head of Network Services, said 'there was no "political" input into the decision. At the time, the involvement (or indeed knowledge) of politicians and Civil Servants in the development of the mobile business was minimal.'[14] Some might argue that not much has changed since 1992.

The advent of GSM removed the need for such bespoke technical solutions, but created a new issue of cost. Roaming charges were, and in some cases still are, often a multiple of the cost of using a phone at home. The issue came to the fore with the arrival of data services with mobile phone users often unaware of how much data they used. From the late noughties, there were regular media stories about users who had racked up huge bills when abroad. For example, in 2014 the *Irish Independent* was contacted by an unfortunate visitor to the US who had been charged €7,500.[107] The issue of roaming fees was taken up by Vivienne Reding, the European Commissioner for

Information Society and Media, who championed a number of regulations to curb the costs within EU and EEA countries. Her work was continued by her successors, who saw it as not only 'a market distortion with no rational place in a single market',[108] but a popular cause with MEPs, the public and even Eurosceptic elements of the media.[109] The regulations successfully removed much of the fear of using mobile devices elsewhere in Europe: between 2007 and 2016 the volume of data roaming increased by 630 per cent.[110]

Demand for mobile data at home too continued to rise. The primary reason was the popularity of smartphones with 91 per cent of mobile users in Ireland possessing one by 2019,[111] and a majority of the traffic to Irish websites coming from smartphones rather than from laptop or desktop computers.[112] Per capita mobile data usage in Ireland in 2018 was higher than in Norway. We had become more Norwegian than the Norwegians themselves.

Just as the low penetration of landlines had stimulated demand for mobiles, the poor state of Ireland's fixed-line broadband infrastructure was another reason for this high rate of mobile data traffic. On top of this human-generated traffic was an increasing amount of data used by machine-to-machine (M2M) communications, with 1.1m such devices on mobile networks in 2019, representing 17 per cent of all mobile subscriptions.[65] Such M2M communication was being used, for example, to track delivery vehicles, monitor house alarms and check water levels, all part of the increasingly important Internet of Things (IOT).

The Internet of Things was one of the main drivers for the adoption of 5G mobile technologies. Like its predecessors 4G and 3G, 5G was really a marketing term for a suite of technologies to deliver faster data speeds with much lower latency. Latency is the delay before a transfer of data begins following a request for its transfer: low latency is better for playing video games, video conferencing and, potentially, self-driving cars. In yet another licence allocation round, in 2017 ComReg auctioned off spectrum space in the 3.6 GHz range to five operators for 5G services. The mobile operators Vodafone, Three Ireland and Eir were joined by neutral host provider Dense Air and one-time MVNO, now fixed wireless access provider, Imagine. They stumped up a combined €78m, 78 per cent of which had to be paid upfront.[113] It brought the total paid by telecoms networks to the exchequer for spectrum licences to almost €1.4bn (Table 11.1). Commercial 5G services were launched in Dublin, Cork, Limerick, Galway and Waterford in August 2019. At the Dublin launch, the *Irish Independent* reported data speeds of 630 Mbps[114], about 43,750 times faster than its ill-fated WAP predecessor.

Licence	Year	Eircell/ Vodafone	Digifone/ O2	Meteor/ Eir	Three	Imag-ine	Dense Air	Total
2G	1995–7	12.7	19	14.6				46.3
3G	2002	114	114	114	50.7			392.7
4G	2012	280.64	224.57	244.42	105.01			854.64
5G (3.6 Ghz)	2017	22.73	(now part of Three)	15.66	20.37	9.77	9.64	78.17
Total		430.07	357.57	388.68	176.08	9.77	9.64	1,371.81

Table 11.1: Amounts paid (in millions of euro) for licences by telecoms operators in Ireland.

In Chapter Four, we saw Nicola Tesla's prediction that wireless would enable us to 'communicate with one another instantly, irrespective of distance.' In the same interview in 1926, Tesla also predicted that through:

> telephony we shall see and hear one another as perfectly as though we were face to face, despite intervening distances of thousands of miles, and the instruments through which we shall be able to do all of this, will fit in our vest pockets.[115]

Thanks to the advent of fast mobile data technologies, such as 4G and 5G, by the early twenty-first century we carried 'instruments' in our vest pockets or handbags which enabled us, amongst many other things, to see and hear one another over distances of thousands of miles.

While we still referred to these instruments as 'mobile phones', they were actually vest-pocket-sized computers connected to the internet. Phones had become computers and voice was merely one of many types of data. We had entered the Second Information Age.

CHAPTER TWELVE

ONLINE AND OFFLINE

Howth, Co Dublin, December 1999. John Beckett should have been studying for his Leaving Cert in June, or thinking about a big bash to celebrate the impending new millennium. Instead he was holed up in his bedroom for days on end, putting the finishing touches to his new website design – one, it turned out, that would help change the course of telecommunications in Ireland.

Only two years before, when he was just 15 and in his Transition Year, Beckett had scored a work-experience gig at the Dublin plant of the big computer manufacturer Gateway. There he helped to build the network for an extranet system that allowed Gateway to connect its staff across the globe. Such a system was still a big deal in 1999. As part of the project Beckett had to meet all the department heads including Eddie Wilson, Gateway's HR director. Wilson was impressed by the teenager's understanding of emergent internet technologies and his ability to communicate with senior people.

Soon after Wilson left Gateway and moved a few miles out the M1 to a new job at Ryanair, where he quickly discovered that the budget airline was in need of an internet whizz. For although at this time the web had not really caught on in Ireland, Ryanair's UK rival EasyJet was selling tickets 'online' and Ryanair thought they should give it a go too.

Wilson remembered the whizz kid John Beckett from Gateway and got him on the phone. Would he be interested in building an online booking system for Ryanair? Beckett leapt at the idea but knew he would need help. He rang his mate Tom Lenihan, two years his senior and a dentistry student in Trinity, and gave him a run-down. Lenihan likewise jumped at the opportunity and together they drafted a proposal.

A few days later, Beckett came home from school, changed from his uniform into jeans and a t-shirt and drove over to Dublin Airport.[1] When he arrived, the 17-year-old was surprised to find himself standing in front of Ryanair's board of management. He started talking. Within a few minutes he landed the contract to build Ryanair's online booking system for a fee of £16,000 (*c.* €20,000).[2]

It was a huge sum of money for a 17-year-old. What the Ryanair board did not mention to Beckett, however, was that the established technology consultancies had quoted up to £3.5m (€3.85m) for the same job – figures that did not impress Ryanair's famously miserly CEO, Michael O'Leary, who did not even have a computer in his office.[3] Although O'Leary was not inclined to spend millions on a toy just for the amusement of a few computer nerds, he and others at Ryanair HQ had witnessed EasyJet and other airlines around the world using the web as a supplement to their traditional sales channel of travel agents. There was not much to lose by giving this schoolkid John Beckett a shot at it – £16,000 was less than Ryanair's fuel bill for a day.

Beckett went to work on it. Only a few months later, his mocks out of the way, as the rest of the world planned parties and fretted about the Y2K bug, he did the final checks of the Ryanair online booking system. On 17 January 2000, with all the hangovers over, the new site (Figure 12.1) went live. It proved to be the best £16,000 Michael O'Leary ever spent.

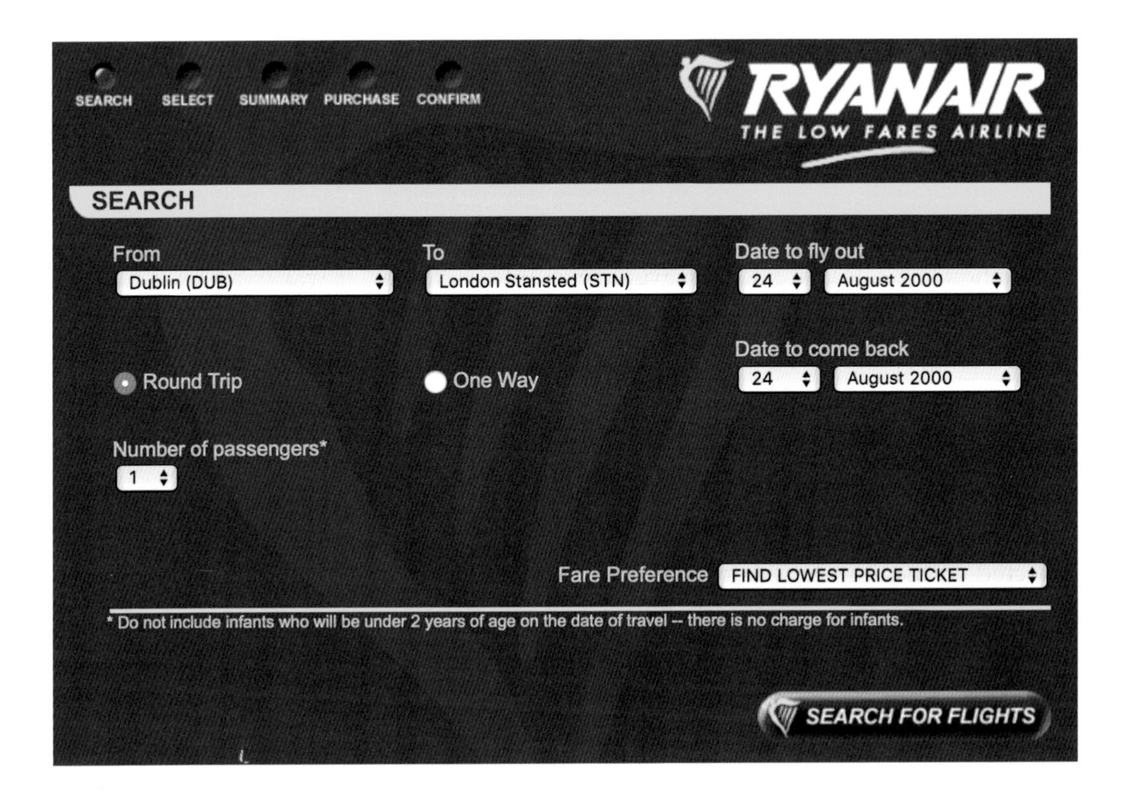

Figure 12.1: The Ryanair online booking site, soon after its launch in 2000. *(Courtesy of Ryanair)*

Offering a discount of £5 (€6.35) for every ticket booked online, within three months www.ryanair.com was taking over 50,000 bookings per week.[4] A year later, over 65 per cent of the company's total bookings were made using the website[5] developed by the two young lads from Dublin. As a result, Ryanair was able to slash its commissions and marketing costs by 65 per cent and ditch the Galileo airline reservation system used by travel agents.[6] By 2003 Ryanair's passenger figures had grown to 21.4m[2], 75 per cent of which were booked online. Soon almost every other airline would study how this aggressive upstart from Ireland had managed to slash its costs by selling online, and getting the passenger to do for free the work that traditionally had been performed by a travel agent at the airline's expense. The Ryanair booking system proved to be a turning point for the aviation industry in Europe.

But Ryanair's system also proved to be a turning point for the internet in Ireland. For while the internet by 1999 had become a commercial juggernaut in countries like the US, its uptake in Ireland remained relatively low.

Across the world the dot-com bubble was in full swing. Online global fashion retailer boo.com, for example, had just been launched after investment of £60m (*c.* €67m) on a swish website with a pixelated shop assistant called Ms Boo to help shoppers.[7] By 1999 online bookstore Amazon had expanded its product range to include CDs and electronic goods, resulting in global annual sales of $1.64 billion. Matters were different in Ireland, however, with the Irish public largely unswayed by this internet malarkey. Research in 1999 found that only 1.6 per cent of Irish people had made a purchase online[8] and only 8 per cent of Irish businesses had an email address.[9] With most Irish people still preferring to phone or call into a local travel agent to book flights, it was hardly surprising that neither Aer Lingus nor Ryanair had bothered with online bookings – until, that is, Ryanair's online booking system went live.

Soon after, Irish consumers overcame their scepticism of the internet and became converts to the joys of booking a cheap flight online. Mary Murphy in Ballygobackwards learnt that she could pop over to her sister in London for just £19 – if she booked online. With a compelling reason to go online, 20.4 per cent of Irish households were connected to the web by the end of 2000, four times the rate in 1998.[10] This rapid growth in internet use has continued unabated ever since, affecting society, culture and, naturally, telecommunications, with the main purpose of the telecoms network mutating from transmitting voice to carrying internet traffic.

Unsurprisingly, given what we've seen with other telecoms developments, this shift was not without pain, a pain exacerbated by structural issues within the telecoms sector in Ireland with the result that the country fell behind its peers again. As a result, the quality of internet access became a major public issue, spilling into the political arena.

But before we delve into the twenty-first-century internet, we need to go back to the twentieth century and the story of data communications, the internet and world wide web. The three terms are sometimes used interchangeably but have distinct meanings and emerged at different times.

The oldest, and most encompassing, term of the three is data communications. This term simply means the electronic transmission of information between different computers. Data communications have been used since the early years of electronic computers in the 1950s as users connected them together over the telecoms network in order to exchange data. Such was the case in Ireland, where one of the earliest networks developed around the country's first computer. This was installed by the semi-state Irish Sugar Company at its Thurles plant in 1957 in order to calculate the annual payments that it made to beet farmers.[11] Through the 1960s the Sugar Company installed three other computers and by 1970 it had linked them together using dial-up modems.[12] Short for 'Modulator-Demodulator', a modem converted digital data from a computer into an analogue signal that could be sent over a standard telephone line. It was a twentieth-century way to send data in digital format over a network designed in the late-nineteenth century to carry the human voice. The Irish Sugar modems, cutting-edge at the time, operated at 600 or 1,200 bps, about 0.0000006 times the speed of a typical 100 Mbps domestic internet connection in 2015.

The airlines were also early adopters of data communications, with Aer Lingus rolling out a network in 1969–71 to connect its new ASTRAL reservation system (Figure 12.2) to its offices in Ireland, the UK and New York.[13] The system also had an interface to the global SITA reservations network and was used by several other airlines around the world to manage their reservations – provided, of course, that P&T were not on strike. Likewise, in 1975, all ten branches of the Dublin Saving Bank (one of the precursors to PTSB) were linked by leased lines to the bank's main computer[14] so that customers could obtain their up-to-date balance and transaction data in their own branch.

These early forms of data communications were based on a method called circuit switching. This treated data communications like a telephone call with a dedicated circuit over a leased line or dial-up connection used for the entire duration of the session. Thus the computer operator at Irish Sugar's Dublin office had to dial the respective numbers for the computers in Mallow and Carlow, and deal with telephone operators to connect to the computer in Thurles through the town's manual switchboard.[12] This communications method made sense for a phone call, where someone is almost always talking, but was quite wasteful for computer communication, which typically happens in short bursts. The necessity for a leased line or dial-up connection between each computer made circuit switched data communications expensive and inflexible.

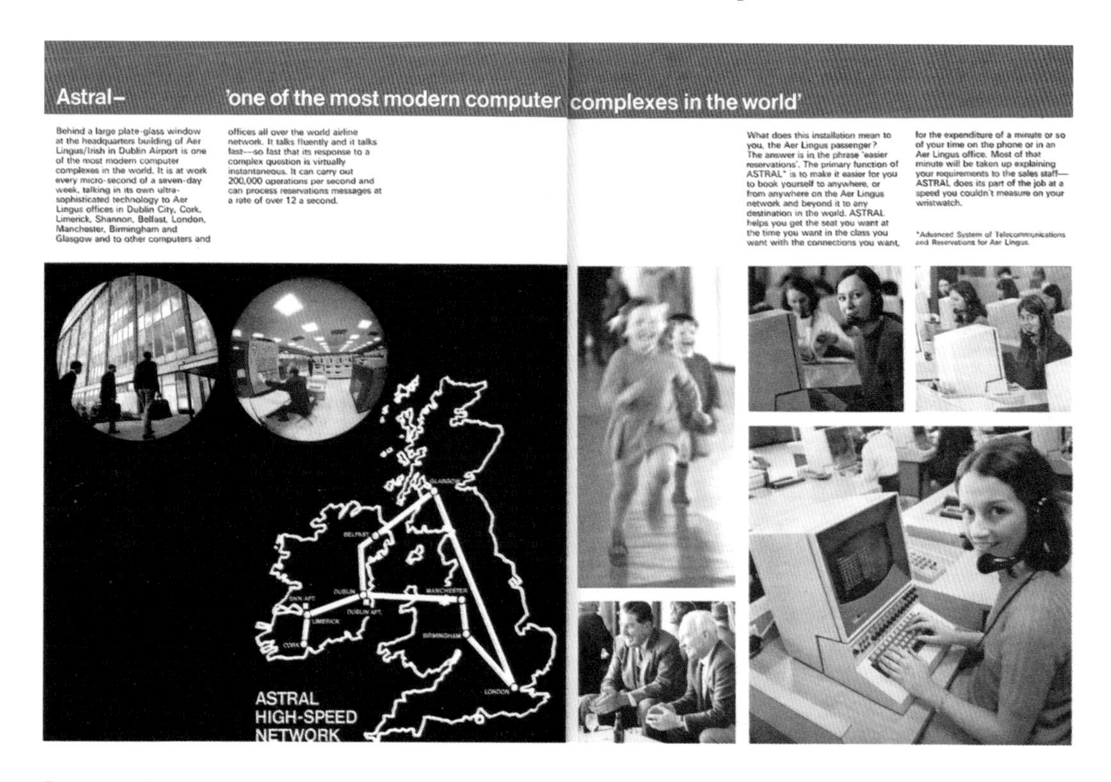

Astral– 'one of the most modern computer complexes in the world'
Behind a large plate-glass window at the headquarters building of Aer Lingus/Irish in Dublin Airport is one of the most modern computer complexes in the world. It is at work every micro-second of a seven-day week, talking in its own ultra-sophisticated technology to Aer Lingus offices in Dublin City, Cork, Limerick, Shannon, Belfast, London, Manchester, Birmingham and Glasgow and to other computers and offices all over the world airline network. It talks fluently and it talks fast—so fast that its response to a complex question is virtually instantaneous. It can carry out 200,000 operations per second and can process reservations messages at a rate of over 12 a second.
What does this installation mean to you, the Aer Lingus passenger? The answer is in the phrase 'easier reservations'. The primary function of ASTRAL* is to make it easier for you to book yourself to anywhere, or from anywhere on the Aer Lingus network and beyond it to any destination in the world. ASTRAL helps you get the seat you want at the time you want in the class you want with the connections you want, for the expenditure of a minute or so of your time on the phone or in an Aer Lingus office. Most of that minute will be taken up explaining your requirements to the sales staff—ASTRAL does its part of the job at a speed you couldn't measure on your wristwatch.
*Advanced System of Telecommunications and Reservations for Aer Lingus.
BELFAST
DUBLIN
DUBLIN APT.
MANCHESTER
LIMERICK
BIRMINGHAM
CORK
LONDON
ASTRAL
HIGH-SPEED
NETWORK

Figure 12.2: Publicity material for Aer Lingus ASTRAL system and accompanying data network. *(Courtesy of Aer Lingus)*

An alternative concept for data communications emerged in the 1960s. Called packet switching, it borrowed a concept from a communications network even older than the telegraph or telephone – the postal system. Packed switching divides data into discrete chunks called packets. Like a packet sent by post, each of these contains the destination, sender's address and priority. As with the post, it's possible that different packets sent from the same origin to the same destination may travel different routes and arrive in a different sequence to the way they were sent (Figure 12.3). And, in the same way that most postal networks comprise multiple sorting offices, a packet switched network typically has multiple nodes[15] and can reroute data as needed.

Packet switching made data communications cheaper and, thanks to its ability to reroute data, created a more resilient network. This was an important consideration for a concept born at the height of the Cold War. Possibly as a result, the first packet switched network, called ARPANET, was sponsored by the US Department of Defense, establishing a pattern of US pre-eminence in data communications which it has largely retained. Access to ARPANET was not limited to the military, however, with universities across the US also connected.

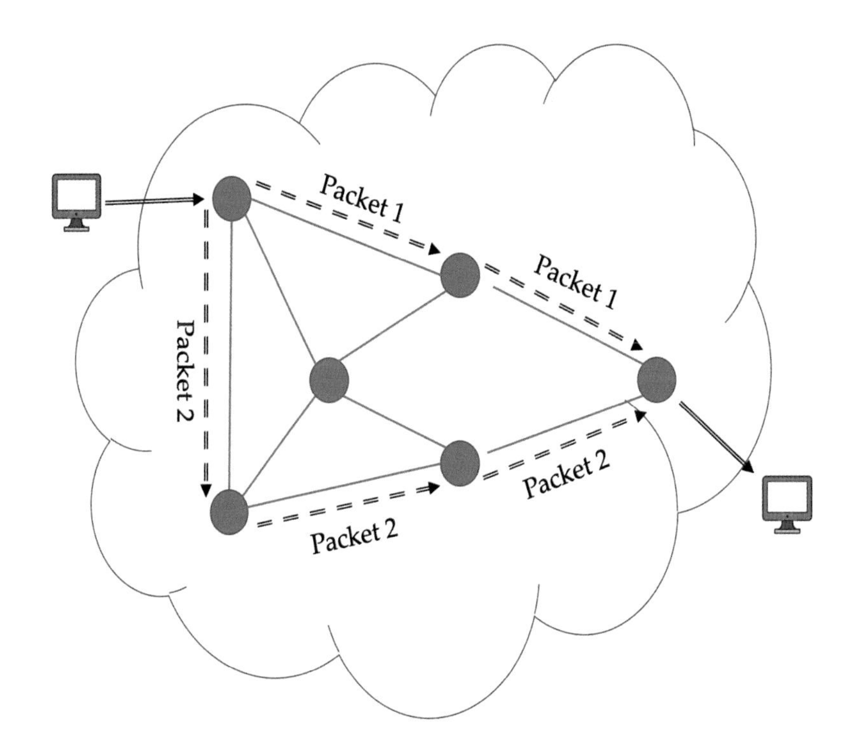

Figure 12.3: With a packet-switched network, different packets sent from the same origin to the same destination may travel along different routes.

Linking those various computers from a plethora of manufacturers, each with a different operating system, was a challenge. To help overcome this, during the 1970s a pair of protocols was developed by researchers in the US that enabled data to travel across different networks, like a digital lingua franca. One was called Transmission Control Protocol, the other Internetworking Protocol, or Internet Protocol for short. Collectively referred to as TCP/IP, the protocols were seen by many as a stop-gap while a complete solution known as OSI (for Open Systems Interconnection) was being worked on by the international community. The development of TCP/IP proved to be a pivotal event, creating the lingua franca still used today for computer communications.

This 'stop-gap' TCP/IP protocol was at the heart of another network linking academic institutions – one with an Irish flavour. NSFnet was funded by the National

Science Foundation in the US, and headed by Irishman Dennis Jennings, on leave of absence from University College Dublin (UCD). Its aim was to allow academics from all over the US to use five supercomputers. Jennings decided on a three-tiered model: a backbone network connecting the five supercomputers, a second-level connecting these to other campuses, with the third level connecting the users' computers provided by campus networks. It was a devolved model, designed to provide access for all.[16] Since the US government was footing the bill, there was no need for a complex set of charges, or indeed any charges at all. To connect to the internet, the institution just needed to pay for a telecommunications link; after that it could exchange any data it wanted with any US research and education establishment that was also connected.

With NSFnet connected to other packet switched networks such as ARPANET, a network of networks developed, becoming known as the 'internet' (Figure 12.4) after the TCP/IP protocol it used. Jennings, who was inducted into the Internet Hall of Fame in 2014 for his role in creating the internet, recalled the plans of the time: 'Our ambition was to link all 300 universities in the US to create a national centre for research. While it was ambitious at the time we had no concept that the internet would grow the way it has.'[17]

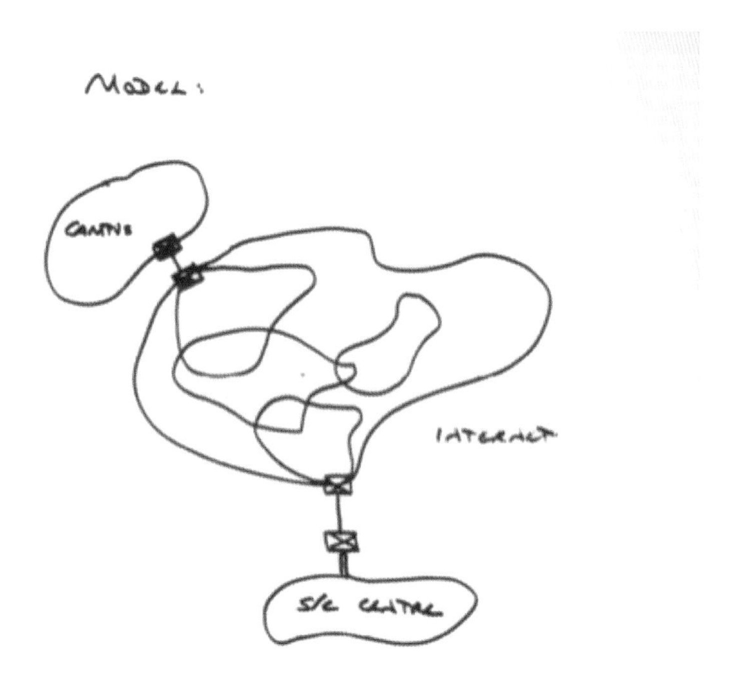

Figure 12.4: Dennis Jenning's conceptual design for NSFnet, connecting the supercomputer centres to the campus networks via a set of networks collectively called the internet. *(Courtesy of Dennis Jennings)*

One of the most popular services on this early internet was, as today, email. Email had been around since the late 1960s, allowing users of a mainframe computer to send each other messages. Its utility increased in 1971 when an American computer engineer called Ray Tomlinson developed the ability to send messages to users on another computer using data communications. His decision about how to separate the username from the name of the computer using the '@' symbol catapulted the once-humble keyboard character to fame.

By the mid-1980s email had become a popular service for the relatively small number of confirmed geeks working in academic institutes and multinational technology companies, such as computer firm Digital Equipment Corporation. In 1987 an employee of Digital in Galway, Liam Ferrie, had the bright idea of using the corporate network to send news from home to Irish people working for the company scattered around the world. His first *Irish Emigrant* email newsletter was sent from his desk in Galway via a satellite dish atop Telecom's Mervue exchange to Digital's head office in the US and then to 15 employees across the globe.[18] It was a small taste of how the internet could connect people around the world, at practically zero cost.

However, outside of universities and tech companies, the potential of email as a communications tool faced problems, both technical and human. For example, Dennis Jennings, while head of Computing Services in UCD, recalls phoning the Department of Posts and Telegraphs around 1983 to request their approval for an administrative level email domain for the EuroKom conferencing service run by UCD for one of the European Commission's research programmes. He was eventually put through to an official who had never heard of email. After Jennings explained the concept to him, the official replied that his role was 'to protect the jobs of P&T workers', before aggressively demanding 'Who did you say you were?' At that point Jennings quietly hung up the phone.[16] It was a small but striking example of how the traditional telecommunications industry viewed the nascent internet.

A few months later, with P&T now carved up between Telecom Éireann and An Post, a row erupted over the issue of which company would be allowed to provide a public email service. In a decision that displayed a strange concept of the potential use of email, the *Irish Times* reported that the Minister, Jim Mitchell, came down on the side of An Post because it already had a national network of offices,[19] as if email was just a fancy telegram that could be sent from a post office. In the end, Telecom Éireann launched its Eirmail service in 1985, though it was not a rip-roaring success, garnering only 660 customers in three years.[20]

To be fair on Telecom Éireann, email had remained a niche product everywhere. Part of the reason for this slow take-up was that the early decades of email were like a replay of Belfast's telephone wars in 1880, but on global scale, with users of one network unable to send messages to users of a different network. This considerably reduced the available pool of recipients and thus the attractiveness of the service.

That all changed, however, from 1989 when the various commercial and corporate networks began to be connected to the embryonic internet. The phenomenal growth of email that followed is a fine example of Metcalfe's Law, which states the effect of a telecommunications network is proportional to the square of the number of connected users of the system. For example, once Digital's corporate network was linked to the internet, Liam Ferrie's *Irish Emigrant* emails could be sent for free to anyone with an email account, not just Digital staff using their own computer network. Thanks to Ireland's huge emigrant community, the initial list of 15 email addresses eventually grew to reach 23,000 across the world. In 2019, 247 billion emails were sent around the world every day – 32 for every person on the planet that year. 'I knew exactly what I was doing. I just had no notion whatsoever of what the ultimate impact would be', Raymond Tomlinson, inventor of inter-computer email, said of his creation in 2012.[21]

The development of the internet was watched with interest, and with a touch of alarm, from across the Atlantic by the European Commission. The Commission was aware of the value of linking together computers in member states to enable exchange of information, foster collaboration and develop European technological prowess. It threw its weight behind the OSI set of network standards that were – slowly – being agreed by international agencies. This multi-national, collaborative, top-down approach, where much of the decision making was in the hands of state-run PTTs, was close to the way the Europe Commission and other EEC institutions operated themselves. The process had produced the GSM standard adopted all over the world for mobile telephony. Surely it would also work for data communications?

These European initiatives had been responsible for much of the development of data communications in Ireland. Back in 1980 Ireland's first public packet switched data communications network called Euronet had been sponsored by the Commission. Responsibility for packet switched networking was then handed over to the telecommunications authority in each member country, with Telecom Éireann launching its Eirpac service in 1985.

Eirpac was aimed at business users who were willing to pay to access specific information. On top of equipment costs and quarterly fees, Eirpac charged an access cost of 2p per minute and 0.03p per byte.[22] If online content download services such as iTunes had been around in 1989, a typical HD movie of about 4 Gb such as that year's Oscar-winning *My Left Foot* would have taken 93 hours to download on a 1,200 bps connection – at a cost of €153,000. In parallel with this public packet switched data network, networks specifically for academic purposes were sponsored by the EEC, such as the Europe-wide EARN research network that used leased lines to provide e-mail and file transfer services. In Ireland the Higher Education Authority Network (HEAnet) linked the country's campuses and provided a gateway to these European networks.

These EEC-led projects were aimed at industry and academia rather than consumers. But here Europe – or rather, one European country – also led the way. As the French telephone network expanded and improved rapidly in the late 1970s, the expense of printing paper telephone directories and employing directory enquiry operators was mushrooming. The French government was also anxious to showcase the country's technological prowess. Thus in 1980 Minitel was born.[23] It consisted of a single unit containing a small screen, keyboard and modem, with a lead to plug into a phone line (Figure 12.5). As well as

Figure 12.5: The Minitel, a success in France but not in Ireland. *(Rama, CC BY-SA 3.0 FR, via Wikimedia Commons)*

searching the telephone directory, users could check the news, make train reservations and carry out bank transactions[23], many years before the advent of what became known as the web. A large part of its success was the fact that Minitel terminals were given free to any telephone subscriber in France who wanted one.[24] At the height of its popularity, 20m users had access to 26,000 services[24] over a data communications network completely separate to the internet. It was a shining example of French technology.

As we saw in a previous chapter, French telecommunications policies and technology had a huge influence in Ireland. Such was also the case with Minitel so that by September 1988 moves were afoot to bring the service to Ireland through the formation of a consortium that included France Télécom, Telecom Éireann, AIB Bank and the Quinnsworth supermarket chain.[25] The end goal for France Télécom was to use Ireland as a pilot country for the Anglophone world before moving on to the bigger markets of the UK and US.[24]

Despite a high-profile consumer launch with Gay Byrne on *The Late Late Show* on 18 October 1991, Minitel proved to be a complete flop in Ireland, with only a few thousand terminals installed out of the forecasted 150,000.[26] Some of the initial investors, such as Quinnsworth, dropped out early on and the service was quietly put to sleep in 1998. Interviewed by the BBC in 2012, Gary Jermyn, Minitel Ireland's former finance director, recalled why:

> First of all, unlike in France, we were selling the terminals, not giving them away. That was a huge handicap … Also by the early 1990s the terminal itself was the clunkiest piece of desk manure you could imagine. It was embarrassing … And then the internet was arriving, and that was the death knell.[24]

But why did the internet, a network designed for American academics, kill off Minitel, a network purpose-built for European consumers? The answer was not technology or money, but governance and control. In networks such as Euronet, EuroKom and Minitel, the PTTs had complete control over how and what kind of data moved through the network. When Bank of Ireland wanted to allow its customers to access their accounts over Minitel, it first had to be approved by the company who operated the service, whose shareholders included Telecom Éireann and two rival banks. By contrast, the internet was designed to do a simple task: take in data packets at one end and do its best to deliver them to their destination, regardless of their contents or purpose. The internet was 'permissionless': if a group of organisations wanted to build their own network and connect it to the internet, they could do so by following some simple rules.[16] If someone had an idea that could be realised using data packets, then the internet would do it for them with no questions asked.[27] The internet had no gatekeepers, no civil servants barking down the phone because they did not like the idea they heard.

The result was that by the late 1980s, European academic institutions were beginning to abandon the carefully curated European networks and connect to the decentralised utopian anarchy of the American-founded internet. By 1990 CERN, the particle physics laboratory based in Geneva, had become the European gateway to the internet with a connection to NSFnet at Cornell University.[28] But much bigger things than a 1.5 Mbps leased line were to emerge from CERN in 1990.

As well as sending email, the early internet allowed users to access information held on other computers. The services used to access this, such as Telnet, Gopher and FTP, required technical knowledge, confining usage largely to confirmed geeks. That all changed with a new concept, the brainchild of two scientists at CERN, and based on a principal called 'hypertext' – text with embedded links that allowed the reader to jump instantly from one electronic document to another. In order to display this information, the reader's computer needed a programme called a browser. The first browser, called 'WorldWideWeb', was quite crude and was soon superseded by a more sophisticated one called Mosaic. But the original browser left its legacy in the form of the term 'world wide web' to mean this subset of the internet consisting of pages containing hypertext. The world wide web, or simply 'the web', quickly exploded in popularity, breaking out of the confines of academia. And since the web was accessed via the internet, it turned the latter into a steam-roller, crushing other forms of data communications such as Minitel and Eirpac.

With no gatekeepers to control admission to the web, anyone could contribute and the amount of content available mushroomed. UCC launched Ireland's first web server which hosted machine-readable copies of Irish manuscript texts[29] at http://curia.ucc.ie in 1991. Meanwhile commercial interests recognised the potential of this 'permissionless' network. Soon, a new form of business, called an Internet Service Provider (ISP), developed across the world to allow users outside of academia or multinational technology companies to use the web, exploiting a massive surge in the demand for internet connections among the general public.[27]

Customers of ISPs, like Ireland OnLine, Indigo and Tinet in Ireland, used a modem that dialled the phone number of the ISP and, after some ear-splitting noise as the two modems established a connection, the user's computer was connected to the internet. No specialist knowledge was required to view information, while someone with a bit of technical know-how could create their own website or add content – with no permission required.

When restrictions about commercial use were relaxed in the early 1990s, the internet became an unstoppable force. The top-down European approach, based on standards agreed through many years of careful international collaboration, and with state-owned PTTs controlling content and access, ultimately failed. The US-developed 'stop-gap' TCP/IP protocol conquered the world, and the internet that resulted had a distinctly American aura. In the ultimate irony, the decentralised anarchy of the internet, designed to suit academics, proved perfect for unbridled capitalism. Amazon launched in 1995 as an online bookstore. A few months later, Alaska Airlines and British Midland Airways started to sell tickets online,[30] eventually to be joined by Ryanair in 2000, thanks to John Beckett and Tom Lenihan.

The popularity of the internet also revolutionised the sphere of communications. The potential of the internet to transcend distance and allow families and friends to easily keep in touch was recognised as early as 1995 by President Mary Robinson, who had made a policy to reach out to the Irish diaspora:

> The magic of E-mail surmounts time and distance and cost. And the splendid and relatively recent technology of the World Wide Web means that local energies and powerful opportunities of access are being made available on the information highway ... The shadow of departure will never be lifted ... But we can make their lives easier if we use this new technology to bring the news from home.[31]

She was probably thinking of services like Liam Ferrie's *Irish Emigrant* email newsletter, which continued publication until 2012. Its demise was partly due to the multitude of competing news sources such as the *Irish Times,* which had become the first newspaper in Ireland or the UK to have a website when it went online in 1994.[32] But there was also another reason for the demise of a newsletter aimed at emigrants. From the mid-1990s, the 'shadow of departure' became less common. More people were moving to Ireland than leaving – in part because of the internet.

For a small island with few resources or traditional industries, information networks such as the internet provided an opportunity for economic development that required neither mineral resources nor geographical proximity. The hope was that Ireland could become a key player in this new Information Age, even more than during the first Information Age of the late nineteenth century, when telegraph cables fanned out from Kerry. The new Information Age, however, still required resources, not of coal or iron ore, but rather modern telecommunications infrastructure and experience in data communications. As we have seen, Ireland's record on telecommunications infrastructure was, to be kind, mixed, and investment in data communications in the 1990s lagged behind her competitors.

Perhaps to divert attention from its lack of investment, state-owned Telecom Éireann announced a national competition in 1997 to select and fund Ireland's 'Information Age Town'. Dubbed 'the largest community technology project in Europe',[33] the competition garnered huge interest, with 50 towns across the State competing for £19m (€24m) in funding. The excitement continued after Ennis in Co Clare was selected as winner, with the *Irish Independent* predicting that 'home shopping and electronic banking will be commonplace ... Renewing driving licences and applying for visas will be a matter of pressing a few keys.'[34] Telecom's CEO Alfie Kane foresaw Ennis as 'a showplace where people will be able to see the future happening right now'.[34]

The most visible impact for most of the town's 15,000 citizens was a new home PC, available at the subsidised price of £260 (€330).[35] A surprisingly small proportion of the programme's capital was devoted to the telecoms network in Ennis. Residential telephone users got free voicemail but they accessed the internet using dial-up modems, just like the rest of the country. Not only was this slow, but thanks to Telecom's controversial changes six years before to charge for local calls based on duration, it was expensive. A letter writer to the *Clare Champion* calculated that 10 hours of internet access per week would result in telephone charges equivalent to €67.36 a month.[36] Despite these limitations, many people clearly wanted to be part of this Information Age: five months after the announcement it was reported that house prices in Ennis had increased by £5,000 (€6,350) as a direct result of the competition win.[37]

Not everyone believed the hype. In a letter to the *Irish Times*, Professor Vance Gledhill of Trinity College's Computer Science department described the Information Age Town initiative as 'an expensive public relations campaign which obscures the real issues for Ireland entering the Information Age. These are the cost of telecommunications in Ireland and the appallingly poor penetration of computing into Irish schools.'[38]

Professor Gledhill's scepticism was borne out with the Information Age Town ultimately proving to be a technological cul-de-sac just like Minitel. While the concept was well-intentioned, it was a case of 'too little, too early'. It was too little because a small town on its own had very limited scope to change what was by definition a world wide web. Online shopping would not become a reality in Ennis unless the national (and, increasingly, international) retail chains invested in systems to allow it. It was not economic for them just to do this for Ennis. When Tesco, successors to Quinnsworth, one of the original partners in Ireland's ill-fated Minitel project, launched its online grocery store in 2000, deliveries were available in Cork city and parts of Dublin – but not Ennis.[39] Bank of Ireland launched online banking just days before Ennis won the title[40], but its service was available to all its personal customers in Ireland, not just the lucky burghers of Ennis.

Confining the project to a single town faced particular handicaps in a small country with a highly centralised government such as Ireland. For example, even if the local council had taken heed of the hype and allowed Ennis residents to renew their driving licences online, it could not have done so without the help of the Department of the Environment which operated the national database of licences. In any case, the problem was somewhat hypothetical since four years after Ennis became an Information Age Town, its council still had no functioning website of its own.[41]

Four years after the project started, a survey of townspeople found disappointing results. Only 11 per cent of Ennis folk shopped online, no greater than elsewhere in the country. A local travel agency, despite winning an Information Age Town award for its embracing of the project, reckoned that online bookings comprised just 5 per cent of its business.[41] The important decisions affecting online commerce were being made by the likes of Ryanair, Tesco or Amazon and there was little that businesses and residents of Ennis could do about it. The Information Age Town was just too little.

The Information Age Town project was also a bit too early. In 1997 there were only three methods to access the internet: by dialling up over a regular phone line, through an ISDN connection, or via an expensive leased line. A year into the Information Age Town project, Telecom Éireann launched a genuine revolution in internet access in the form of a trial of DSL (Digital Subscriber Line) technology, permitting speeds of up to 8 Mbps.[42] However the trial involved not the townspeople of Ennis, but 40 Telecom Éireann staff members, 200 km away in the Dublin suburb of Malahide.

By the time the programme in Ennis wound up quietly in 2002, the term 'Information Age' better described the length of time it took an Irish person to connect to the internet. For most domestic users, this still involved the discordant squealing noise of a dial-up connection. The solution was broadband, but apart from those 40 lucky households in Malahide, this was not an option available in Ireland.

Broadband is loosely defined as internet access that is always on and faster than dial-up access. The two main delivery methods were – and for many people still are – DSL and cable broadband. DSL technology uses the traditional telephone line to deliver broadband by transmitting data over the line at higher frequencies than those required for regular voice phone calls. Thus a telephone line consisting of a pair of copper wires, similar to those laid down by the National Telephone Company in the 1890s, can carry data at relatively high speed at the same time as a phone call.

By contrast, cable broadband uses the coaxial cables of a cable television network to carry high-speed data as well as television stations. Since traditional television required only one-way communication to the customer's premises, cable networks required upgrading to allow the two-way traffic required for internet connectivity. While DSL became the most popular wired broadband delivery channel in the twenty-first century, cable was the first broadband delivery channel in many parts of the world. Such was the case in Canada and Belgium, where, like Ireland, cities and towns had extensive cable television systems originally built mainly to relay television from a larger neighbour.[43, 44] In both countries, the telephone operators quickly reacted and introduced DSL broadband[45] to compete with the cable companies.

Ostensibly Ireland appeared ripe for similar innovation. In 1993, 40 per cent of Irish households had cable television[46] but, as we saw in Chapter Nine, it was a moribund market. During the 1990s, cable operators became distracted by MMDS and cable customers began to migrate to satellite television. It was, however, the boardroom battles about ownership of the country's cable networks that had the most malevolent effect. Thanks to Ray Burke's ministerial meddling in 1989, the country's largest cable system was controlled by Telecom Éireann. Telecom was not going to spend money upgrading Cablelink to carry other forms of telecommunications, only to see it be spun off to become a rival. A Forfás report in 1996 described the Dublin cable system as the 'jewel in the crown' of potential advanced communications infrastructure[47], but by this stage its sparkle was distinctly faded following years of neglect. The service provided by Cablelink in the 1990s was little different to that offered in 1958 on the country's first cable TV system in Mespil House.

In 1998 Telecom's concerns about a divestiture were realised when, in line with a European directive, the government put Cablelink up for sale. Since his failed attempt at an Irish satellite channel, Denis O'Brien had been busy creating a telecommunications network, including a large stake in mobile phone operator Esat Digifone. With the dot-com bubble inflating rapidly, O'Brien offered £450m (€571m) for Cablelink and its ready access to virtually every home in Dublin, Waterford and Galway. He was thwarted by the even more cash-rich NTL, which simply offered to pay 15 per cent more than the highest bid.[48] Within a year, however, the dot.com bubble burst and NTL's parent entered Chapter 11 bankruptcy protection.[49]

This pattern of mergers and acquisitions went further than just the networks in Dublin, Galway and Waterford. By 2000 almost all of the State's other cable operators had been merged to form Chorus. The company struggled to consolidate its motley hotch-potch of networks and platforms and, inexplicably, terminated a broadband service in Swords and Malahide, only to relaunch broadband two years later in the towns of Kilkenny, Clonmel and Thurles. With logic like this it was perhaps unsurprising that Chorus was awarded the dubious honour of the worst customer service of any telecoms company in the world in a 2004 global survey.[51]

Clearly sensing that their investment was a bad news story, Independent News and Media sold its shares in Chorus that same year, exiting the television market it had been so keen to join. The country's cable infrastructure was in the hands of two underfunded companies, both haemorrhaging customers to Sky's satellite tv platform.

Meanwhile, thanks to John Beckett and his Ryanair website, consumer demand for always-on, high-speed internet was beginning to soar.

In a twist of fate that was to have ruinous consequences for Ireland's infrastructure, the lack of investment in the cable network was mirrored by the goings-on at the incumbent telco. By the mid-1990s investment in the telecommunications network had dropped from the heady days of the 1980s and Ireland was lagging behind in terms of data communications.[52] It was widely believed that the flotation of Telecom Éireann would cure this by leading to injections of competition – and capital – from the private sector. Unfortunately the opposite was the case. The share price of Eircom, as it was now called, plummeted as a result of the dot-com bubble burst that had also crippled NTL, falling from €4.77 soon after its flotation in July 1999 to €2.45 in August 2000.[53] This drop led to a barrage of criticism from the 500,000 or so small shareholders who had not cashed in immediately after flotation. It also left the company vulnerable to a takeover.

The ensuing sequence of buyouts and sales continued for years, further hindering Ireland's broadband progress. The first move came in October 2000 and involved two familiar players. First up was Denis O'Brien who, with the €300m from the sale of Esat to BT still warm in his pocket, offered €2.25bn for the former state-owned telco.[54] Media baron Tony O'Reilly reacted by assembling a group of investors called Valentia, which included Hungarian-American billionaire George Soros. Any bidder had to win the approval of the ESOT, which owned 14.9 per cent of the shares. With the ESOT expected to vote in line with the recommendations of the Communication Workers Union (CWU) both bidders set about wooing the union. O'Reilly was more generally favoured but the clincher came when he persuaded Finance Minister Charlie McCreevy to amend the tax laws to favour his deal by permitting Valentia to issue preference shares that would benefit Eircom employees without giving them extra representation on the board.[55] McCreevy subsequently admitted that O'Reilly could not have won the battle without the tax changes he introduced, justifying this in the Dáil on the grounds that 'if this change had not been made, it would have favoured the Denis O'Brien consortium and the deputy can imagine the hullabaloo that would have caused.'[54]

Within 30 months, the Valentia consortium had re-floated Eircom on the stock exchange, netting themselves around €950m[55] in the process. That was only the start of it. The former state-owned utility went through ten changes of ownership between 2004 and 2017, bought and sold by a series of rapacious private equity owners solely interested in extracting cash. *Irish Times* writer Fintan O'Toole expressed a view held by many when he wrote 'instead of the genius-level private-sector management that was going to make Eircom a dynamic national champion, the company has been passed around like a joint at a student party'.[56] The company, which was debt-free when privatised, ended up in examinership in 2012 having racked up €4bn in debt,[57] despite having flogged off its network of 340 mobile phone masts in 2007 for €155m.[58] Current and former staff who were members of the ESOT did rather nicely out of the arrangement, receiving an average total pay-out of €67,142 each, tax free[59], though as the leader of the CWU pointed out, 'You compare what they got to Tony O'Reilly or George Soros. They're in the halfpenny place'.[59] The net result was that Eircom's capital investment in the five years between 2002 and 2006 was only €1.18bn.[60] Coupled with the parallel lack of investment in the cable network, this under-investment was catastrophic for the development of the country's broadband infrastructure.

Having performed so well for most of the 1980s, Ireland was back at the bottom of the telecoms league tables again in the 2000s (Figure 12.6), along with Greece, Slovakia and Turkey. Even Hungary, which had studied Telecom Éireann's progress of the 80s with envy, was doing better than Ireland in the broadband race. In 2001 the *Irish Times* asked 'are we in danger of entering a millennial equivalent of the days when Ireland had wind-up telephones and three-year waiting lists while the rest of the world enjoyed international dialling?'[61]

Finally, in May 2002, DSL was launched to the public by Eircom in Dublin and Esat-BT in Limerick. For download speeds of up to 512 kbps, Eircom charged a whopping €89 per month – plus €165 for installation.[62] In the US, Verizon charged half as much for a service that was 50 per cent faster. With such high charges it was not surprising that Eircom managed to notch up just 3,500 users in 12 months. A slightly cheaper service aimed at residential customers was launched in April 2003[63], though it was still a pricey €54.45 per month. Of course, with virtually no competition from cable, Eircom could charge what it liked. It could also largely govern the prices of other providers offering DSL, such as BT, since they were dependent on the wholesale rates it charged for access to the local loop and its telephone exchanges.

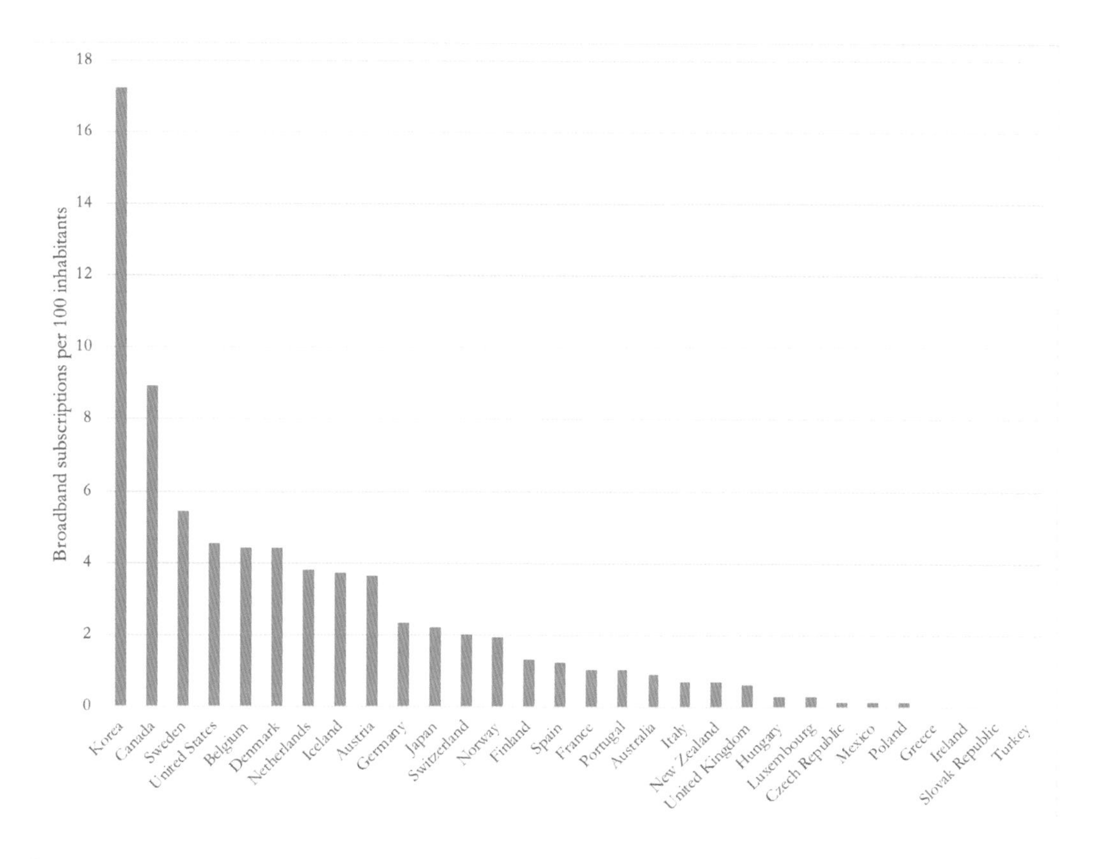

Figure 12.6: Broadband subscribers per 100 in 2001. Ireland was back at the bottom of the class. Source: OECD, *The Development of Broadband Access in the OECD Countries* (Paris, 2001).

The initial version of DSL to be deployed, called ADSL, was far from perfect. The problem of attenuation, which affected the early transatlantic cables, reared its head again, so that the maximum speed possible with ADSL declined with distance from the exchange. This posed a particular problem in Ireland where only 16 per cent of all premises in rural areas were within a 1.5 km radius of village centres.[64] A further problem was that Telecom Éireann, largely to avoid the costs of laying new lines, had from 1996 employed carrier systems to allow several customers to share the same pair of wires back to the exchange.[65] This was fine for voice or dial-up internet but it prevented such customers being able to use ADSL. All of these factors conspired to make ADSL service availability patchy in urban areas – and almost unknown in sparsely populated areas.

A report in July 2003 found that Ireland had the second lowest proportion of households with broadband access of the 19 countries then in the EU.[66] Partly thanks to those Vikings from Telenor, Ireland had become Scandinavian in its adoption of mobile technologies but the proportion of homes in Denmark with a broadband connection was 42 times higher than in Ireland. The report urged the government to deliver an investment programme, noting that 'Broadband connectivity should be treated as a priority basic infrastructure which can deliver very tangible benefits to the economy in the short to medium term.'[66] In 2004 Ryanair was the most profitable airline in the world,[67] soaring above its competitors thanks to the savings it made by getting its passengers to book online, while broadband in its home country was barely airborne.

The *Irish Independent* highlighted the gap between the image of Ireland as a high-tech haven, portrayed by agencies like the IDA, and the reality for most consumers: 'In a country that pitched itself as a European technology hub, Ireland's broadband infrastructure has been akin to the dotty family relative that should not really be mentioned at the dinner table.'[68] Despite its supposed high-tech economy, only 3.4 per cent of Irish GDP came from the internet economy, little more than half the European average.[69] While millions of cable customers from Canada to Belgium enjoyed high-speed internet, a mere 18,000 of these were in Ireland. One of these was presumably not Prisoner 33791, Ray Burke, who was serving time in Arbour Hill Prison for tax evasion.[70] His intervention while Minister in 1989 to ensure that Cablelink came under the control of Telecom Éireann was one of the factors contributing to the sorry state of broadband.

Meanwhile, new online services fuelled even more public demand for better internet access. Ryanair launched web check-in in 2006,[71] a year after the introduction of RIP.ie, a website dedicated to publishing death notices.[72] There were now more reasons than ever for Mary from Ballygobackwards to use the internet, except that Ballygobackwards had no cable and no DSL.

This vacuum of wired broadband service was partly filled from 2006 when mobile phone companies began to offer broadband service over their new 3G networks. For example, a SIM card could be inserted into a wireless router which then acted as a WiFi hotspot to which a laptop could be connected. Mobile broadband was generally considered inferior to wired broadband as typically data capacity was shared with everyone else using the same site, so speeds dropped when lots of people in the same area went online. As a result, in most countries mobile broadband was a niche product used by a businessperson on the move, or in a holiday home. In Ireland, however, the paucity of DSL and cable broadband made mobile the only option for many and in

2011 35.5 per cent of all broadband users utilised mobile broadband.[73] As availability was dependent on the proximity to a 3G (or later, 4G) mobile phone mast and the topography of the land in between, many rural areas were excluded from the footprint. Often these were the same areas that also had no DSL or cable broadband.

Over the coming years fixed broadband services in urban Ireland finally began to catch up with those in other countries, but this only widened the urban–rural divide. The cable networks in towns and cities across Ireland were consolidated under the control of multinational Liberty Global. Rebranded as UPC, during 2006–8 it undertook a long-overdue upgrade of the cable network, adding a fibre-optic backbone to create a hybrid fibre-coaxial (HFC) network which could provide digital television, telephony and broadband to almost all its customers. Meanwhile Eircom expanded the number of DSL enabled exchanges and, thanks to competition from cable and mobile operators, prices dropped. Broadband take-up began to soar from 2013 when Eircom, having exited examinership shorn of much of its accumulated debt, commenced an aggressive rollout of fibre-to-the-cabinet (FTTC) (Figure 12.7).[74] This brought fibre-optic cables as far as street-side cabinets, with existing copper lines used solely for the last leg into the home or business. Coupled with two innovative new technologies known as VDSL2 and vectoring, this provided speeds of 100 Mbps in the cities, towns and suburbs. This was an impressive speed to achieve over a traditional phone line.

While a vast improvement on the previous ADSL technology, it still could not match the potential capacity of bringing a fibre-optic cable directly into a home or business. For years this concept, known as fibre-to-the-premises (FTTP), had been the holy grail for communications. With much more bandwidth than the traditional copper pair used by telephones networks or even the coaxial cable used for cable TV, fibre could replace both – and supply much faster internet connection speeds. Japan and South Korea were leaders in the field: with 50 per cent of Koreans living in apartments[75], it was relatively cheap to connect them up to a brand-new network of fibres. For the same reason, it was apartment-dwellers in Ireland who were the first domestic users to benefit from FTTP in 2005, after US-owned telco Magnet signed a deal with the developers of a new housing development at Clongriffin in north Dublin to provide broadband, telephone and TV over fibre-optic cables.[76] However, fibre provision remained limited to new housing complexes so that six years later only 9,700 of Irish homes were connected to fibre. By contrast Lithuania had 343,400.[77]

In the 2010s fibre rollout finally commenced in earnest, thanks to a number of factors. One was the insatiable demand for faster speeds as more and more services went online. Early in 2012, video-streaming service Netflix expanded into Ireland[78], growing so quickly that within 15 months the State's largest DVD rental chain, Xtravision, had gone into receivership.[79] The other factor was a decline in the cost of fibre technologies, making it economic to wire up cities, towns and suburbs – though not the many houses dotting the Irish countryside.

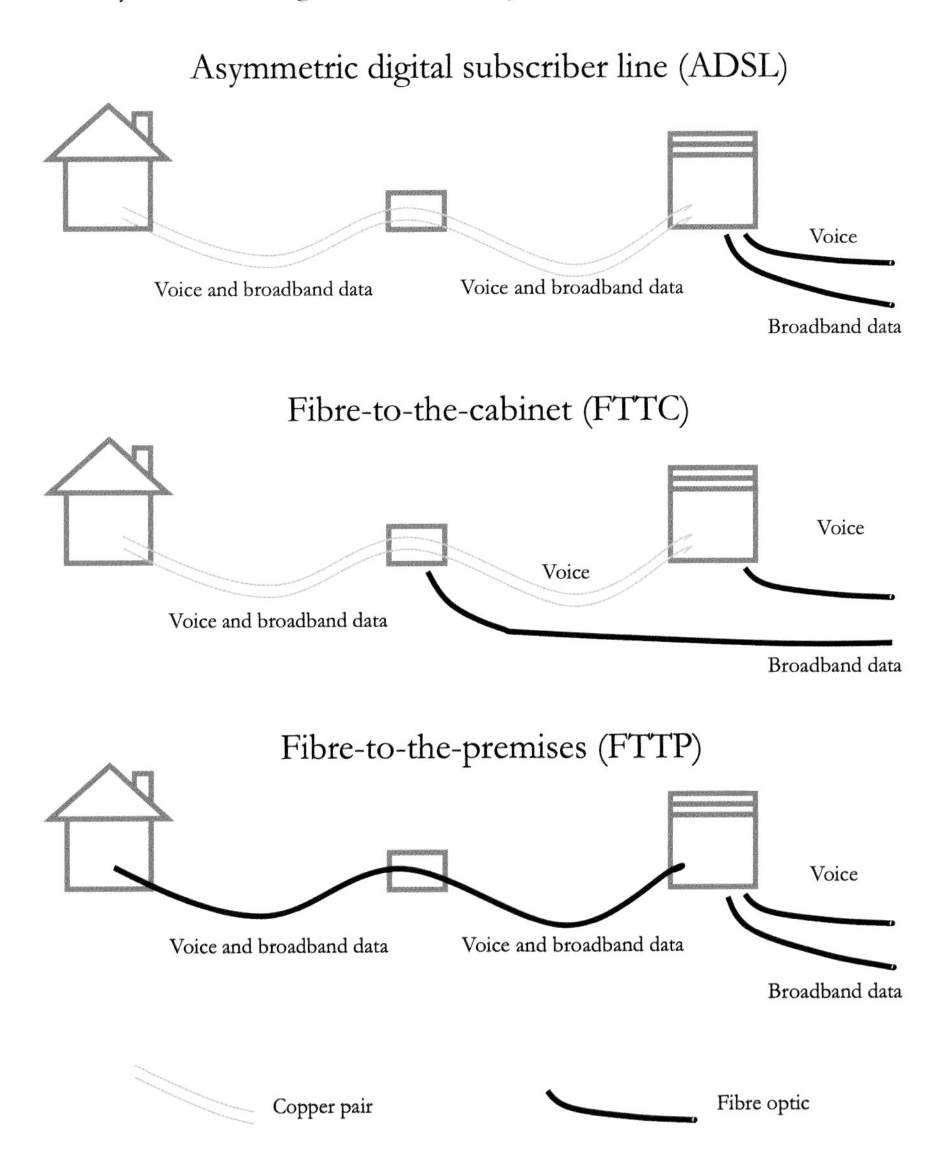

Figure 12.7: The evolution of broadband data networks has brought fibre-optic cables from the exchange firstly to street-side cabinets and, more recently, direct to homes and other premises.

Eircom started a FTTP pilot in 2011 with Wexford town and Sandyford in south Dublin the first to be lit up.[80] Later in the decade, progress ramped up as the former monopoly was joined by Siro, a joint venture formed by the state-owned ESB and mobile operator Vodafone. With a budget of €450m, Siro aimed to pass over 450,000 homes with its fibre-optic network, focusing mainly on provincial towns and suburban areas, with Carrigaline in Cork the first town to be lit in 2017.[81] By 2019 residents of many parts of urban Ireland had a choice of two high-speed networks based on either fibre or coaxial cable to the premises, offering speeds of 500 Mbps or 1 Gbps. Meanwhile most of rural Ireland could get a small fraction of that speed through DSL – if they were lucky. The urban–rural divide could not have been starker.

History seemed to be repeating itself. For decades residents of the larger towns and cities had enjoyed an automatic telephone service available 24 hours a day, while many rural dwellers had a part-time manual service. In the late 1980s the disparity evaporated: the standard of telecoms service was almost identical nationwide, while MMDS, cable and satellite brought multi-channel television to almost everyone. But now this inequality had returned. Part of the reason was, as before, financial. It simply was not economic to cable a rural area, or equip a roadside cabinet with fibre to serve just a handful of customers.

The result was that the potential for broadband access to spread economic development more evenly across the State went unrealised. For centuries, innovation had largely taken place in towns and cities, with new industries locating in major population centres to take advantage of the density of potential customers, suppliers, experts and workers in order to create new technologies, products and services. The internet provided a virtual city, connecting previously isolated local and regional economies into a more closely integrated global economy. A 2011 study of OECD countries found that a ten percentage-point increase in broadband penetration raised annual per capita economic growth by 0.9 to 1.5 percentage points.[82] But rural Ireland seemed stuck, yet again, in a catch-22, unable to access the technology that could, if deployed more evenly, benefit them the most.

The poor quality of internet access in much of the country had more profound impacts than just slowing down the likes of Mary Murphy trying to book her Ryanair flight. Apple advertised in 2012 for customer support agents to work from home – provided their broadband connection was faster than 5 Mbps.[83] It was as good as saying 'rural dwellers need not apply.'

The government had identified an urban–rural divide as a potential problem back in 1999, before the privatisation of Eircom. The *National Development Plan 2000–2006* was realistic about the limits of market economics:

> There is now evidence that a competitive market alone will not ensure the provision of advanced communications networks ... Under the current investment plans of the communications companies, large areas of the midlands will not have access and much of the west coast will not be served ... This deficiency has clear implications for regional development.[84]

The *National Development Plan* was devised just before John Beckett and Ryanair brought the internet to the mass market. The emphasis was thus mainly on ensuring broadband access across the country for business users, with less attention paid to how residential customers would access the internet. One of the main outcomes of the plan was the development of the Metropolitan Area Networks (MANs). These were open-access fibre-based networks that could be used by any broadband provider to link its customers to its backbone network. The MANs eventually grew to connect 94 towns and cities. Being largely used by Eircom's competitors, the investment was criticised by the incumbent as unnecessary duplication though analysis from the IDA suggested that the towns connected to the MANs gained considerably in terms of foreign investment.[60]

A variety of other companies also busied themselves building their own fibre-optic connections between the towns and cities, often along the same routes that had been used by the earlier telegraph and telephone networks. For example, back in the mid-1800s, the Magnetic telegraph company signed deals with the railway companies to erect lines along the tracks, a pattern revived in 1997 when Esat Telecom signed a deal with state transport holding company CIÉ to lay fibre-optic cables along the railway.[85] As a result by 2003 BT, which had acquired Esat in 2000, possessed 2,000 km of fibre,[86] largely following the railway network. In another evocation of the early days of telegraphy, other companies reached agreements with Waterways Ireland[87] and the roads authority to install their own fibre networks along the canals and roadways. Though Eircom had been privatised in 1999, the government retained involvement in this infrastructure through cables laid by state-owned companies such as the ESB and Bord Gáis.[86] In most cases, the companies involved were selling capacity on their fibres to others, rather than providing telecoms services directly themselves.

Of course, this frantic laying of cables along the main roads, railways and canals was to create a backbone network, not bridge the 'last mile' that linked the customer's home or business to the telecoms network. Ireland's dispersed settlement pattern meant that the 'last mile' was often many miles long, and thus technically unsuitable for DSL. Solving this issue was the prize that eluded several governments.

The first attempt at a solution was the County and Group Broadband Scheme. Launched in 2004, this was a self-help initiative to assist rural communities with the

State funding up to 55 per cent of the capital expenditure required to provide broadband service. The plan was that such initiatives would cover 575 communities with a population of 420,000 people. It was a nice theory, but in practice there was 'a lot of foot dragging and quibbling. It was badly handled and planned'.[88] Two years after the scheme's launch, only 6,000 subscribers were served.[88]

In addition to lacking wired broadband services, rural areas often had patchy mobile phone reception. To help address this, in 2009 the government signed a €220m contract with mobile network Three to extend broadband services throughout the country.[89] Under the plan, the mobile operator committed to expand its network of 3G infrastructure. While this brought easy internet access to many, it was no magic bullet. The speeds offered by 3G were inadequate for many purposes and availability was constrained by Ireland's hilly topography.

The private sector itself responded to this rural 'last mile' problem with a wireless solution known as Fixed Wireless Access (FWA), which used technologies similar to WiFi to connect homes to the internet. For those with a clear line-of-sight to the provider's mast, FWA could provide reasonable service and found a sizeable niche in rural areas unserved by cable or DSL. Early providers such as Irish Broadband and Leap were joined by a gaggle of smaller players and local community organisations. In the auction for the 3.6 GHz radio spectrum in May 2017, a company that had once been Ireland's first MVNO, Imagine, made a successful bid to enable it to enhance its FWA service using 5G. It planned to deploy 325 sites to cover up to 400,000 homes largely in rural areas and promising speeds of up to 300 Mbps.[90]

However, providing internet access by wireless means to all premises in Ireland was a huge challenge, despite the advent of improved mobile data technologies such as 5G. A 2018 study commissioned by Comreg found that providing internet speeds of 30 Mbps or more to 99.5 per cent of homes and businesses by wireless means would require a 65 per cent increase in the number of cell sites[91], largely due to the hilly landscape. Furthermore, it found that modern building insulation materials tended to impair reception and degrade speed. There seemed to be no getting away from the need to extend the network of fibre-optic cables deeper into rural Ireland.

Continued political pressure from rural residents led to the third, and most ambitious, of the government-backed programmes. Announced in September 2012, the National Broadband Plan (NBP) aimed to provide broadband speeds of 30 Mbps for all by 2020[52] with the State contributing €125m. Even by the low standards most Irish people were used to in 2012, 30 Mbps was not exactly ambitious. By April 2014 the plan had become bolder with an aim to bring fibre to 1,000 rural towns and villages[92] at an estimated cost of €512m. There was still little detail, though, on how the problematic 'last mile', would be bridged, or who would pay for it.

The muddled government policies were perhaps unsurprising in view of the limited understanding of many politicians of the complexities of the internet. A 2013 Oireachtas Joint Committee where members competed to display their deep ignorance of digital technology, prompted the *Sunday Independent* to comment 'the belief of second division Irish politicians that they can control a global phenomenon like the internet is almost touching.'[93]

Eventually, in December 2015, the NBP project went to tender, seeking bidders to build a network where commercial operators did not offer an adequate service. The decision process proved to be even slower and more controversial than the mobile phone licence awards of the 1990s. Even the number of premises included in what was called the Intervention Area kept changing. The initial figure of 757,000 was increased to 927,000 in 2016[94] and then cut to 542,000 in 2017. The reason for the latter change was an announcement by Eir, as Eircom was now known, that it was expanding the number of properties it was planning to serve with fibre by 300,000. This was widely seen as a cynical move by Eir to thwart its rivals[95] as it left the NBP with only the most remote and expensive of premises to provide with high-speed broadband. In addition the minimum speed requirement was increased to 150 Mbps.[64] One by one, bidders withdrew leaving only one in the race – and the Government effectively over a barrel on price.

Months of uncertainty ensued, partly as a result of changes within the consortium and partly due to political wrangling. In the end, the contract was awarded to the sole remaining bidder, now called National Broadband Ireland (NBI) and backed by US investor Granahan McCourt. The contract, with an estimated cost to the State of up to €3 billion over 25 years, was the most expensive State contract ever awarded to a third party[96] and worked out at over €18,000 per household assuming that 30 per cent of premises passed would take up the service. Part of the reason for the high cost was the need to lease poles and ducts from an existing utility – in this case Eir's infrastructure division, Open Eir – to reach homes and businesses in the intervention area.

Understandably, the high projected cost of the programme, and the fact that the bid-winner would own the network outright after 25 years, created a political and media rumpus. This proved relatively short-lived as, within four months of the contract being awarded, Ireland, like much of the world, went into lockdown thanks to the Covid-19 pandemic. Broadband became a critical tool, not for booking Ryanair flights, but to allow employees to work from home and help 'cocooning' grandparents keep in touch with their grandchildren. With many city dwellers pondering a move to the country, opposition to the cost of the NBP melted away.

2020 also saw the introduction of another potential solution to the rural broadband problem, one with no need for thousands of miles of fibre optic cable. While many Irish households received their television signals from a satellite in geo-stationary orbit 36,000 km above the earth, very few used satellite broadband. Just as telephone calls routed via a satellite in geo-stationery orbit suffer from annoying lags in speech, satellite broadband suffered from high latency, making video calls difficult and online gaming well-nigh impossible. However, late in 2020 entrepreneur Elon Musk inaugurated Starlink, a network of satellites that orbit at only 550–1,100 km above the earth instead of 36,000 km. It is planned to launch 12,000 satellites[97] to provide almost global coverage with latency and speed comparable with that provided by FTTC. Amongst the first Irish users were residents of the Black Valley in Co Kerry,[98] a telecoms blackspot for decades. Meanwhile the National Space Centre at the former Telecom Éireann satellite site at Elfordsdown in Co Cork became a base station to link the orbiting satellites to the internet.[99]

That base station, along with the fibre networks being laid by Open Eir, Siro and NBI, connected Ireland to the world over a network of fibre cables that stretched across the country and under the seas. As noted earlier, the fibre backbone largely followed the railway and road and canal networks. A map of the various fibre networks in the mid-noughties, if anyone had drawn one, would have looked rather like the map of the telegraph networks from 1870 – but with one notable addition. This was a 44-km-long snake around the west of Dublin containing multiple fibres. Dubbed the T50, after the M50 orbital motorway that it roughly followed, the cables held in this duct connected the many submarine cables[100] landing on the east coast with the sprawling industrial estates and business parks around Dublin. The reason for this was the growth of a new type of business activity – the data centre.

While media reports about Ireland's multinational tech industry were invariably accompanied by images of glass-walled office blocks in Dublin's 'Silicon Docks',[101] in reality those Facebook updates, email messages and Ryanair bookings were being processed by acres of servers housed in data centres occupying low-slung anonymous windowless buildings in unglamorous industrial estates around the periphery of Dublin. It became a huge business, partly because Ireland's cool climate lessened the need for expensive cooling. Even so, the power demand of these vast server farms was considerable: a 2019 report about the electricity grid forecast that by 2027 data centres would consume 31 per cent of Ireland's total electricity demand.[101]

These data centres needed to be connected with internet users all over the world, driving a flurry of new submarine cables not seen since the heady days of the 1850s. Though Ireland in the early twenty-first century was not the global telecommunications hub it had been in the late 1800s, it was remarkably well-connected for a small island. The government became involved in this flurry, negotiating a public–private partnership in 1999 with US-based fibre provider Global Crossing to lease capacity on a pair of new cables landing in Wexford[102] (see Appendix 1). The deal was signed 14 months after the well-publicised Microsoft decision not to locate a data centre in Ireland owing to its poor data connectivity, and, more curiously, two days before the floatation of Eircom. It was almost as if the government had doubts of the abilities of the free market to provide Ireland with adequate telecommunications. Howanever, by 2019, 14 subsea cables linked the Republic to the rest of the world; by contrast New Zealand, with a very similar population size, had four.[103]

While fibre-optics and data centres were new concepts, much of the business being conducted along the cables was essentially unchanged since the days of the *Great Eastern*. In the late nineteenth century, multiple cables laid by different companies from Co Kerry across the Atlantic provided capacity and redundancy, allowing a stockbroker in New York who telegraphed an order to London to receive a reply within five minutes.[104] In a similar move, twenty-first-century cable companies competed to provide the route with the lowest latency, allowing financial trades to arrive a few milliseconds ahead of the competition.

It was for that reason that a new cable was planned in 2011, its route remarkably similar to the first telegraph cable between Howth and Holyhead in 1852. Indeed the city terminus of the new CeltixConnect cable was to be in the IFSC, only a stone's throw from Connolly station, the terminus of the original cable. The reason for this was that, like in 1852, it made sense to be close to potential customers to reduce latency. In 1852 this was to reduce the distance a telegram boy had to carry the message to the recipient's premises; in 2012 this was to reduce the total length of cable between financial traders at Dublin's IFSC and their counterparts in London.

On a gloomy Friday morning at the tail-end of 2011, some of the hardy souls braving the elements on Dollymount Strand must have noticed their usual view of Howth Head obscured by a ship moored close to shore (Figure 12.8). The ship was the *Cable Innovator* and it had arrived to lay the CeltixConnect cable under the Irish Sea to Holyhead, 123 km away.[105] Unlike its predecessor, 159 years before, this new cable comprised not of strands of copper, but 72 pairs of optical fibre, giving it a total potential capacity of 1,440Tbps – 863,000,000,000,000 times greater than the capacity of that first, ill-fated, telegraph cable.

And while the arrival of the *Britannia* and *Prospero* on that warm summer's evening in Howth attracted huge crowds and media attention, in 2011 the arrival of the *Cable Innovator* went virtually unnoticed save for a few dog-walkers and joggers. It was just another submarine cable, joining the many already linking Ireland to the outside world. It was just another unremarkable piece of fibre-optics in a mosaic of copper and coaxial, routers and exchanges, poles and masts.

Just another cable, connecting a nation.

Figure 12.8: The CeltixConnect cable being taken ashore at Bull Island. *(Courtesy of Aqua Comms)*

CHAPTER THIRTEEN

CONCLUSION

I was in London on my way to visit the BT Archives. It was the start of my research for this book and I felt a growing sense of excitement as I made my way there. On the tube my fellow-passengers were reading newspapers with stories about Northern Ireland, 'soft' borders, smartphone apps and customs drones. This unusual fare for the English media had come about as a result of the seemingly intractable Brexit negotiations between the EU and the UK, and a recent UK election that had given Northern Ireland's Democratic Unionist Party (DUP) and its leader Arlene Foster a leading role in decision making. It turned out that the themes of borders, politics and technology were to dominate my day long beyond that tube journey.

I got off the Central line at Holborn and, using the Google Maps app on my phone, soon reached a severe-looking 1920s edifice, home of the BT Archives. I pressed a buzzer and was brought up to the second floor.

I quickly realised that I was the only reader and that the trolley I could spy laden with neatly arranged piles of boxes and envelopes contained the material about Ireland I had requested to see. I sat down at the desk beside the trolley, picked up a manilla envelope from the top of the pile and carefully opened the flap.

A large map, hand-drawn on tracing paper, fell out. *Diagram of Routes crossing the Nor. Ireland–Free State* border was written at the top, with the year 1929 added in pencil below. The map was disintegrating with age but the colours were still vivid and revealed that this was not just another map of the border in Ireland (Figure 6.3).

For this map had *two* borders: a *green* line indicating the familiar political border and a *yellow* line showing the border agreed on by the two post office authorities now sharing the island. Intriguingly, the two lines sometimes diverged, creating a type of 'soft border' for telecoms. For example, the yellow telecommunications border bulged out into Co Donegal so that two of its telephone exchanges were served from Derry on the other side of the green political border while, conversely, two exchanges on the northern side were maintained by the Free State.

The map blew my mind. I began to realise that factors other than just technology were involved in the story of telecommunications in Ireland. There on this old map was Roslea in Co Fermanagh on the northern side of the green political border – but on the *southern* side of the yellow telecoms border. In the midst of all the news coverage of 'soft' borders, it seemed ironic that 90 years earlier Roslea, home town of DUP leader Arlene Foster, had been in the *south*, telephonically speaking.

That map from 1929 shows that the story of telecommunications in Ireland is more than just a one-man-show of technological advances, dizzying though they are. In fact, as we've seen, the story is one that involves a cast of other actors with politics and geography playing critical roles, along with socio-economic factors.

The political dimension to the story of Irish telecommunications is clear. An early example occurred when in 1804 the government, fearing a French invasion, commissioned Richard Lovell Edgeworth to construct an optical telegraph link between Dublin and Galway. 120 years later, politically motivated sabotage of the transatlantic cable stations during the Civil War sowed an image of Ireland as a hostile environment for transatlantic communications. Soon after that, politically inspired antipathy towards the British-dominated Marconi company contributed to the company winding down its Irish operations in the early years of the Free State. This all left Ireland as a global telecommunications backwater.

Domestically, it was a political decision to establish a semi-state company in 1927 to develop the electricity network but to leave the telecommunications service under direct government control. For decades the telecommunications service was accorded scant priority by successive governments and crippled by the lack of investment.

With many competing demands for capital investment, perhaps the reason for this consistent pattern of underinvestment in telecoms was the length of time such investment took to deliver benefits. There was little to be gained by approving a big project when, by the time the new exchange was planned, designed, built, equipped and commissioned, the Minister who made the investment decision had been replaced, leaving their successor to gain the kudos.

This pattern of short-termism and underinvestment eventually blew up in the faces of politicians in the 1970s when public outrage demanded a change in direction. The result was two pivotal political decisions in 1979: first, to throw a huge wad of taxpayers cash at the network, and second, to establish it as an autonomous organisation. The huge influence of politics on telecommunications was not just an Irish phenomenon: political decisions about governance and investment had effects on telecoms around the world. Even in the free-market US, it was government-funded programmes such as NSFnet in the 1970s and 80s that led to the creation of the internet.

But as we saw in Chapter Ten, by the 1990s the pendulum had swung the other way as governments all over Europe and further afield jumped on the privatisation bandwagon. With even Cuba selling off a 49 per cent share of its telecoms network to a Mexican company in 1993[1], the privatisation of Telecom Éireann had a certain inevitability. The timing of its sale in 1999 was unfortunate, coming just before the bursting of the dot-com bubble but even so there were some aspects that seem ill-judged. First was the decision by the government to sell its entire stake at once rather than in tranches over a number of years. Second, unlike in many other countries, the government did not seek to retain a 'golden share' granting them certain veto rights over decisions in relation to changes in company ownership.[2] Thirdly, there was no attempt to divide Telecom Éireann into separate network and retail entities – it was just a simple sale as a job lot.

When the gas and electricity sectors were deregulated a little later, a different approach was adopted. Since its formation, the ESB had controlled all aspects of electricity supply from generating and distributing power to billing customers in what is called a vertically integrated utility. A similar model was employed for gas supply. As part of deregulation, the ESB and Bord Gáis were divided into separate network and retail entities, with the State retaining the distribution networks while allowing competition for retail customers. Thus the network of gas pipes was vested in state-owned Gas Networks Ireland while customers could choose between a number of competing suppliers, including the now-privatised Bord Gáis Energy.

Potentially if a similar separation had been applied to telecoms, with the State retaining control of infrastructure, the problems caused by Eircom being flipped by a series of asset-stripping owners would not have occurred. But separating a telecoms network into network and retail entities is more complex than separating a gas or electricity utility. Gas and electricity are relatively generic products: an electricity customer is pretty unlikely to contact their supplier looking for current at 60 Hz instead of 50 Hz, but a telecoms customer is very likely to look for a higher broadband speed. Probably for that reason, while many other countries divided their gas and electricity utilities in the same manner as Ireland, none divided up ownership of their telecoms into separate network and retail entities prior to privatisation.

While the liberalisation of gas and electricity provoked relatively little controversy, the reverberations from the privatisation of Eircom in 1999 continued for decades, with regular calls for either the network, or the entire company, to be renationalised. For example, during discussions of the National Broadband Plan, Taoiseach Leo Varadkar, not known as a radical leftist, told the Dáil that 'one of the things we gave consideration to was ... the renationalisation of Eircom so that we could get our hands on the real infrastructure which is the poles and the ducts' but the government baulked at an estimated cost of €6bn.[3] Even more extraordinary was the opinion of Eir's CEO Carolan Lennon in a 2019 interview that privatisation in 2000 was 'probably [a] bad decision'.[4]

Problems with telecoms infrastructure post-privatisation were not unique to Ireland. While discussion about political policies and telecoms is often reduced to a binary public vs private trope, in reality the form of ownership seems to be one of the least important factors. Historically this was the case in 1912 when the Post Office takeover of the private telephone companies failed to improve the quality of service as much as proponents of nationalisation had hoped. More recently, countries as different as Australia and Venezuela renationalised their telecoms infrastructure, though neither has become a broadband trailblazer as a result.[5] In 2020 that top position was held by Singapore, where the dominant provider, SingTel, remained government owned. In part, the relative success of Singapore is indeed due to government policy, but more in areas such as governance and regulation rather than its ownership of a key player.

The example of Singapore, where 85 per cent of residents live in apartments, shows the second main force impacting telecoms: geography. Weaving in and out of this book is a narrative about the disparity in telecommunications access between urban and rural parts of Ireland. The problem of providing telecommunications economically to rural areas is a widespread one but it is more acute in Ireland for reasons that go back to Co Mayo during the 1880s. In Chapter One we learnt about the campaign for a telegraph connection to Blacksod involving philanthropist James Hack Tuke. This was not the extent of Tuke's lobbying to improve conditions in the west of Ireland, however, as he went on to help establish a government agency called the Congested Districts Board. The Board began to acquire and sub-divide large parcels of land and distribute them amongst former tenant farmers, in doing so creating a striking new settlement pattern in rural Ireland with traditional clusters of homes replaced by a dispersed system of farmhouses. The Board and the Land Commission, its post-independence successor, built thousands of new dwellings, each surrounded by regular enclosed fields, radically reshaping the countryside.[6]

During the era of the Congested Districts Board, services such as piped water, electricity and telephones were novelties of little concern to ordinary people either in town or country and so the impact of this new settlement pattern on providing such conveniences was not considered. However as such utilities became considered normal in the twentieth century, the problems of providing them to scattered household manifested themselves. For example, the Rural Electrification Scheme, which successfully brought light and power to the countryside, was only possible thanks to government subsidies from its inception in 1946. A confidential 1976 report estimated that widely scattered housing cost the State between three and five times more to service than closely knit dwellings.[7] Despite the extra costs imposed, this pattern of isolated homes became the norm throughout rural Ireland and, assisted by the advent of mass motoring, accounted for 72 per cent of rural housing in 2016.[8]

The impact of this dispersed distribution on telecommunications was even more pronounced than on the supply of water or electricity. While there are many countries with even lower average population densities than Ireland, in most cases their rural communities are clustered in the same manner as in Ireland before the late nineteenth century. In such places, a group of houses could be connected to a local telephone exchange with a relatively small amount of cable. By contrast in Ireland only 16 per cent of premises in rural areas lay within a 1.5 km radius of a village centre[9] making it much more expensive to serve the same number of people due to the miles of cable required.

As we saw in Chapters Six and Eight, for decades P&T sought to recoup the extra expense entailed by applying extra installation and rental charges where a telephone was located more than a few miles from an exchange. This created a vicious circle where the service in rural areas was expensive because it failed to attract sufficient customers and failed to attract customers because it was expensive. By the 1980s, with this policy becoming a political liability, Telecom Éireann began to apply the same connection and rental charges to all customers, so that those living in remote locations were effectively cross-subsidised by urban customers. Coupled with the replacement of the last manual exchanges and flattening of call costs, by the early 1990s rural dwellers finally enjoyed service and charges similar, on fixed-lines at least, to their urban cousins. That equalisation proved brief, however.

At privatisation Eircom retained this obligation to charge the same rental and, within certain limits, connection fees regardless of location. However under this Universal Service Obligation it was only obliged to provide a fixed line capable of voice, fax and dial-up internet, not broadband data. The technical limitations of DSL and enormous cost of laying new fibre cables would have made such an obligation financially crippling in a country with such a scattered housing pattern. As a result many rural dwellers were virtually cut off from the internet. The urban–rural divide was back.

The political pressure this generated led eventually to the government-funded National Broadband Plan which aimed to provide broadband to 542,000 premises, the vast majority of which were the isolated houses so favoured by the Congested Districts Board. The forecast cost of €18,000 per house served was not just due to the scattered housing but also the fact that a completely separate local fibre network was required, often running alongside fibres owned by commercial operators Eir and Siro. Expensive as it may be, however, the NBP and investment by commercial operators should lead to almost every home in the State being provided with a fibre connection by 2027. If delivered, not only will this remove the urban–rural divide but it is also forecast to move Ireland to the top of the telecoms class, a position it has rarely held, in terms of fibre rollout.[10] Of course no-one knows if this will happen or not. Aside from the technological challenges and the difficult geography, forecasting politics is a hazardous undertaking.

One chilly evening recently I took a trip to Howth, back to the spot where the first telegraph cable was taken ashore on a more pleasant Wednesday evening in the early summer of 1852. Unlike then, however, there were no crowds and no cable ship. As I walked back towards the railway station, I noticed the Martello tower, which had housed equipment for some of those early cables. The charming little museum of radio that now occupies the tower had closed up for the evening, leaving the building in darkness.

My train journey back to Connolly Station was also very different to that experienced by the day-trippers of 1852, with steam supplanted by electric traction and overhead electric wires replacing the telegraph poles. There are still telecommunications cables alongside the track, but made of fibre-optic instead of iron and hidden away in a conduit beside the rails. And, unlike that evening in 1852 when extra trains were laid on to carry the excited masses back from Howth, my DART carriage was quite empty, at least until a group of teenagers boarded at Kilbarrack.

The teenagers immediately started showing each other videos and pictures on their phones, the digital cacophony only momentarily interrupted by a faltering signal as the train entered a cutting after Killester. I smiled as I thought about how when I was a teenager telephones were tethered by a wire to the wall and music videos were a weekly television treat on *Top of the Pops*. I suspected that these teenagers would react to such a description with a mixture of derision and disbelief.

But then I wondered that maybe they'd be taken down a peg or two if they knew that their hyper-connectivity was foreseen nearly a 100 years ago by Nicola Tesla. As we saw, Tesla predicted that wireless would enable us to 'communicate with one another instantly, irrespective of distance' and he boldly stated that through:

> telephony we shall see and hear one another as perfectly as though we were face to face, despite intervening distances of thousands of miles, and the instruments through which we shall be able to do all of this, will fit in our vest pockets.[11]

Yet we also saw how Heinrich Hertz, who probably deserves the accolade 'father of radio' more than Marconi, reportedly said 'I do not think that the wireless waves I have discovered will have any practical application.'[12] It all goes to show the difficulties in predicting the next stage in the story of telecoms in Ireland, or anywhere.

One thing is certain. A future generation of teenagers will hear a description of our era of 5G networks, satellite television, data centres, the Internet of Things, smart-phone apps, online gaming, virtual reality goggles and the rest and will react with the same mixture of derision and disbelief.

As my train slowed to approach Connolly station, I took my own phone out of my coat pocket and clicked on the real time travel app to plan the rest of my journey from Connolly. Perfect: a connecting bus was due in a few minutes. With telegraphic brevity, I texted my husband. 'On way home. C u soon.'

APPENDIX 1: LIST OF SUBMARINE TELECOMMUNICATIONS CABLES LANDING IN IRELAND, 1852–2022

Date	Irish landing point	Overseas landing point	Length (km)	Company	Capacity (see note)	Name	Notes
1852, June 1	Howth, Co Dublin	Holyhead, Wales	126	Irish Sub-Marine Telegraph Company (bought out by the Electric)	1 telegraph circuit		Failed after 3 days. Cable weighed 1 ton per mile.
1852, July	Donaghadee, Co Down	Portpatrick, Scotland	50	Electric Telegraph Company of Ireland	2 telegraph circuits		Completion abandoned due to bad weather. Company wound up in 1856.
1852, October	Donaghadee, Co Down	Port Mora, Portpatrick, Scotland	50	English & Irish Magnetic Telegraph Company (merged into the 'Magnetic')	6 telegraph circuits		Completion abandoned due to bad weather.
1853, May 23	Donaghadee, Co Down	Port Mora, Portpatrick, Scotland	50	English & Irish Magnetic Telegraph Company (merged into the 'Magnetic')	6 telegraph circuits		First successful cable. Cost was £13,000. Cable taken over by the Post Office in 1870.
1853, November	Donaghadee, Co Down	Port of Spittal Bay, Portpatrick, Scotland	50	Electric Telegraph Company of Ireland	1 telegraph circuit		Cable broke during paying out and was abandoned.. Company wound up in 1856.
1854, June	Whitehesd, Co Antrim	Portpatrick, Scotland	50	British Electric Telegraph Company (merged into the 'Magnetic')	6 telegraph circuits		Successful. Taken over by the Post Office in 1870.
1854, September	Howth, Co Dublin	Holyhead, Wales	126	International Telegraph Company (subsidiary of the 'Electric')	1 telegraph circuit		First successful cable from Dublin area. Failed in 1859.
1855, June	Howth, Co Dublin	Holyhead, Wales	126	International Telegraph Company (subsidiary of the 'Electric')	1 telegraph circuit		Laid due to volume of business on previous cable. Failed in 1859.
1857	White Strand, Ballycarbery, Co Kerry	Bay Bull arm, Trinity Bay, Newfoundland	2,315	Atlantic Telegraph Co.	1 telegraph circuit		Cable lost.
1858	White Strand, Ballycarbery, Co Kerry	Bay Bull arm, Trinity Bay, Newfoundland	4,074	Atlantic Telegraph Co.	1 telegraph circuit		Cable lost.

1858	Knightstown, Valentia, Co Kerry	Bay Bull arm, Trinity Bay, Newfoundland	4,074	Atlantic Telegraph Co.	1 telegraph circuit		Failed within weeks.
1861	Howth, Co Dublin	Porth Trecastell, Holyhead, Wales	126	Electric Telegraph Company (the 'Electric')	1 telegraph circuit		Failed in 1865. Cable recycled from an Anglo-Dutch cable. Landing place at Howth was Balscaddan Bay.
1862	Greenore Point, Co Wexford	Abermawr, Wales	115	The London & South-of-Ireland Direct Telegraph Company (subsidiary of the 'Electric')	4 telegraph circuits		First cable from Wexford. Taken over by the Post Office in 1870.
1865	Foilhommerum Bay, Valentia, Co Kerry	Heart's Content, Trinity Bay, Newfoundland	3,511	Atlantic Telegraph Co. (later Anglo-American Telegraph Co)	1 telegraph circuit		Cable lost, but recovered in 1866. Abandoned in 1877.
1866	Whitehead, Co Antrim	Port Kale, Portpatrick, Scotland	50	Electric & International Telegraph Company (subsidiary of the 'Electric')	6 telegraph circuits		Last cable laid across the Irish Sea by private company for 123 years.
1866	Foilhommerum Bay, Valentia, Co Kerry	Heart's Content, Trinity Bay, Newfoundland	3,430	Anglo-American Telegraph Co	1 telegraph circuit		First working cable. Abandoned in 1872
1870	Donaghadee, Co Down	Port Kale, Portpatrick, Scotland	43	GPO	4 telegraph circuits		Still in use in 1938.
1871	Howth, Co Dublin	Porth Trecastell, Holyhead, Wales	120	GPO	7 telegraph circuits	Cable No 1	Included a private circuit beween Dublin Castle and London.
1873	Knightstown, Valentia, Co Kerry	Heart's Content, Trinity Bay, Newfoundland	3,476	Anglo-American Telegraph Co (later Western Union)	1 telegraph circuit	1VA.	Submarine repeaters installed in the 1950s.
1874	Ballinskelligs, Co Kerry	Tor Bay, NS, Canada	4,126	Direct United States Telegraph Company	1 telegraph circuit		First competitor to the Anglo-American Telegraph Co. Bought by the Post Office in 1920 and diverted to Mousehole in Cornwall.

Date	Irish landing point	Overseas landing point	Length (km)	Company	Capacity (see note)	Name	Notes
1875	Cooscroom pier, Cooncrome, Co Kerry	Heart's Content, Trinity Bay, Newfoundland	3,476	Anglo-American Telegraph Co. (later Western Union)	1 telegraph circuit	2VA.	Submarine Repeaters installed 1957. Worked until 1966.
1879	Whitehead, Co Antrim	Knock Bay, Portpatrick, Scotland	43	GPO	4 telegraph circuits		Still in use in 1938.
1880	Blackwater, Co Wexford	Abermawr, Wales	104	GPO	4 telegraph circuits		Cable abandoned 1922/3.
1880	Knightstown, Valentia, Co Kerry	Heart's Content, Trinity Bay, Newfoundland	3,430	Anglo-American Telegraph Co. (later Western Union)	1 telegraph circuit	3VA	A near complete renewal of the 1866 cable: only the shore ends were not replaced. Abandoned in 1949.
1882	Foilhommerum Bay, Valentia, Co Kerry	Greetsiel, Emden, Germany	1,585	German Union Telegraph Co	1 telegraph circuit		First connection to mainland Europe. This cable was often damaged by shipping and was diverted to Brest in 1911 to form an Emden–Brest link.
1883	Blackwater, Co Wexford	Fishguard, Wales	115	GPO	4 telegraph circuits		Repaired 1912. Cable abandoned 1922/3.
1884	Waterville, Co Kerry	Canso, NS, Canada	3,356	Commercial Cable Co	1 telegraph circuit	MAIN – 1	
1884	Waterville, Co Kerry	Canso, NS, Canada	3,408	Commercial Cable Co	1 telegraph circuit	MAIN – 2.	Onward connections from Dover Bay to Duxbury, MA and Far Rockaway, NY.
1885	Waterville, Co Kerry	Le Havre, France	952	Commercial Cable Co	1 telegraph circuit	HR, HAVRE – 1	
1885	Waterville, Co Kerry	Weston-super-Mare, England	611	Commercial Cable Co	2 telegraph circuits	SN1	

1886	Newcastle, Co Wicklow	Aber Geirch, Nefyn, Wales	102	GPO	4 telegraph circuits	Aber-geirch-New-castle : No. 1	Still in use in 1922.
1888	Whitehead, Co Antrim	Port Mora, Portpat-rick, Scotand	48	GPO	4 telegraph circuits		Still in use in 1938.
1892	Newcastle, Co Wicklow	Aber Geirch, Nefyn, Wales	102	GPO	4 telegraph circuits	Probably Aber-geirch-New-castle : No. 3	Still in use in 1922.
1893	Donaghadee, Co Down	Port Kale, Portpat-rick, Scotand	44	GPO	2 telephone circuits		First telephone cable connecting Ireland. Four seven-wire conductors. Still in use in 1938.
1894	Waterville, Co Kerry	Canso, NS, Canada	3,274	Commercial Cable Co	1 telegraph circuits	MAIN – 3	In 1926 this cable was diverted to St John's in Newfoundland.
1894	Waterville, Co Kerry	Weston-super-Mare, England	611	Commercial Cable Co	1 telegraph circuit	MAIN – 3	
1894	Knightstown, Valentia, Co Kerry	Heart's Content, Trinity Bay, New-foundland	3,432	Anglo-American Telegraph Co. (later Western Union)	1 telegraph circuit	4VA.	Submarine Repeater installed 1957.
1899	Newcastle, Co Wicklow	Aber Geirch, Nefyn, Wales	115	GPO	4 telegraph circuits		Four air-spaced copper conductors. Intended for telephone use but found unsuitable and used for telegraph purposes instead. WFU 1935.
1901	Waterville, Co Kerry	Horta, Azores	2,230	Commercial Cable Co	1 telegraph circuit	MAIN - 4A	
1901	Waterville, Co Kerry	Weston super Mare, England	611	Commercial Cable Co	1 telegraph circuit	SN2, WESTON - 2	

Date	Irish landing point	Overseas landing point	Length (km)	Company	Capacity (see note)	Name	Notes
1902	Howth, Co Dublin	Porth Trecastell, Holyhead, Wales	126	GPO	4 telegraph circuits	No 2	
1905	Waterville, Co Kerry	Canso, NS, Canada	3,513	Commercial Cable Co.	1 telegraph circuit	MAIN – 5	In 1926 the western landing point was moved to St John's in Newfoundland.
1910	Waterville, Co Kerry	Weston super Mare, England	626	Commercial Cable Co.	1 telegraph circuit	SN3	
1913	Howth, Co Dublin	Aber Geirch, Nefyn, Wales	117	GPO	2 telephone circuits	Aber Geirch - Howth 1	Abandoned in 1938.
1918	Valentia, Co Kerry	Sennen Cove, Penzance, England	509	Western Union	1 telegraph circuit	1PZ.	Connected to one of the transatlantic cables during the Civil War to allow continued service.
1920	Valentia, Co Kerry	Sennen Cove, Penzance, England	530	Western Union	1 telegraph circuit	2PZ	
1921	Donaghadee, Co Down	Port Kale, Portpat-rick, Scotland	41	GPO	2 telephone circuits		
1923	Waterville, Co Kerry	Le Havre, France	948	Commercial Cable Co.	1 telegraph circuit	HAVRE2, MAIN 6	
1923	Waterville, Co Kerry	Horta, Azores	2,263	Commercial Cable Co.	1 telegraph circuit	MAIN – 6(A)	
1923	Valentia, Co Kerry	Sennen Cove, Penzance, England	537	Western Union	2 telegraph circuits	3PZ	
1923	Waterville, Co Kerry	Weston super Mare, England	578	Commercial Cable Co	1 telegraph circuit	SN4, WESTON - 4	

1929	Ballyhornan, Co Down	Port Erin, IOM	57	GPO	2 telephone circuits		Continued to Blackpool, England. Continuously loaded paper insulated telephone cable.
1937	Donaghadee, Co Down	Port Kale, Portpatrick, Scotland	44	GPO	16 telephone circuits		Coaxial pair.
1938	Howth, Co Dublin	Aber Geirch, Nefyn, Wales	119	GPO	16 telephone circuits.		Coaxial pair. One telephone circuit used for 18 channel telegraphy. Upgraded in 1946 to 24 telephone circuits plus two 18-channel telegraph circuits. Still in use in 1961.
1942	Ballyhornan, Co Down	Port Erin, IOM	57	GPO	Unknown		Continued to Cemaes Bay, Wales, via a cable with the first submerged repeater. WFU before 1980.
1947	Dollymount, Dublin	Holyhead, Wales	115	GPO	24 telephone circuits		Coaxial Pair. Two submarine amplifiers were inserted in each cable in 1953, increasing the combined circuit capacity to 120. WFU 1984.
1983	Drains Bay, Larne, Co Antrim	Port Mora, Portpatrick, Scotand	44	BT	480 telephone circuits	UK Inland 7	Coaxial cable recycled from a recovered England-Netherlands cable laid in 1967.
1988	Portmarmock, Co Dublin	Porth Dafarch, Holyhead, Wales	130	BT, TE	0.84 Gbps	BT - TE 1	First fibre-optic cable. 6 x 140 Mb/s. WFU 2007.
1989	Ballinspittle, Co Cork	Manasquan, NJ, US, Bream, England, and Bermuda	Unknown	C&W - Sprint	0.42 Gbps	PTAT 1	420 Mb/s. WFU 2004.
1989	Donaghadee, Co Down	Portpatrick, Scotland	35	BT	3.36 Gbps	Scotland-Northern Ireland 1	6 x 560 Mb/s.

Date	Irish landing point	Overseas landing point	Length (km)	Company	Capacity (see note)	Name	Notes
1990	Ballyhornan, Co Down	Peel, IOM	155 (total)	BT, Manx Telecom	480 Gbps	BT-MT-1	Onwards to Silecroft Beach, Cumbria, England. Upgraded in 2000 to 12 x 40 Gbps.
1992	Whitehead, Co Antrim	Troon, Scotland	122	Vodafone (was Mercury Communications)	3.39 Gbps	Lanis-3	6 x 565 Mb/s.
1992	Ballywalter, Co Down	Peel, IOM	67	Vodafone (was Mercury Communications)	3.39 Gbps	Lanis-2	6 x 565 Mb/s.
1993	Larne, Co Antrim	Girvan, Scotland	83	BT	0.565 Gbps	Scotland-Northern Ireland 2	
1994	Kilmore Quay, Co Wexford	Sennen Cove, England	275	BT, eir, Orange	2.5 Gbps	Celtic	WFU before 2014.
1999	Portmarnock, Co Dublin	Lytham St Annes, England	219	Virgin Media (was NTL)	480 Gbps	Sirius South	
1999	Carrickfergus, Co Antrim	Saltcoats, Scotland	147	Virgin Media (was NTL)	480 Gbps	Sirius North	
1999	Kilmore Quay, Co Wexford	Sennen Cove, England	259	Esat BT	80Gbps	ESAT-1	Upgraded to 12Tb/s.
2000	Sandymount, Dublin	Southport, England	240	Esat BT	80Gbps	ESAT-2	Upgraded to 100Tb/s.
2000	Kilmore Quay, Co Wexford	Oxwich Bay, Wales	140	eir, Vodafone	5Gbps	Solas	2 x STM-16.
2000	Ballinesker, Co Wexford	Bude, Cornwall, England	134	Lumen/Level 3	40Gbps	Pan European Crossing (UK-Ireland 2) (Global Crossing 1 & 2)	Public/Private Partnership.

2000	Ballygrangans, Kilmore Quay, Co Wexford	Whitesand Bay, Sennen Cove, England	161	Lumen/Level 3	40Gbps	Pan European Crossing (UK-Ireland 1) (Global Crossing 1 & 2)	Public/Private Partnership.
2001	Baldoyle, Co Dublin	Southport, England	Unknown	GTT	100Gbps	GTT South (was Hibernia Atlantic) (Segment C)	Upgraded to 100 Gb/s in 2011.
2001	Sutton, Dublin	Herring Cove, NS, Canada	Unknown	GTT	100Gbps	GTT South (was Hibernia Atlantic) (Segment D)	Upgraded to 100 Gb/s in 2011.
2009	Portrush, Co Antrim	Herring Cove, NS, Canada & Southport, England	Unknown	GTT	100Gbps	GTT North (was Hibernia Atlantic) (NI 6) (Segment B)	
2012	Bull Island, Dublin	Porth Dafarch, Holyhead, Wales	131	Sea Fibre Networks, later Aqua Comms	1,440Tbps	Celtix-Connect-1 (CC-1)	72 fibre pairs.
2012	Portmarnock, Co Dublin	Porth Dafarch, Holyhead, Wales	120	Zayo Group, ESB Telecoms	100Gbps	Emerald Bridge Fibres	State owned. 48 fibre pairs.
2012	Lusk, Co Dublin	Deeside, Wales	187	Zayo Group, Eirgrid	700Gbps	Geo-Eirgrid	Part of electricity interconnector. Not carrying traffic.
2012	Garretstown, Co Cork	Halifax NS, Canada & Brean, England	Unknown	GTT	53,000Gbps	GTT Express (was Hibernia Express)	

Date	Irish landing point	Overseas landing point	Length (km)	Company	Capacity (see note)	Name	Notes
2016	Killala, Co Mayo	Shirley, NY, USA	5,536	Aqua Comms	78,000Gbps	AEC-1 (was AEConnect)	6 fibre pairs: 130 wave-lengths at 100 Gb/s on each fibre pair.
2019	Portrane, Co Dublin	Cleveleys, Southport, England	221	euNetworks	Unknown	Rockabill	
Planned for 2020	Dublin	Blackpool, England via IOM	1750	Aqua Comms	300Tbps	Havhing-sten/Celtix-Connect-2 (CC-2)	
Planned for 2022	Killala, Co Mayo	Norway, Scotland, Iceland	Un-known	Eidsiva Energi, NTE, TrønderEn-erg	Unknown	Celtic Norse	
Planned for 2022	Galway	Molvik, Iceland	Un-known	Farice	Unknown	IRIS	
Planned for 2022	Lacanvey, Co Mayo	Wall Township, NJ, US, Denmark & Norway	Un-known	Aqua Comms, Facebook, Bulk Infrastructure, Google	108,000Gbps	Havfrue/ AEC-2	
Planned for 2022	Cork	Lannion, Brittany, France	Un-known	Ireland-France Subsea Cable Limited	120,000Gbps	IFC-1	

Note: Capacity figures are based on the best information available but may not be completely accurate. In particular, terminal equipment is often upgraded after a fibre-optic cable has been in use for some time, resulting in a huge capacity increase, but such upgrades may not be widely publicised.

SOURCES:

AquaComms, personal communication, 13 August 2021.
Bill Burns, *Atlantic Cable*, https://atlantic-cable.com (various pages). Accessed on: 05/11/2020.
Infrapedia, https://www.infrapedia.com/app. Accessed on: 08/11/2020.
TeleGeography, *The Submarine Cable Map*, https://www.submarinecablemap.com. Accessed on: 07/11/2020.

GLOSSARY

ADP	Accelerated Development Programme: a major investment programme in telecommunications from 1979 to 1984
ADSL	Asynchronous Digital Subscriber Line: one of the first types of DSL (qv) technology
AMPS	Advanced Mobile Phone System: an analogue mobile phone system originally developed by Bell Labs in the US
AT&T	American Telephone and Telegraph Company: a company that had a virtual monopoly over telecommunications in the US until its forced breakup in 1984
ATM	Asynchronous Transfer Mode: a telecommunications standard for digital transmission of multiple types of traffic, including voice, data and video signals in one network
AXE	a model of digital telephone exchange built by Ericsson
BMH	Bureau of Military History (in bibliography): a collection of 1,773 witness statements and other material concerning the revolutionary period in Ireland from 1913 to 1921
bps	bits per second: a measure of speed of data transmission
BT	originally British Telecom: formerly owned by the UK government, now a global communications provider
Comreg	Commission for Communications Regulation
Crossbar	a type of automatic telephone exchange
CW	Continuous Wave: a radio wave of constant amplitude and frequency which replaced earlier forms of radio transmission based on spark-gap
DBS	Direct Broadcast Satellite
DDI	Direct Dial In
DECC	Department of the Environment, Climate and Communications
DMP	Dublin Metropolitan Police
DTT	Digital Terrestrial Television
DSL	Digital Subscriber Line: a type of broadband internet technology that transmits data over a traditional telephone line comprised of a pair of copper wires
EEA	European Economic Area: in this context Iceland, Liechtenstein and Norway

EBU	European Broadcasting Union
ESB	Electricity Supply Board
ESOT	Employee Share Ownership Trust
FDM	Frequency-division multiplexing
FM	Frequency Modulation
FTTC	fibre-to-the-cabinet
FTTP	fibre-to-the-premises
GAA	Gaelic Athletic Association
Gbps	Gigabits per second: equal to 1,000 Mbps
GHz	Gigahertz: equal to 1,000 MHz
GPO	General Post Office: a building on Dublin's O'Connell Street
GPRS	General Packet Radio Service: a packet-switched mobile data standard sometimes called 2.5G
GSM	Global System for Mobile communications (originally *Groupe Speciale Mobile*)
ICA	Irish Countrywomen's Association
ICT	Information and Communications Technology
IDA	Industrial Development Authority: government agency responsible for attracting foreign direct investment (FDI) to Ireland
IP	Internet Protocol
IRB	Irish Republic Brotherhood
ISD	International Subscriber Dialling
ISDN	Integrated Services Digital Network: a set of communication standards for simultaneous digital transmission of voice, fax, data and other network services
ISEQ	Ireland Stock Exchange Overall Index
ITT	International Telephone & Telegraph: a US-based multinational telecoms company active from the 1920s to the 1980s. Subsidiary companies included STC (qv) in the UK and CGCT in France.
ITU	International Telecommunication Union: a United Nations agency for information and communication technologies
kbps	Kilobits per second: equal to 1,000 bps
kHz	Kilohertz: equal to 1,000 Hertz (oscillations per second)
LOP	Look Out Post
Magneto	An early form of manual exchange where each subscribers' telephone has a handle that must be wound to call the operator
Mbps	Megabits per second: equal to 1,000 kbps
MHz	Megahertz: equal to 1,000kHz
Microwave	Part of the radio spectrum typically defined as 3GHz to 30GHz (see Figure 4.7)
MMDS	Multipoint Microwave Distribution Service
MP	Member of Parliament
MVNO	Mobile Virtual Network Operator
NI	Northern Ireland

NMT	Nordic Mobile Telephone
NTC	National Telephone Company
NTL	originally National Transcommunications Limited: a former cable television provider
PABX	Private automatic branch exchange
PRONI	Public Records Office of Northern Ireland
OECD	Organisation for Economic Co-operation and Development
OPW	Office of Public Works
RAX	Rural automatic exchange: a small step-by-step telephone exchange used in rural areas
RDS	Royal Dublin Society
RIC	Royal Irish Constabulary
RTÉ	Raidió Teilifís Éireann
SIM	Subscriber Identity Module: a chip, typically embedded on a plastic card, that identifies a subscriber on a communications network
SITA	Société Internationale de Télécommunications Aéronautiques: a company founded in 1949 by 11 airlines in 1949 that provides IT and telecommunication services to the air transport industry
SMS	Short Message Service
SMSC	Short Message Service Centre
STC	Standard Cable and Telephone: a British-based supplier of telecoms equipment from 1883 to 1991
STD	Subscriber Trunk Dialling
STM	Synchronous Transport Module: a fiber-optic network standard
TACS	Total Access Communication System: a variant of AMPS (qv) developed for the UK and used in Ireland by Eircell for the '088' network
TCP	Transmission Control Protocol
TD	Teachta Dála
TDM	Time-division multiplexing
UAX	Unit automatic exchange: see RAX
UCD	University College Dublin
UHF	Ultra High Frequency: part of the radio spectrum typically defined as 300Mhz to 3GHz (see Figure 4.7)
URL	Uniform Resource Locator: the address of a specific webpage or file on the Internet
UTC	United Telephone Company
UTV	Ulster Television
VHF	Very High Frequency: part of the radio spectrum typically defined as 30Mhz to 300MHz (see Figure 4.7)
VDSL	Very high-speed Digital Subscriber Line
WAP	Wireless Application Protocol: an early form of data access from mobile phones
WFU	Withdrawn from use

ENDNOTES

Author's Note: This book spans the somewhat artificial divide between science and humanities, highlighting a difference of approach between the two disciplines. Coming from a science background, I have used the AMA citation style. With AMA, as with its close cousin the Vancouver style, a reference which is cited more than once in a chapter retains the same original endnote number. This is different to the Chicago notes and bibliography system, widely used in the humanities, which allocates a unique number for each endnote, even those which cite a reference used before. I hope this does not prove too disconcerting for readers with a background in humanities!

Chapter 1: The Information Age

1. *Illustrated London News*, 12 June 1852.
2. *Dublin Evening Mail*, 2 June 1852.
3. *Evening Packet and Correspondent*, 5 June 1852.
4. *Catholic Telegraph*, 19 June 1852.
5. John Kennedy, 'US$15m UK–Ireland fibre network begins rollout today', www.siliconrepublic.com/comms/us15m-uk-ireland-fibre-network-begins-rollout-today, 25 Oct. 2020.
6. Anton A. Huurdeman, *The Worldwide History of Telecommunications* (Hoboken NJ, 2003).
7. *Derry Journal*, 15 Aug. 2015.
8. A. J. Kirwan, 'R. L. Edgeworth and optical telegraphy in Ireland, c. 1790–1805', in *Proceedings of the Royal Irish Academy, Section C: Archaeology, Celtic Studies, History, Linguistics and Literature* 117C (2017) pp 209–35.
9. Richard Lovell Edgeworth, 'An essay on the art of conveying secret and swift intelligence', in *The Transactions of the Royal Irish Academy* 6 (1797) pp 95–139.
10. Charles Mollan, *It's Part of What We Are – Volume 1* (Dublin, 2007).
11. Russell W. Burns, *Communications: An International History of the Formative Years* (Stevenage, 2004).

12. Desmond John Clarke and Richard Lovell Edgeworth, *The Ingenious Mr Edgeworth* (London, 1965).
13. *Saunders's News-Letter*, 6 July 1804.
14. 'Cappagh', www.dgsys.co.uk/btmicrowave/sites/90.php, 28 Feb. 2018.
15. William H. Clements, *Billy Pitt Had Them Built: The Napoleonic Towers in Ireland* (Stamford, 2013).
16. Tom Standage, *The Victorian Internet: The Remarkable Story of the Telegraph and the Nineteenth Century's On-Line Pioneers* (New York, 1998).
17. Jamie Johnston, *Victorian Belfast* (Belfast, 1993).
18. Ken Beauchamp and Institution of Electrical Engineers, *History of Telegraphy*, IEE history of technology series (London, 2001).
19. Simone Fari, *Victorian Telegraphy Before Nationalization* (Basingstoke, 2015).
20. Christopher Morash, *A History of the Media in Ireland* (Cambridge and New York, 2010).
21. 'Nurses in the Crimea', www.nationalarchives.gov.uk/womeninuniform/crimea_intro.htm, 28 Feb. 2018.
22. Samuel Finley Breese Morse, *Foreign Conspiracy Against the Liberties of the United States: The Numbers of Brutus, Originally Published in the New-York Observer* (New York, 1835).
23. Michael B. Schiffer, *Power Struggles: Scientific Authority and the Creation of Practical Electricity Before Edison* (Cambridge MA and London, 2011).
24. John Steele Gordon, *A Thread Across the Ocean: The Heroic Story of the Transatlantic Cable* (New York, 2003).
25. Oliver Doyle and Stephen Hirsch, *Railways in Ireland 1834–1984* (Dublin, 1983).
26. Garth Pedler, *Rail Operations Viewed from South Devon* (Leicester, 2017).
27. Jeffrey L. Kieve, *The Electric Telegraph: A Social and Economic History* (England, 1973).
28. Thomas F. Wall, 'Railways & telecommunications – 1', in *Journal of the Irish Railway Record Society* 20142 (2000) pp 479–85.
29. Steven Roberts, 'Distant Writing – 8 Non-Competitors', http://distantwriting.co.uk/noncompetitors.html, 23 Jan. 2021.
30. Stephen Ferguson, *The Post Office in Ireland: An Illustrated History* (Newbridge, 2016).
31. Plastics Historical Society, 'William Montgomerie', http://plastiquarian.com/?page_id=14213, 27 Feb. 2019.
32. *Freeman's Journal*, 22 Dec. 1851.
33. Steven Roberts, 'Distant Writing – 5 Competitors and Allies', http://distantwriting.co.uk/competitorsallies.html, 2 Mar. 2019.
34. Steven Roberts, 'Anglo-Irish Cables', http://atlantic-cable.com//Cables/Anglo-Irish/index.htm, 5 Mar. 2019.
35. Steven Roberts, *Distant Writing: A History of the Telegraph Companies in Britain between 1838 and 1868*, http://distantwriting.co.uk/Documents/Distant%20Writing%202012.pdf, 5 Mar. 2019.
36. *Dublin Evening Mail*, 19 Aug. 1853.
37. R. N. Barton, 'Brief lives: Three British telegraph companies 1850–56', in *The International Journal for the History of Engineering & Technology*, 802 (2010) pp 183–98.

38. *The Times (London)*, 18 Jan. 1854.
39. *Irish Times*, 31 July 2000.
40. *Evening Freeman*, 13 Nov. 1861.
41. *Warder and Dublin Weekly Mail*, 16 Nov. 1861.
42. *Saunders's News-Letter*, 4 Apr. 1862.
43. Jimmy O'Connor, 'Aspects of Galway postal history 1638–1984', in *Journal of the Galway Archaeological and Historical Society* 44 (1992) pp 119–94.
44. Arthur Maltby and Jean Maltby, *Ireland in the 19th Century: A Breviate of Official Publications* (Elkins Park PA, 1994).
45. *The Advocate: or, Irish Industrial Journal*, 14 Sept. 1853.
46. Seija-Riitta Laakso, *Across the Oceans: Development of Overseas Business Information Transmission, 1815–1875,* (Helsinki, 2007).
47. Bob Clarke, *From Grub Street to Fleet Street: An Illustrated History of English Newspapers to 1899* (London, 2017).
48. Eamonn G. Hall, *The Electronic Age* (Dublin, 1993).
49. K A. Murray, 'The Armagh Collision, 1889', in *Journal of the Irish Railway Record Society* 1054 (1971) pp 10–28.
50. Damian Woods, *The Fateful Day: A Commemorative Book of the Armagh Railway Disaster June 12th 1889* (Armagh, 1989).
51. Charles R. Perry, 'Frank Ives Scudamore and the Post Office telegraphs', in *Albion: A Quarterly Journal Concerned with British Studies* 124 (1980) pp 350–67.
52. The National Archives (UK), 'Ireland: Extension of telegraph service to Belmullet', T 1/14952.
53. Gerard Moran, 'James Hack Tuke and his schemes for assisted emigration from the west of Ireland', in *History Ireland* 213 (2013).
54. Charles R. Perry, *The Victorian Post Office: The Growth of a Bureaucracy* (London, 1992).
55. Mícheál Ó Bréartúin, *Sreangscéal: Forbairt na Teileagrafaíochta Leictrí* (Baile Átha Cliath, 2005).
56. *Longford Journal*, 11 Nov. 1909.
57. *Thom's Official Directory of the United Kingdom of Great Britain and Ireland* (Dublin, 1898).
58. Silvanus Phillips Thompson, *The Life of William Thomson, Baron Kelvin of Largs – Vol. II* (Cambridge, 1910).
59. John Coakley and Michael Gallagher, *Politics in the Republic of Ireland* (Abingdon, 2017).
60. BT Archives, 'Telegrams in languages other than those authorised', TCB 2/208.
61. Jean-Francois Fava-Verde, 'A tale of two telegraphs: Cooke and Wheatstone's differing visions of electric telegraphy', http://journal.sciencemuseum.ac.uk/browse/issue-08/a-tale-of-two-telegraphs, 23 Jan. 2021.
62. *Evening Herald*, 6 Jan. 1894.

Chapter 2: Kerry, communications hub of the world

1. Ainissa Ramirez, 'A wire across the ocean', www.americanscientist.org/article/a-wire-across-the-ocean, 2 Aug. 2018.
2. John Steele Gordon, *A Thread Across the Ocean: The Heroic Story of the Transatlantic Cable* (New York, 2003).
3. Tom Standage, *The Victorian Internet: The Remarkable Story of the Telegraph and the Nineteenth Century's On-Line Pioneers* (New York, 1998).
4. Jill Hills, *The Struggle for Control of Global Communications: The Formative Century* (Chicago, 2002).
5. Aaron Stanton, 'The periphery at the center: Valentia Island and the transatlantic telegraph', in *Brown Journal of History* Spring (2008) pp 127–49.
6. Donard De Cogan, 'Background to the 1858 telegraph cable', in *Institution of Engineering and Technology Seminar on the Story of Transatlantic Communications 2008* (Manchester, 2008).
7. *Warder and Dublin Weekly Mail*, 8 Aug. 1857.
8. *Liverpool Daily Post*, 8 Aug. 1857.
9. Cornelia Connolly, 'The transatlantic cable – An Irish perspective', https://corneliathinks.wordpress.com/2017/06/05/the-transatlantic-cable-an-irish-perspective, 7 Aug. 2018.
10. Gillian Cookson, *The Cable: The Wire That Changed the World* (England, 2012).
11. H. M. Field, *History of the Atlantic Telegraph* (New York, 1867).
12. *Daily Express (Dublin)*, 24 Aug. 1858.
13. Bill Burns, 'The Curious Story of the Tiffany Cables', http://atlantic-cable.com/Article/Lanello/index.htm, 8 Aug. 2018.
14. *London Evening Standard*, 4 Sept. 1858.
15. *Kings County Chronicle*, 8 Sept. 1858.
16. Eamonn G. Hall, *The Electronic Age* (Dublin, 1993).
17. Bill Burns, 'Messages carried by the 1858 Atlantic Telegraph Cable', https://atlantic-cable.com/Article/1858Messages/index.htm, 7 Aug. 2018.
18. Jeffrey L. Kieve, *The Electric Telegraph: A Social and Economic History* (England, 1973).
19. Allan Green, 'A Meeting of Minds? Dr E. O. Wildman Whitehouse, Electrician to The Atlantic Telegraph Company, and his relationship with Professor William Thomson during the development and construction of the first transatlantic telegraph cable 1857/8', https://atlantic-cable.com/Books/Whitehouse/AG/amom.htm, 18 Oct. 2019.
20. Anton A. Huurdeman, *The Worldwide History of Telecommunications* (Hoboken NJ, 2003).
21. Bernard S. Finn and Daqing Yang, *Communications Under the Seas: The Evolving Cable Network and its Implications* (Cambridge MA and London, 2009).
22. Bruce J. Hunt, 'Scientists, engineers and Wildman Whitehouse: Measurement and credibility in early cable telegraphy', http://atlantic-cable.com/Books/Whitehouse/BJH/index.htm, 8 Aug. 2018.
23. Arthur C. Clarke, *Voices Across the Sea* (New York, 1974).
24. Charles Bright, *The Story of the Atlantic Cable* (New York, 1903).

25. Mark McCartney and Andrew Whitaker, *Physicists of Ireland: Passion and Precision* (Bristol and Philadelphia, 2003).
26. *Evening Freeman*, 31 Aug. 1857.
27. David and Julia Bart, 'Sir William Thomson, on the 150th Anniversary of the Atlantic Cable', http://atlantic-cable.com/CablePioneers/Kelvin, 5 Aug. 2018.
28. Donard de Cogan, 'Dr. E. O. W. Whitehouse and the 1858 trans-Atlantic cable', http://atlantic-cable.com/Books/Whitehouse/DDC/index.htm, 1 Oct. 2019.
29. Donard de Cogan, 'The Irish submarine cable stations: A technological history', https://dandadec.files.wordpress.com/2013/07/valentia-technology1.pdf, 6 Aug. 2018.
30. Andrew R. Holmes, 'Professor James Thomson Sr. and Lord Kelvin: Religion, science, and Liberal Unionism in Ulster and Scotland', in *Journal of British Studies* 501 (2011) pp 100–24.
31. Simone Fari, *Victorian telegraphy before nationalization* (Basingstoke, 2015).
32. Massimo Guarnieri, 'The Conquest of the Atlantic', in *IEEE Industrial Electronics Magazine* 81 (2014) pp 53–67.
33. BT Archives, 'Abstracts of minutes', TGF 1/1.
34. Don Fox, 'James Graves (1833–1911), his Descendants, the Electric Telegraph and Valentia Island', http://donaldpfox.blogspot.com/2018/01/james-graves-1833-1911-his-descendants.html, 9 Aug. 2018.
35. William Howard Russell, *The Atlantic Telegraph* (London, 1865).
36. IET Archive, 'Sir Peter FitzGerald's album of telegraph material titled "Atlantic Telegraph 1865"', SC MSS 254.
37. Dwayne R. Winseck and Robert M. Pike, 'Communication and empire: Media, markets, and globalization, 1860–1930', in *Business History Review*, 82 (2008) pp 138–78.
38. *Dublin Evening Mail*, 6 June 1866.
39. *Belfast Morning News*, 4 June 1866.
40. Daniel Farrell, 'Robert Halpin: The Irishman who revolutionized global communications', https://coastmonkey.ie/robert-halpin-irishman-global-communication/, 10 Aug. 2018.
41. Stephen Ferguson, *The Post Office in Ireland: An Illustrated History* (Newbridge, 2016).
42. 'History of Tinakilly House', https://tinakilly.ie/wp-content/uploads/2017/12/TK_History.pdf, 21 Oct. 2019.
43. Ken Mitchell, Robert Halpin – the Wicklow master mariner who connected the world, https://web.archive.org/web/20180329231854/www.engineersjournal.ie/2017/12/05/robert-halpin-transatlantic-cable-wicklow/, 10 Aug. 2018.
44. Donard De Cogan, *Thirty Six years in the Telegraphic Service, 1852 To 1888: Being a Brief Autobiography of James Graves MSTE* (Norwich, 2016).
45. Donard de Cogan, 'Harbour Grace and the "Direct": The amazing story of a telegraph cable', https://dandadec.files.wordpress.com/2013/07/direct-us-telegraph-company.pdf, 11 Aug. 2018.
46. Peter J. Hugill, *Global Communications Since 1844: Geopolitics and Technology* (Baltimore, 1999).
47. Daniel R. Headrick and Pascal Griset, 'Submarine telegraph cables: business and politics, 1838–1939', in *The Business History Review* 753 (2001) pp 543–78.

48. Donard De Cogan and Bill Burns, *They Talk Along the Deep: A Global History of the Valentia Island Telegraph Cables* (Norwich, 2016).
49. The 'Story of Waterville Cable Station' Exhibition, Tech Amergin, Waterville, Co. Kerry, Ireland.
50. 'Registration of the United Kingdom', in *Journal of the Statistical Society of London* 33 (1870) pp 153–7.
51. Stephen Kern, *The Culture of Time and Space, 1880–1918: With a New Preface* (Cambridge MA and London, 2003).
52. *Telephone Directory Vol. 5 (Belfast, Dublin and Cork)* (London, 1921).
53. The Anglo-American Telegraph Company's Station, Valentia, Ireland, The Telegraphist, 1 Dec. 1885, available at https://atlantic-cable.com/CableCos/Valentia/index.htm, 13 Sept. 2021
54. Des Lavelle, interviewed in 'Valentia Lecture and Fireside Chat', https://valentiacable.com/events/valentia-lecture-2020, 9 Oct. 2020.
55. The National Archives of Ireland, *Census of Ireland 1901.*
56. Alan Leon Varney, 'NetValley: A New Home for the Mind?', www.netvalley.com/archives/mirrors/telegraph__radio_timeline-3.htm, 13 Aug. 2018.
57. Donard De Cogan, 'Ireland, telecommunications and international politics', in *History Ireland* 2 (1993) .
58. BT Archives, 'Landing licences for transatlantic cable companies for cables between England and Ireland', POST 30/4473B.
59. BT Archives, 'Irish Free State: Malin Head and Brow Head old wireless stations and Ballinskelligs cable station', in POST 33/2678.
60. Ivan Stoddart Coggeshall, et al., *An Annotated History of Submarine Cables and Overseas Radiotelegraphs, 1851–1934: With Special Reference to the Western Union Telegraph Company* (Norwich, 1993).
61. *Irish Times*, 30 Sept. 1961.
62. *Irish Independent* 4 Apr. 1966.

Chapter 3: Miss Agnes Duggan, telephone operator and revolutionary

1. Thomas F. Wall, 'Some notes towards a history of telecommunications with particular reference to Ireland' (2005), Dublin City Archive.
2. *Thom's Official Directory of the United Kingdom of Great Britain and Ireland* (Dublin, 1881).
3. A. J. Litton, 'The growth and development of the Irish telephone system', in *Journal of the Statistical and Social Inquiry Society of Ireland* XX:V (1962) pp 79–115.
4. Agnes Duggan, 'Operating reminiscences', in *National Telephone Journal* 4 (1909) pp 96–7.
5. Agnes Duggan, 'Telephone women', in *National Telephone Journal* 4 (1909) p. 99.
6. J. H. Robertson, *The Story of the Telephone: A History of the Telecommunications Industry of Britain* (London, 1948).
7. *Irish Times*, 15 July 1881.

8. Anton A. Huurdeman, *The Worldwide History of Telecommunications* (Hoboken NJ, 2003).
9. Silvanus P. Thompson, *Philipp Reis, Inventor of the Telephone: A Biographical Sketch* (London, 1883).
10. Daniel P McVeigh, 'Stephen Mitchell Yeates perfects Reis's invention', http://oceanofk.org/telephone/html/part10.html, 19 June 2020.
11. 'Bell, Gray and the invention of the telephone', www.ericsson.com/en/about-us/history/communication/early-developments/bell-gray-and-the-invention-of-the-telephone, 8 June 2020.
12. *Irish Times*, 11 Mar. 1976.
13. *Northern Whig*, 21 Nov. 1877.
14. *Irish Independent*, 3 Jan. 2017.
15. *Cork Examiner*, 22 Feb. 1877.
16. R. J. Chapuis and A. E. Joel, *100 Years of Telephone Switching* (Amsterdam, 2003).
17. *Freeman's Journal*, 30 Mar. 1880.
18. BT Archives, Private and municipal telephone companies, available at https://web.archive.org/web/20170924141940/https://www.btplc.com/Thegroup/BTsHistory/BTgrouparchives/Informationsheetsandtimelines/Info_sheet_Private_Telephone_Co_Aug_2008.pdf, 19 June 2020.
19. *Northern Whig*, 1 May 1880.
20. Ibid.
21. Roddy Flynn, *The Development of Universal Telephone Service in Ireland 1880–1993*, (1998) DCU Doctor of Arts thesis, http://doras.dcu.ie/18734/1/Roddy_Fynn_20130509140820.pdf, 3 May 2020.
22. F. J. Cullen, *Local government and the management of urban space: a comparative study of Belfast and Dublin, 1830–1922* (2005) NUIM PhD thesis, www.jstor.org/stable/24338495.
23. PRONI, 'The National Telephone Company Limited', D2194/106/1.
24. The Anglo-American Telegraph Company's Station, Valentia, Ireland, The Telegraphist, 1 Dec. 1885, available at https://atlantic-cable.com/CableCos/Valentia/index.htm, 13 Sept. 2021 https://atlantic-cable.com/CableCos/Valentia/index.htm.
25. Telephone Directory: Scotland/North Western/Northern/Ireland (London, 1904).
26. *Northern Whig*, 19 Feb. 1880.
27. *Irish Times*, 22 June 1998.
28. James Joyce, *Ulysses* (New York, 1986).
29. *Freeman's Journal*, 30 June 1883.
30. Ibid., 26 July 1883.
31. *Evening Herald*, 6 Jan 1894.
32. *Belfast News-Letter*, 6 Apr. 1892.
33. Eli M. Noam, *Telecommunications in Europe* (New York and Oxford, 1992).
34. *Irish News*, 7 Apr. 1893.
35. Hansard, HC Debates, 3 Feb. 1902.
36. W. E. Vaughan and A. J. Fitzpatrick (eds), *Irish Historical Statistics: Population, 1821–1971* (Dublin, 1978).

37. *Belfast News-Letter*, 13 Dec. 1913.
38. Charles R. Perry, *The Victorian Post Office: The Growth of a Bureaucracy* (London, 1992).
39. BT Archives, 'The history of telecommunications 1912–1968', www.bt.com/about/bt/our-history/history-of-telecommunications/1912-to-1968#history-1960, 22 July 2020.
40. BT Archives, 'Supervising allowances to telephonists in the London Telephone Service', POST 30/2642B.
41. *Irish telephone directory – Eolaí telefóin na hÉireann* (Dublin, 1977).
42. *Daily Chronicle*, 23 Apr. 1915.
43. BT Archives, 'Organisation charts for the National Telephone Company', TPF 2/4.
44. *Daily Express (Dublin)*, 27 July 1912.
45. Agnes Duggan, 'Operators in training', in *National Telephone Journal* 2 (1907) pp 142–3.
46. BT Archives, 'Telephone exchange operating statistics – Volume 1', TPF 3/2/1.

Chapter 4: The Italian and the hungry hawks

1. Augustine McCurdy, 'Marconi and Rathlin', www.culturenorthernireland.org/article/1316/marconi-and-rathlin, 5 Nov. 2018.
2. *Irish Times*, 10 Aug. 2017.
3. Stanley Warren, 'Montrose House and the Jameson family in Dublin and Wexford: A personal reminiscence', in *The Past: The Organ of the Uí Cinsealaigh Historical Society*, 28 (2007) pp 87–97.
4. Anton A. Huurdeman, *The Worldwide History of Telecommunications* (Hoboken NJ, 2003).
5. Peter A. Davidson, *An Introduction to Electrodynamics* (Oxford, 2019).
6. Gregory Malanowski, *The Race for Wireless: How Radio was Invented (or Discovered)* (Bloomington, IN, 2011).
7. Inez Hunt, *Lightning in his Hand: The Life Story of Nikola Tesla* (Hawthorne CA, 1981).
8. Marc Raboy, *Marconi: The Man Who Networked the World* (Oxford, 2016).
9. Peter Homer, 'Malin Head Marconi Radio Station', www.malinhead.net/marconi/Marconi.htm, 28 Nov. 2018.
10. Hugh Alexander Boyd, 'Marconi and Ballycastle', http://antrimhistory.net/marconi-and-ballycastle, 5 Nov. 2018.
11. Michael Sexton, *Marconi: The Irish Connection* (Dublin, 2005).
12. The National Archives of Ireland, *Census of Ireland 1901*.
13. *Belfast News-Letter*, 24 Aug. 1898.
14. *Daily Express (Dublin)*, 21 July 1898.
15. *Irish Times*, 11 May 2011.
16. Kenneth Mitchell, Marconi and Ireland – the small country that played a big role in the radio age, www.engineersireland.ie/Engineers-Journal/News/marconi-and-ireland-the-small-country-that-played-a-big-role-in-the-radio-age, 22 Dec. 2019.
17. William John Baker, *A History of the Marconi Company* (London, 1970).

18. *Irish Times*, 25 Aug. 1898.
19. *Freeman's Journal*, 24 Aug. 1898.
20. Petri Launiainen, *A Brief History of Everything Wireless: How Invisible Waves Have Changed the World* (Switzerland, 2018).
21. *Cork Examiner*, 27 Nov. 1903.
22. Wade Rowland, *Spirit of the Web: The Age of Information from Telegraph to Internet* (Vancouver, BC, 2007).
23. BT Archives, 'Malin Head wireless station', POST 31/61B.
24. BT Archives, 'Valencia Island wireless station: General papers', POST 31/62 (2).
25. Russell W. Burns, *Communications: An International History of the Formative Years* (Stevenage, 2004).
26. Shane Joyce, 'The Marconi Station', www.connemara.net/the-marconi-station, 14 July 2018.
27. Dwayne R. Winseck and Robert M. Pike, 'Communication and Empire: Media, Markets, and Globalization, 1860-1930' in *Business History Review* 82 (2008) pp 138–78.
28. *New York Times*, 29 Jan. 1912.
29. *Marconigrams [advertisment]*, Marconigraph, November 1912 p. iv.
30. Ivan Stoddart Coggeshall, et al., *An Annotated History of Submarine Cables and Overseas Radiotelegraphs, 1851–1934: With Special Reference to the Western Union Telegraph Company* (Norwich, 1993).
31. *Kerry Evening Post*, 14 Sept. 1907.
32. *Daily Express (Dublin)*, 24 Feb. 1908.
33. BT Archives, 'Universal Radio Syndicate wireless trial at Ballybunion and New Brunswick stations', POST 30/3895.
34. 'Radio Signal from Kerry to Canada, 1965', www.rte.ie/archives/2015/0518/701871-hello-canada-hello-canada/, 6 Nov. 2018.
35. *Midland Counties Tribune*, 21 Sept. 1919.
36. *Irish Times*, 14 Aug. 1923.
37. Ibid., 17 Aug. 1923.
38. *Belfast Telegraph*, 11 July 1925.
39. Eric Eastwood, 'Marconi: His wireless telegraphy and the modern world', in *Journal of the Royal Society of Arts* 122 (1974) pp 599–611.
40. *Londonderry Sentinel*, 5 Sept. 1935.
41. Peter Young, *Power of Speech: A History of Standard Telephones and Cables, 1883–1983* (London and Boston, 1983).
42. 'Ballygomartin', www.dgsys.co.uk/btmicrowave/sites/86.php, 21 Jan. 2020.
43. Giancarlo Masini, *Marconi* (New York, 1998).
44. Brian Friel, *Dancing at Lughnasa* (London and Boston, 1990).
45. Andrew Robinson, 'Marconi forged today's interconnected world of communication', *New Scientist*, www.newscientist.com/article/mg23130862-900-marconi-forged-todays-interconnected-world-of-communication, 10 Aug. 2016.
46. *Irish Independent*, 22 July 1937.

Chapter 5: 'Get back, son, get back, the British are in the Telephone Exchange!'

1. Oxford, Bodleian Libraries, MS. Nathan 476.
2. Stephen Ferguson, *GPO Staff in 1916: Business as Usual* (Cork, 2012).
3. Keith Jeffery, *The GPO and the Easter Rising* (Dublin and Portland OR, 2006).
4. Working on the Railway in Dublin, 1900–1925 (Transcript), https://www.dublincity.ie/library/blog/working-railway-dublin-1900-1925-transcript, 6 Dec. 2018.
5. Michael McNally and Peter Dennis, *Easter Rising 1916: Birth of the Irish Republic* (Oxford and New York, 2007).
6. Max Caulfield, *The Easter Rebellion* (London, 1965).
7. E. Gomersall, 'The impact of the 1916 rebellion on the refurbished GPO', in *The Post Office Electrical Engineers' Journal* IX (1916).
8. Jonathan Davis, *Historical Dictionary of the Russian Revolution* (Lanham MD, 2020).
9. '"Rozmowy kontrolowane": satyra na stan wojenny', www.polskieradio.pl/39/156/Artykul/2420638,Rozmowy-kontrolowane-satyra-na-stan-wojenny, 9 Jan. 2020.
10. Military Archives, Bureau of Military History Witness Statements (BMH WS) 543.
11. Ibid., BMH WS 1686.
12. Ibid., BMH WS 188.
13. Ibid., BMH WS 579.
14. Eddie Bohan, *Rebel Radio: Ireland's First International Radio Station 1916* (Dublin, 2016).
15. 'The Sinn Féin Rising', in *The Telephone and Telegraph Journal* 221 (1916) p. 177.
16. Neil Richardson, *According to their Lights* (Cork, 2015).
17. Military Archives, BMH WS 510.
18. Keith Jeffery, *The Sinn Féin Rebellion as They Saw It* (Newbridge, 1999).
19. *London Gazette*, 7 Jan. 1918.
20. Stephen Ferguson, *The Post Office in Ireland: An Illustrated History* (Newbridge, 2016).
21. Hansard, HC Debates, 31 July 1916.
22. Bridget McAuliffe, Mary McAuliffe and Owen O'Shea, *Kerry 1916: Histories and Legacies of the Easter Rising: A Centenary Record* (Listowel, 2016).
23. Donard De Cogan, 'Ireland, telecommunications and international politics', in *History Ireland* 2 (1993).
24. Generational Change in Rising Memories, https://www.irishexaminer.com/news/arid-20385885.html 14 Sept. 2021.
25. Military Archives, BMH WS 804.
26. *The Kerryman*, 31 Jan. 2015
27. Colum Ring, personal communication, 15 Jan. 2021.
28. Cathal Brennan, 'Radio and the 1916 Rising', https://timeline.ie/radio-and-the-1916-rising, 27 Nov. 2018.
29. *Irish Independent*, 15 Mar. 2016.
30. Francis Devine, *Connecting Communities: A Pictorial History of the Communications Workers' Union* (Dublin, 2013).

31. *Irish Independent*, 26 May 1916.
32. Ibid., 5 July 1916.
33. Military Service Pensions Collection, Kevin O'Reilly, MSP34REF50604.
34. *Belfast News-Letter*, 27 July 1920.
35. *Ballymena Observer*, 10 June 1921.
36. Military Archives, BMH WS 1773.
37. BT Archives, 'Irish Free State: Wireless telegraph services', POST 33/1109.

Chapter 6: An Roinn Puist agus Telegrafa

1. *Irish Times*, 12 July 1929.
2. Tim Carey, *Dublin Since 1922* (Dublin, 2016).
3. Stephen Ferguson, *The Post Office in Ireland: An Illustrated History* (Newbridge, 2016).
4. P. Abercrombie, S. Kelly, and A. Kelly, *Dublin of the Future* (Liverpool, 1922).
5. Dáil Debates, 21 June 1923.
6. Cormac Ó Gráda, *A Rocky Road: The Irish Economy since the 1920s* (Manchester, 1997).
7. Joseph Lee, *Ireland 1912–1985: Politics and Society* (Cambridge, 2006).
8. Telecho, vol. 5, Dec. 1988/Jan 1989.
9. The National Archives of Ireland, 'Blacksod – Telephonic (sic) Interruption', TAOIS/ S 2030.
10. Military Archives, 'Patrick Lawson', BMH WS 667.
11. *The Derry Journal*, 31 Jan. 1923.
12. Bernard Share, *In Time of Civil War: The Conflict on the Irish Railways, 1922–1923* (Cork, 2006).
13. John Crowley, Donal Ó Drisceoil, and John Borgonovo, *Atlas of the Irish Revolution* (Cork, 2018).
14. The National Archives of Ireland, 'Anglo-Irish Treaty – 6 December 1921', available at https://web.archive.org/web/20190916102036/http://treaty.nationalarchives.ie/document-gallery/anglo-irish-treaty-6-december-1921/anglo-irish-treaty-6-december-1921-page-1/, 14 Sept. 2021.
15. The National Archives of Ireland, 'Protection of Commercial Cable Company, Waterville, Co. Kerry, and Western Union Telegraph Company, Valentia Island, Co. Kerry', TSCH/ S1658.
16. Donard De Cogan and Bill Burns, *They Talk along the Deep: A Global History of the Valentia Island Telegraph Cables* (Norwich, 2016).
17. A. J. Litton, 'The growth and development of the Irish telephone system', in *Journal of the Statistical and Social Inquiry Society of Ireland* XX:V (1962) pp 79–115.
18. Thomas F. Wall, 'Railways & telecommunications – 4', in *Journal of the Irish Railway Record Society* 21147 (2002) pp 194–203.
19. Marc Raboy, *Marconi: The Man Who Networked the World* (Oxford, 2016).
20. *Manchester Guardian*, 2 Aug. 1922.
21. Richard Pine, *2RN and the Origins of Irish Radio* (Dublin, 2002).

22. Oxford, Bodleian Libraries, MS. Marconi 193.
23. Michael Sexton, *Marconi: The Irish Connection* (Dublin, 2005).
24. PRONI, 'Postal and telegraph arrangement on border', CAB9F/44/1s.
25. Roddy Flynn, *The Development of Universal Telephone Service in Ireland 1880–1993* (1998) DCU Doctor of Arts thesis, http://doras.dcu.ie/18734/1/Roddy_Fynn_20130509140820.pdf, 14 Mar. 2019.
26. PRONI, 'Post Office telegraphic and telephonic communications dependent upon Free State Post Offices', HA/32/1/244.
27. The National Archives of Ireland, 'Cross-border telegraph & telephone circuits 1951', DFA/5/305/179.
28. Cathal Brennan, 'The Postal Strike of 1922', www.theirishstory.com/2012 June 08/the-postal-strike-of-1922/#.XBlp5qecau4, 11 Dec. 2018.
29. Francis Devine, *Connecting Communities: A Pictorial History of the Communications Workers' Union* (Dublin, 2013).
30. *Kilkenny Moderator*, 6 May 1922.
31. *Telephone Directory Vol. 5 (Belfast, Dublin and Cork)* (London, 1921).
32. *Leabhar Seolta Telefona Saorstat Éireann* (Dublin, 1929).
33. Donal P. Corcoran, *Freedom to Achieve Freedom: The Irish Free State 1922–1932* (Dublin, 2013).
34. Seanad Debates, 23 Apr. 1931.
35. *Larne Times*, 17 Nov. 1923.
36. *Freeman's Journal*, 13 Sept. 1923.
37. Andrew Emmerson, 'From Strowger to System X', http://strowger-net.telefoniemuseum.nl/tel_hist_edgelane.html, 17 July 2018.
38. Thomas F. Wall, 'Some notes towards a history of telecommunications with particular reference to Ireland' (2005), Dublin City Archive.
39. *Irish Times*, 23 July 1927.
40. 'Timeline of telephony in Sweden', https://timelines.issarice.com/wiki/Timeline_of_telephony_in_Sweden, 20 July 2018.
41. The UAX Project, 'UAX5 General Features', www.uax.me.uk/uax5/uax5general.html, 2 Aug. 2018.
42. *Portadown Times*, 14 July 1929.
43. *Belfast News-Letter*, 13 Sept. 1929.
44. Ibid., 1 Nov. 1935.
45. *Derry Journal*, 4 Oct. 1937.
46. Lars Kleberg, 'Silence And Surveillance: A History of Culture and Communication', http://balticworlds.com/a-history-of-culture-and-communication, 10 July 2018.
47. Diarmaid Ferriter, *The Transformation of Ireland: 1900–2000* (London, 2004).
48. *Irish Times*, 28 July 1927.
49. The National Archives of Ireland, 'Dep Post and Telegraphs: Special reports 1931', TAOIS/ S 2221.

50. *Frankfurter Allgemeine*, 1 June 2016.
51. Jeremy Leon Stein, *Ideology and the Telephone* (1996) UCL PhD thesis, http://discovery.ucl.ac.uk/1381834/1/389060.pdf.
52. D. A. Levistone Cooney, 'Switzer's of Grafton Street', in *Dublin Historical Record* 552 (2002) pp 154–65.
53. *Waterford Standard*, 4 May 1935.
54. *Committee on Finance Debates*, 8 July 1927
55. Donal Fallon 'Dublin's first public telephone', https://comeheretome.com/2012/04/06/dublins-first-public-telephone/, 12 May 2019.
56. *Irish Times*, 26 Feb. 1935.
57. Hansard, HC Debates, 17 July 1939.
58. Dáil Debates, 25 Feb. 1926.
59. J. Atkinson, *Telephony Vol. 1* (London, 1948).
60. Dáil Debates, 16 May 1945.
61. *Irish Times*, 16 Apr. 1934.
62. *Belfast News-letter*, 24 Oct. 1929.
63. *Waterford Standard*, 25 Oct. 1952.
64. *Irish Times*, 28 May 1925.
65. BT Archives, 'Transatlantic telephone services and a new transatlantic submarine telephone cable', POST 122/53.
66. John L. O'Sullivan, *From Morse to Mobile* (Cork, 2000).
67. Peter Young, *Power of Speech: A History of Standard Telephones and Cables, 1883–1983* (London and Boston, 1983).
68. *The Liberator (Tralee)*, 24 Aug. 1929.
69. BT Archives, 'Irish Free State: telephone communication with Great Britain and abroad', POST 33/2623 (1).
70. Keith Ward, 'A short history of telecommunications transmission in the UK', in *Journal of the Communications Network* 5:1 (2006) pp 30–41.
71. W. G. Radley, 'Telephony and telegraphy', in *Journal of the Institution of Electrical Engineers* 84507 (1939) pp 359–67.
72. *Irish Times*, 26 Nov. 1938.
73. Ibid., 2 Dec. 1938.
74. Dáil Debates, 27 Mar. 1931.
75. Johnlepo, 'The price of a pint from 1928–2015 in todays money', http://publin.ie/2015/the-price-of-a-pint-from-1928-2015-in-todays-money, 3 Aug. 2018.
76. Dáil Debates, 30 Mar. 1938.
77. *Irish Times*, 5 Dec. 1929.
78. Maurice Manning, Moore McDowell, and Electricity Supply Board, *Electricity Supply in Ireland: The History of the ESB* (Dublin, 1985).
79. 'An overview of the Shannon Scheme', https://esbarchives.ie/2016/02/13/shannon-scheme, 7 Nov. 2018.

80. Michael J. Shiel, *The Quiet Revolution: The Electrification of Rural Ireland, 1946–1976* (Dublin, 2005).
81. The National Archives of Ireland, 'Proposed purchase of the Saorstat Telephone Organisation', DFA/1/GR/1363.
82. *Freeman's Journal*, 15 Feb. 1924.
83. Eamonn G. Hall, *The Electronic Age* (Dublin, 1993).
84. Gerald Condon, 'Telecommunications', in R. C. Cox (ed.), *Engineering Ireland* (Cork, 2006).
85. *The Kerryman*, 21 Mar. 1936.
86. Clair Wills, *That Neutral Island: A Cultural History of Ireland during the Second World War* (London, 2007).
87. 'Volledige Lijst der Europeesche Omroepzenders', www.radioheritage.net/europe/images/lists/frequentieplan1933.pdf, 11 May 2020.
88. Paddy Clark, *Dublin Calling* (Dublin, 1986).
89. Jonathan Marks, 'MN.20.March.1987 – Ireland Calling on Shortwave', *The Media Network Vintage Vault* (2010), http://jonathanmarks.libsyn.com/mn_20_march_1987_ireland_calling_on_shortwave, 8 Aug. 2018.

Chapter 7: Neutral and Allied

1. Vincent Sweeney, personal communication, 14 Feb. 2020.
2. *Irish Independent*, 31 May 2014.
3. *Station Weather Record – Blacksod Point*, Department of Industry and Commerce, Meteorological Service, p. 3.
4. John Bull, 'The Forecaster: The Man Who Decided D-Day', https://medium.com/p/the-weatherman-the-man-who-decided-d-day-d0eb5cad3f7e, 6 Oct. 2019.
5. Gerard Fleming, interviewed in *Storm Front in Mayo: The Story of the D-Day Forecast*, (2019).
6. A. J. Litton, 'The growth and development of the Irish telephone system', in *Journal of the Statistical and Social Inquiry Society of Ireland* XX:V (1962) pp 79–115.
7. Confidential report from John W. Dulanty to Joseph P. Walshe, www.difp.ie/books/?volume=6&docid=3351, 11 Jan. 2020.
8. Joseph Lee, *Ireland 1912–1985: Politics and Society* (Cambridge, 2006).
9. Aoibheann Lambe, 'This Kerry peninsula shows how the Irish have communicated for thousands of years', http://jrnl.ie/2975807, 21 Feb. 2019.
10. Michael Kennedy, *Guarding Neutral Ireland: The Coast Watching Service and Military Intelligence, 1939–1945* (Dublin, 2008).
11. Seanad Debates, 6 Dec. 1939.
12. The National Archives (UK), 'Liaison and exchange of information with Eire authorities', KV 4/280.
13. Eamonn G. Hall, *The Electronic Age* (Dublin, 1993).

14. The National Archives of Ireland, 'Communications (external and internal): Emergency measures in event of invasion or internal attack', TAOIS/ S 11992.
15. Eunan O'Halpin, 'Intelligence and security in Ireland, 1922–45', in *Intelligence and National Security* 51 (1990) pp 50–83.
16. Paul McMahon, *British Spies and Irish Rebels: British Intelligence and Ireland, 1916–1945* (England, 2011).
17. Eunan O'Halpin, *Spying on Ireland: British Intelligence and Irish Neutrality during the Second World War* (Oxford, 2008).
18. The National Archives (UK), 'Liddell Diaries Volume 1', KV 4/185.
19. Ibid., 'Papers: 71–194, JIC (40) 108', CAB 81/97.
20. *Belfast News-letter*, 6 Apr. 1944.
21. The National Archives of Ireland, 'Telephone communications with GB: Censorship of 1940', TAOIS/ S 11963.
22. Donal Ó Drisceoil, *Censorship in Ireland, 1939–1945: Neutrality, Politics and Society* (Cork, 1996).
23. Eunan O'Halpin, *Defending Ireland: The Irish State and its Enemies since 1922* (Oxford, 2002).
24. Darragh Biddlecombe, *Colonel Dan Bryan and the evolution of Irish Military Intelligence, 1919–1945* (1999) NUIM MA thesis, http://mural.maynoothuniversity.ie/5176/1/Darragh_Biddlecombe_20140708161420.pdf, 20 Oct. 2019.
25. Brian Barton, *The Belfast Blitz: The City in the War Years* (Belfast, 2015).
26. The National Archives of Ireland, 'Commercial Cable Company London, Loan of cable channel for use of USA Officer of War importation', 1943, DFA/233/192.
27. Telegram from Winston Churchill to Eamon de Valera 8 Dec. 1941, www.difp.ie/volume-6/1941/-now-or-never-a-nation-once-again-/3577, 28 Feb. 2019.
28. Ian S. Wood, *Ireland during the Second World War* (London, 2002).
29. Roddy Flynn, *The Development of Universal Telephone Service in Ireland 1880–1993* (1998) DCU Doctor of Arts thesis, http://doras.dcu.ie/18734/1/Roddy_Fynn_20130509140820.pdf, 14 Mar. 2019.
30. Dáil Debates, 16 May 1945.
31. Ibid., 26 Mar. 1942.
32. Ibid., 30 Mar. 1938.
33. *Irish Independent*, 20 Jan. 1944.
34. *Irish Times*, 20 Sept. 1938.
35. 'Chronologie von Fernsprechwesen und Telefonbuch in Berlin', www.schwender.in-berlin.de/Fernsprechwesen.html, 26 Feb. 2019.
36. 'Fernamt Berlin', https://dewiki.de/Lexikon/Fernamt_Berlin, 26 Feb. 2019.
37. Michael Kirwan, 'Foynes Aeradio in the years 1936 to 1945', www.limerickcity.ie/media/olj%202013%20p007%20to%20010.pdf, 27 Feb. 2019.
38. Eddie Bohan, 'The Saint: A Shocking Irish Discovery', http://ibhof.blogspot.com/2013/06/the-saint-shocking-irish-discovery.html, 13 Aug. 2018.

39. Clair Wills, *That Neutral Island: A Cultural History of Ireland during the Second World War* (London, 2007).
40. *Irish Examiner*, 9 Mar. 2015.

Chapter 8: A painful case

1. The National Archives of Ireland, 'Central telephone exchange, Crown Alley: General file (plans, specs etc), 1912–14', OPW5 6114/14.
2. John L. O'Sullivan, *From Morse to Mobile* (Cork, 2000).
3. Eamonn G. Hall, *The Electronic Age* (Dublin, 1993).
4. *Irish Independent*, 7 Nov. 1946.
5. The National Archives of Ireland, 'Telephone development: Postwar planning', TSCH/3/S13086 C.
6. Michael J. Shiel, *The Quiet Revolution: The Electrification of Rural Ireland, 1946–1976* (Dublin, 2005).
7. Joseph Brady, *Dublin, 1950–1970: Houses, Flats and High Rise* (Dublin, 2016).
8. León Ó Broin, *Just like Yesterday: An Autobiography* (Dublin, 1986).
9. Tom Garvin, *Preventing the Future: Why Was Ireland so Poor for so Long?* (Dublin, 2005).
10. Diarmaid Ferriter, *The Transformation of Ireland: 1900–2000* (London, 2004).
11. Bernadette Whelan, *Ireland and the Marshall Plan, 1947–57* (Dublin, 2000).
12. Thomas F. Wall, 'Some notes towards a history of telecommunications with particular reference to Ireland' (2005), Dublin City Archive.
13. The National Archives of Ireland, 'Castleisland tel exch', INDC/B 16179/49 Vol 1.
14. *Telecho*, Vol. 4 February 1987.
15. The National Archives of Ireland, 'Standard engineering structures for small exchanges', INDC/B 29545/ 59 Parts 1 & 2.
16. Gary A. Boyd and John McLaughlin, *Infrastructure and the Architectures of Modernity in Ireland 1916–2016* (London, 2015).
17. The National Archives of Ireland, 'Telephone development: Postwar planning', TSCH/3/S13086 A.
18. *Irish Times*, 3 Sept. 1946.
19. *Belfast Newsletter*, 7 Sept. 1946.
20. *Irish Independent*, 1 Mar. 1951.
21. Commission for Communications Regulation, *Emergency Call Answering Services: Call Handling Fee review 2017* (Dublin, 2016).
22. Luca Thanei, 'Please connect me! Automating our telephone exchanges', https://blog.nationalmuseum.ch/en/2019/12/please-connect-me-automating-our-telephone-exchanges/, 12 Apr. 2020.
23. Dáil Debates, 30 Apr. 1953.
24. *Westmeath Independent*, 23 Mar. 1957.

25. International Telecommunication Union, *Annual Report by the Secretary-General of the International Telecommunication Union 1957* (Geneva, 1958).
26. A. J. Litton, 'The growth and development of the irish telephone system', in *Journal of the Statistical and Social Inquiry Society of Ireland* XX:V (1962) pp 79–115.
27. *Irish Independent*, 15 Dec. 1958
28. *Irish Press*, 25 June 1959.
29. International Telecommunication Union, General Telephone Statistics (Geneva, 1958).
30. Joseph Lee, *Ireland 1912–1985: Politics and Society* (Cambridge, 2006).
31. *Programme For Economic Expansion* (Dublin, 1958).
32. Roddy Flynn, *The Development of Universal Telephone Service in Ireland 1880–1993* (1998) DCU Doctor of Arts thesis, http://doras.dcu.ie/18734/1/Roddy_Fynn_20130509140820.pdf, 14 Mar. 2019.
33. The National Archives of Ireland, 'Application for telephone service: Priority classes 1951', TAOIS/ s 15211.
34. Dáil Debates*,* 21 July 1965.
35. Dargan, M. J., *Report of Posts and Telegraphs Review Group: 1978–1979* (Dublin, 1979).
36. *WeekIn:CouldYouHoldAMinutePlease?,* RTÉ(1980)www.rte.ie/archives/2020/0211/1114635-irelands-telephone-system, 23 Apr. 2020.
37. Eli M. Noam, *Telecommunications in Europe* (New York and Oxford, 1992).
38. NESC, *The Importance of Infrastructure to Industrial Development in Ireland* (Dublin, 1981).
39. The National Archives of Ireland, 'Telephone service: Delays Sep 1970 – Jan 1986', 2016/51/320.
40. Dáil Debates*, 5 Feb. 1975.*
41. Terry Kelleher, *The Essential Dublin* (Dublin, 1972).
42. Joe Kearney, personal communication, 14 Apr. 2020.
43. *Cork Examiner*, 23 Oct. 1974.
44. Michelle Ward, personal communication, 11 June 2020.
45. *Donegal Democrat*, 25 July 1986.
46. *Death, Drink and Luke Kelly*, Magill, 27 Dec. 2006.
47. *Donegal Democrat*, 30 Aug. 1963.
48. *Meath Chronicle*, 20 July 1963.
49. *Irish Times*, 27 Aug. 1963.
50. Research and Technology Survey Team, *Science and Irish economic development: report of the Research and Technology Survey Team appointed by the Minister for Industry and Commerce in November 1963 (in association with OECD)* (Dublin, 1966).
51. *Offaly Independent*, 3 June 1961.
52. *Evening Herald*, 21 Nov. 1961.
53. Ibid., 21 Feb. 1962.
54. Stephen Ferguson, *The Post Office in Ireland: An Illustrated History* (Newbridge, 2016).
55. *Longford Leader* 14 Jan. 1956.

56. L. M. Ericsson Holdings, *Ireland's Communication Company* (Dublin, 1985).
57. Cormac Ó Gráda, *A Rocky Road: The Irish Economy since the 1920s* (Manchester, 1997).
58. *Irish Press*, 4 June 1981.
59. International Telecommunication Union, Operational Bulletin No. 765 – 1.VI.2002 (Geneva, 2002).
60. *Cork Examiner*, 4 July 1987.
61. Seanad Debates, 6 Dec. 2007.
62. 'Northern Atlantic communications: History', www.iaa.ie/air-traffic-management/north-atlantic-communications/north-atlantic-communications---history, 12 Apr. 2020.
63. Brian Nagel, 'Shannon Aeradio', in *Monitoring Times* (1988), https://worldradiohistory.com/Archive-Monitoring-TImes/1980s/Monitoring-Times-1988-04.pdf, 12 Sept. 2021.
64. *Irish Times*, 29 July 2003.
65. *Irish Independent*, 10 Jan. 1962.
66. *Irish Times*, 22 Apr. 2017.
67. *Irish Press*, 15 Nov. 1972.
68. *PagusT*, Oct. 1969.
69. Ibid., Feb. 1974.
70. Dáil Debates, 9 Nov. 1972.
71. Ibid., 24 Nov. 1971.
72. *The Kerryman*, 9 June 1973.
73. Dáil Debates, 26 July 1973.
74. *Telecho*, June 1985.
75. Dáil Debates, 19 Apr. 1983.
76. *The Kerryman*, 27 Apr. 1990.
77. Dáil Debates, 3 Feb. 1977.
78. *Belfast Telegraph*, 7 Aug. 1970
79. Ibid., 13 Apr. 1972.
80. *Anglo-Celt*, 1 Sept. 1972.
81. *Belfast Telegraph*, 16 Sept. 1970.
82. *Ballymena Weekly Telegraph*, 27 Nov. 1953.
83. Houses of the Oireachtas - Joint Committee on Justice, Equality, Defence and Women's Rights, *Interim Report on the Report of the Independent Commission of Inquiry into the Dublin Bombings of 1972 and 1973* (2004).
84. Christopher Morash, *A History of the Media in Ireland* (Cambridge and New York, 2010).
85. *Irish Press*, 24 July 1972.
86. 'Cappagh', www.dgsys.co.uk/btmicrowave/sites/90.php, 28 Feb. 2018.
87. 'Standing Stones', www.dgsys.co.uk/btmicrowave/sites/89.php, 23 Apr. 2020.
88. *Irish Independent*, 29 June 1962.
89. *Evening Herald*, 30 June 1962.
90. *Irish Independent* 21 Oct. 1965.
91. *Irish Press*, 1 Jan. 1975.

92. *Cork Examiner*, 11 Mar. 1975.
93. *Le Monde*, 31 May 1956.
94. Diarmaid Ferriter, *Ambiguous Republic: Ireland in the 1970s* (London, 2013).
95. *Cork Examiner*, 12 Mar. 1975.
96. Jimmy O'Connor, 'Aspects of Galway Postal History 1638–1984', in *Journal of the Galway Archaeological and Historical Society* 44 (1992) pp 119–94.
97. *Cork Examiner*, 28 Jan. 1976.
98. eircom, *eircom Reference Interconnect Offer Price List 2.17* (Dublin, 2007).
99. 'At The Third Stroke: The story of the speaking clock in Britain', https://strowger-net.telefoniemuseum.nl/tel_hist_tim.html, 17 May 2020.
100. *Irish Times*, 24 July 1970.
101. 'The End Of The Speaking Clock', *The Ray D'Arcy Show*, RTÉ Radio 1 (2018).
102. Roy F. Foster, *Luck and the Irish: A Brief History of Change from 1970* (Oxford and New York, 2007).
103. Ray Mac Sharry, Padraic White, and Joseph O'Malley, *The Making of the Celtic Tiger: The Inside Story of Ireland's Boom Economy* (Cork, 2001).
104. Louis O'Halloran, interviewed in 'Irish Life and Lore' www.irishlifeandlore.com/product/louis-ohalloran-b-1935-former-lecturer-department-of-electrical-engineering, 20 Apr. 2020.
105. Thomas F. Wall, personal communication, 21 Sept. 2020.
106. *PagusT*, May 1970.
107. Ben Jones, personal communication, 8 July 2020.
108. Liam O'Toole, *Post office rules OK*, Magill, 1 Dec. 1977.
109. *Irish Press*, 15 June 1974.
110. *Irish Times*, 11 Feb. 1978.
111. *Irish Independent*, 15 Feb. 1978.
112. Dáil Debates, 25 May 1978.
113. *Belfast Telegraph*, 19 Apr. 1978.
114. *Irish Independent*, 14 Feb. 1978.
115. Roderick Flynn and Paschal Preston, 'The long-run diffusion and techno-economic performance of national telephone networks: A case study of Ireland, 1922–1998', in *Telecommunications Policy* 235 (1999) pp 437–57.
116. *Irish Farmers Journal*, 7 July 1988.
117. *Irish Press*, 26 Apr. 1978.
118. Basil Chubb, *The Government and Politics of Ireland* (Abingdon, 2014).

CHAPTER 9: 'AN INSTRUMENT SO POWERFUL'

1. *Irish Times*, 17 June 1953.
2. Ireland, The Coronation and Television 1953, https://ibhof.blogspot.com/2019/04/ireland-coronation-and-television-1953.html, 12 Jan. 2020.

3. *Irish Independent*, 9 Feb. 1927.
4. Ibid., 2 Oct. 1929.
5. *Irish Times*, 19 Apr. 1952.
6. Ibid., 20 Mar. 1953.
7. *Thursday Television*, Radio Times, 15 July 1955.
8. *Anglo-Celt*, 13 Aug. 1955.
9. *Irish Times*, 4 May 1956.
10. *Wicklow People*, 24 Nov. 1956.
11. *The Irish Times*, 8 Oct. 1958.
12. Robert J. Savage, *Irish Television: The Political and Social Origins* (Cork, 1996).
13. The National Archives of Ireland, 'TV Aerials', INDC/TW 9821.
14. PRONI, 'Television in the Irish Republic', CAB/9F/165/13/1.
15. Christopher Morash, *A History of the Media in Ireland* (Cambridge and New York, 2010).
16. *Irish Times*, 29 Dec. 2001.
17. *Irish Independent*, 28 Jan. 2016.
18. Gareth Ivory, 'The Provision of Irish Television in Northern Ireland: A slow British–Irish success story', in *Irish Political Studies* 291 (2014) pp 134–53.
19. The National Archives of Ireland, 'Reception of Telefís Éireann programme in the Six Counties, 1964'; AHG/2001/78/77.
20. BBC Engineering Division, Visit to Telefis Eireann 5th and 6th March 1964 (London, 1964).
21. 'President First On New Television Service, 1961', RTÉ (1961) www.rte.ie/archives/exhibitions/681-history-of-rte/704-rte-1960s/139351-opening-night-presidents-address, 3 Nov. 2019.
22. Eamonn Jordan and Eric Weitz, *The Palgrave Handbook of Contemporary Irish Theatre and Performance* (London, 2018).
23. Technical Centre of the European Broadcasting Union, 'Eurovision links', in *EBU Review A, Technical* 89–94 (1965).
24. *Irish Independent*, 26 June 1965.
25. E. G. Aughey, 'Eurovision Song Contest 1971', in *The Post Office Electrical Engineers' Journal* 64:3 (1971) p. 198.
26. John L. O'Sullivan, *From Morse to Mobile* (Cork, 2000).
27. PRONI, 'Television in the Irish Republic', CAB/9F/165/13/2.
28. *Derry Journal*, 23 Feb. 1968.
29. *Belfast Telegraph*, 17 Nov. 1966.
30. The National Archives (UK), 'Distribution of RTE television programmes on broadcast relay stations in NI', HO 256/956.
31. The National Archives of Ireland, 'Report of the Committee to Consider the Question of Getting RTÉ Television signals in to the Six Counties, 3 July 1970', AHG/2001/78/17.
32. The National Archives (UK), 'Irish Republic Broadcasting (1972–80)', CJ 4/2960.
33. *Irish Times*, 16 Aug. 1996.
34. Ofcom, *Communications Market Report: Northern Ireland* (Belfast, 2017).

35. Broadcasting Authority of Ireland, *A Report on Market Structure, Dynamics and Developments in Irish media* (Dublin, 2017).
36. The National Archives (UK), 'RTÉ/Broadcasting in the ROI 1991', FCO 87/3309.
37. The National Archives of Ireland, 'Wireless Telegraphy (Wired Broadcast Relay Service Licence) Regulations', AGO/2004/2/1.
38. *Connacht Sentinel*, 11 May 1971.
39. *Waterford News and Star*, 19 Mar. 1971.
40. *Sunday Independent*, 8 Nov. 1970.
41. The National Archives (UK), 'Relay of UK to foreign countries', FCO 26/1700.
42. Conor Cruise O'Brien, *An Open Television Area for Ireland* (Fortnight, 22 Feb. 1974), pp 7–8.
43. *Orphan of the Storm*, Hibernia, 25 May 1973.
44. The National Archives of Ireland, 'Re-broadcasting rights of BBC1', AGO/2008/17/748.
45. *Cork Examiner*, 16 Mar. 1981.
46. Ibid., 7 May 1982.
47. Ibid., 25 Feb. 1982.
48. *Irish Times*, 5 Aug. 1986.
49. *Evening Echo*, 21 Sept. 1982.
50. The National Archives of Ireland, 'Copyright issues arising from the relay of BBC or ITV programmes', AGO/2016/1/1357.
51. Dáil Debates, 13 Dec. 1988.
52. *Munster Express*, 13 Mar. 1987.
53. *Connacht Sentinel*, 21 Sept. 1982.
54. *Irish Independent*, 4 Mar. 1988.
55. Richard Barbrook, 'Broadcasting and National Identity in Ireland', www.imaginaryfutures.net/2007/01/20/broadcasting-and-national-identity-in-ireland-by-richard-barbrook, 12 Jan. 2020.
56. Vincent Browne, *Beating the System*, Magill, 22 Sept. 2005.
57. *Irish Times*, 16 Apr. 1997.
58. Ibid., 21 Apr. 1997.
59. *Donegal News*, 15 Nov. 1996.
60. Vincent Browne, *Tony O'Reilly's Fitzwilton gave a £30,000 cash cheque to Ray Burke in June 1989*, Magill, 1 June 1998.
61. *Get Things Done Vote No. 1 Gildea Thomas*, Election leaflet (1998).
62. *Irish Independent*, 24 Jan. 1998.
63. *Irish Times*, 2 Jan. 1989.
64. Ibid., 15 Sept. 1989.
65. *Irish Independent*, 28 June 1990.
66. *Connacht Tribune*, 3 Oct. 1986.
67. *Cork Examiner*, 20 Oct. 1986.
68. Jean K. Chalaby, *Transnational Television in Europe: Reconfiguring Global Communications Networks* (London, 2009).
69. Brendan O'Kelly, *The Esat Story* (Dublin, 2001).

70. 'New digital service starts on trial basis', www.rte.ie/news/2010/1029/137461-dtt/, 3 Feb. 2020.
71. *Anglo-Celt*, 22 Dec. 1962.
72. *The Sunday Times*, 27 June 2021.
73. *Irish Independent*, 12 Aug. 2020.

Chapter 10: Where's my bloody phone?!

1. *Irish Press*, 29 May 1987.
2. Florence Bugler, personal communication, 7 May 2020.
3. 'Mountshannon Goes Automatic', RTÉ (1987), www.rte.ie/archives/collections/news/21249852-mountshannon-goes-automatic., 3 July 2021.
4. Roddy Flynn, *The Development of Universal Telephone Service in Ireland 1880–1993* (1998) DCU Doctor of Arts thesis, http://doras.dcu.ie/18734/1/Roddy_Fynn_20130509140820.pdf, 14 Mar. 2019.
5. Dáil Debates, 14 May 1981.
6. *Irish Independent*, 2 Feb. 1987.
7. Michael Smurfit, *A Life Worth Living: The Autobiography* (Cork, 2014).
8. Isolde Goggin, personal communication, 5 Aug. 2020.
9. *PagusT*, Summer 1977.
10. *Irish Times*, 10 May 1983.
11. Ibid., 16 May 1984.
12. Ibid., 24 July 1980.
13. *Sligo Champion,* 29 July 1983.
14. *Fermanagh Herald,* 29 Oct. 1983.
15. Ibid., 3 Mar. 1990.
16. Eli M. Noam, *Telecommunications in Europe* (New York and Oxford, 1992).
17. John Mulrane, 'Early Irish Dial Telephones', in *Telecommunications Heritage Journal* 103 (2018).
18. L. M. Ericsson Holdings, *Ireland's Communication Company* (Dublin, 1985).
19. *Cork Examiner,* 14 Mar. 1980.
20. *Irish Press,* 15 Dec. 1981.
21. *Irish Independent*, 20 Jan. 1982.
22. Eamonn G. Hall, *The Electronic Age* (Dublin, 1993).
23. *Irish Independent*, 17 Sept. 1985.
24. *Irish Times*, 7 May 1983.
25. *Tuam Herald*, 1 June 1974.
26. *Cork Examiner*, 9 Nov. 1982.
27. BT Archives, 'The history of telecommunications 1912–1968', www.bt.com/about/bt/our-history/history-of-telecommunications/1912-to-1968, 22 July 2020.

28. *Irish Independent*, 20 June 1986.
29. 'New telephone fault handling system for Telecom Eireann', in *Telecommunication Journal* 9 (1984) .
30. 'Minister launches new technology: The Payphone', *RTÉ News* (1981), www.rte.ie/archives/2016/0302/772141-introduction-of-payphone, 11 June 2020.
31. Rankin, Nick. 'About Irish Callcards.' https://irishcallcards.net/callcard-info/about-irish-callcards.html, 2 Aug. 2021.
32. Thomas F. Wall, *Some notes towards a history of telecommunications with particular reference to Ireland* (2005), Dublin City Archive.
33. Federal Communications Commission, *FCC Record: A Comprehensive Compilation of Decisions, Reports, Public Notices, and Other Documents of the Federal Communications Commission of the United States* (Washington DC, 1995).
34. *Irish Times*, 6 Mar. 1990.
35. *Telecho*, June 1984.
36. Ibid., March 1986.
37. *Irish Times*, 9 May 1984.
38. *Telecho*, Jul/Aug 1987.
39. *Sunday Tribune*, 29 Nov. 1987.
40. Tom Doyle and Joseph Styles, 'A 126 km unrepeatered optical fibre submarine cable between Ireland and the UK', in *International Journal of Digital & Analog Cabled Systems* 21 (1989) pp 35–48.
41. *Southern Star*, 3 June 1989.
42. *Irish Times*, 19 June 1991.
43. Ibid., 10 May 2011.
44. Marine Licence Vetting Committee, *Report of the Marine Licence Vetting Committee (MLVC) on Foreshore Licence Application by Emerald Bridge Fibres Ltd for Installation of a Subsea Telecommunciations Cable from Portmarnock, Dublin to Port Darfach, Wales* (Dublin, 2012).
45. *Irish Times*, 27 Dec. 2013.
46. J. B. Burnham, 'Why Ireland boomed', in *The Independent Review* 7:4 (2003).
47. *Irish Times*, 14 Nov. 1998.
48. *Telecho*, December 1985.
49. *Irish Press*, 20 Apr. 1989.
50. Thomas F. Wall, personal communication, 21 Sept. 2020.
51. Dargan, M. J., *Report of Posts and Telegraphs Review Group: 1978–1979* (Dublin, 1979).
52. *Cork Examiner*, 12 Feb. 1987.
53. Ibid., 26 Aug. 1994.
54. *Meath Chronicle*, 25 July 1981.
55. *Irish Times*, 24 Mar. 1981.
56. 'Record reductions for DQ' in *BT Journal* 22 (1981) pp 4–6.
57. *Irish Press*, 1 Apr. 1970.

58. 'Ireland's first computers 1956–69', https://techarchives.irish/irelands-first-computers-1956-69, 10 Apr. 2020.
59. *Irish Press*, 4 Jan. 1977.
60. *Irish Independent*, 26 Mar. 1986.
61. *Irish Times*, 2 June 1984.
62. *Telecho*, February 1987.
63. Anton A. Huurdeman, *The Worldwide History of Telecommunications* (Hoboken NJ, 2003).
64. *Irish Press*, 15 Sept. 1947.
65. *Irish Times*, 18 June 1992.
66. *City Tribune*, 9 Feb. 1990.
67. Ray Mac Sharry, Padraic White, and Joseph O'Malley, *The Making of the Celtic Tiger: The Inside Story of Ireland's Boom Economy* (Cork, 2001).
68. *Irish Times*, 5 June 1987.
69. Ibid., 10 Dec. 1987.
70. *Irish Press*, 31 Oct. 1988.
71. Andy Bielenberg and Raymond Ryan, *An Economic History of Ireland Since Independence* (Abingdon, 2013).
72. Andreas Scheibelhut, *The Unprecedented Assimilation of Mobile Telephony in Ireland: A Phenomenon of the Celtic Tiger Era or a Result of Cultural Traits?* (2012) DIT, Masters dissertation, 10.21427/D7WP6F.
73. *Irish Times*, 11 May 1994.
74. Ibid., 16 Dec. 1992.
75. Richard G. Barry, *Ireland as a Haven for International Banking and Financial Services* (Dublin, 1987).
76. *Irish Press*, 22 Feb. 1988.
77. *Irish Times*, 29 Apr. 1982.
78. *Irish Farmers Journal*, 9 July 1988.
79. *Irish Times*, 23 Feb. 1991.
80. Ibid., 14 Aug. 1993.
81. *Telecom faces financial crunch*, Business and Finance, 30 July 1992.
82. BT, *Annual Report and Accounts 1996* (London, 1996).
83. Noel Curran, *Telecom Eireann's uncertain future*, Business and Finance, 30 July 1992.
84. Joseph W. Goodman, *Telecommunications Policy-Making in the European Union* (Cheltenham, 2006).
85. Brendan O'Kelly, *The Esat Story* (Dublin, 2001).
86. *Irish Times*, 17 June 1993.
87. Mary O'Rourke, personal communication, 17 June 2020.
88. *Irish Press*, 5 Aug. 1994.
89. *Cork Examiner*, 7 July 1988.
90. *Irish Independent*, 29 July 1994.

91. Ibid., 7 Nov. 1996.
92. Ibid., 24 Aug. 1990.
93. Howard R. Brown, 'British Telecom's ISDN Experience', in *IEEE Communications Magazine* 2512 (1987) pp 70–3.
94. Gerhard Fuchs, 'ISDN: The telecommunications highway for Europe after 1992?', in *Telecommunications Policy* 168 (1992) pp 635–45.
95. *Xchange*, April 1995.
96. Forfás, Broadband Telecommunications Investment in Ireland (Dublin, 1998).
97. *Sunday Independent*, 17 May 1998.
98. *Irish Independent*, 1 Dec. 1998.
99. *Irish Times*, 5 Feb. 1999.
100. Ibid., 13 Mar. 1999.
101. Ibid., 21 May 1999.
102. Mary O'Rourke, *Just Mary: A Memoir* (Dublin, 2013).
103. *Sunday Independent*, 25 Apr. 1999.
104. *Irish Independent*, 3 Jan. 2000.
105. *Evening Herald*, 21 July 1999.

Chapter 11: A mobile nation

1. Brian Noble, personal communication, 24 June 2020.
2. Sarah Carey, personal communication, 10 Aug. 2020.
3. Siobhan Creaton, *A Mobile Fortune* (London, 2010).
4. Brendan O'Kelly, *The Esat Story* (Dublin, 2001).
5. International Telecommunication Union, *Yearbook of Statistics: Telecommunication Services* (Geneva, 2004).
6. 'Zugfunk 1918 – 1926 – 1940', www.oebl.de/A-Netz/Rest/Zugfunk/Zug1926.html, 16 June 2020.
7. Petri Launiainen, *A Brief History of Everything Wireless: How Invisible Waves Have Changed the World* (Switzerland, 2018).
8. Alexis C. Madrigal, 'The 1947 Paper That First Described a Cell-Phone Network', www.theatlantic.com/technology/archive/2011/09/the-1947-paper-that-first-described-a-cell-phone-network/245222, 21 June 2020.
9. Tomi T Ahonen, 'Celebrating 30 Years of Mobile Phones, Thank You NTT of Japan', https://communities-dominate.blogs.com/brands/2009/11/celebrating-30-years-of-mobile-phones-thank-you-ntt-of-japan.html, 24 June 2020.
10. Svenolof Karlsson and Anders Lugn, 'The launch of NMT', www.ericsson.com/en/about-us/history/changing-the-world/the-nordics-take-charge/the-launch-of-nmt, 28 June 2020.
11. Sara Breselor, 'Gordon Gekko's Cell Phone', https://slate.com/culture/2010/09/gordon-gekkos-cell-phone.html, 1 July 2020.

12. Maggie Shiels, 'A chat with the man behind mobiles', http://news.bbc.co.uk/2/hi/uk_news/2963619.stm, 22 June 2020.
13. *Irish Press*, 16 Nov. 1981.
14. Andreas Scheibelhut, *The Unprecedented Assimilation of Mobile Telephony in Ireland: a Phenomenon of the Celtic Tiger Era or a Result of Cultural Traits?* (2012) DIT Masters dissertation, 10.21427/D7WP6F.
15. Tom Allen, personal communication, 7 July 2020.
16. *Irish Times*, 14 Aug. 1984.
17. *Irish Independent,* 12 Dec. 1985.
18. *Evening Echo*, 22 June 1987.
19. *Irish Independent*, 24 Jan. 1990.
20. Ibid., 28 Feb. 1990.
21. Roddy Flynn, *The Development of Universal Telephone Service in Ireland 1880–1993* (1998) DCU Doctor of Arts thesis, http://doras.dcu.ie/18734/1/Roddy_Fynn_20130509140820.pdf, 14 Mar. 2019.
22. *City Tribune*, 20 Nov. 1992.
23. *Evening Herald*, 26 Jan. 1961.
24. Ibid., 31 Aug. 1972.
25. *Irish Times*, 29 Sept. 1993.
26. Muhammad Haroon, et al., 'Perceptions and attitudes of hospital staff toward paging system and the use of mobile phones', in *International Journal of Technology Assessment in Health Care* 264 (2010) pp 377–81.
27. *Irish Times*, 7 Dec. 2000.
28. Ibid., 1 May 2003.
29. Gerard O'Regan, *A Brief History of Computing* (Berlin, 2008).
30. 'Motorola to close plant in Cork', www.rte.ie/news/2007/0308/86574-motorola, 4 July 2020.
31. Svenolof Karlsson and Anders Lugn, 'The first GSM meeting', www.ericsson.com/en/about-us/history/changing-the-world/world-leadership/the-first-gsm-meeting, 18 July 2020.
32. *Memorandum of Understanding on the Implementation of a Pan European 900 MHz Cellular Mobile Telecommunications Service by 1991* (Copenhagen, 1987).
33. *Sunday Tribune*, 6 Feb. 1994.
34. *Irish Press*, 25 June 1993.
35. *Irish Times*, 3 Mar. 1995.
36. *Irish Independent*, 12 Sept. 1995.
37. *The Kerryman*, 11 Oct. 1996.
38. *Irish Times*, 10 Aug. 1995.
39. *Southern Star*, 6 May 1995.
40. Ibid., 30 Mar. 1996.
41. *Irish Times*, 20 Nov. 1995.

42. Ibid., 5 Aug. 1995.
43. John Sterne, *Adventures in Code: The Story of the Irish Software Industry* (Dublin, 2004).
44. *Irish Times*, 15 Nov. 1996.
45. Ibid., 26 Oct. 1995.
46. Ibid., 27 Mar. 2010.
47. Ibid., 23 Mar. 2011.
48. 'Moriarty Tribunal report: The main points', www.rte.ie/news/2011/0322/298955-moriartymainpoints/, 25 June 2020.
49. *Guardian*, 11 Jan. 2016.
50. *Irish Times,* 17 Dec. 2019.
51. Ibid., 23 Mar. 2011.
52. Ibid., 1 Nov. 2019.
53. *Irish Independent*, 20 Jan. 1997.
54. *Irish Times*, 22 Feb. 1997.
55. *Irish Independent*, 21 Mar. 1997.
56. Margaret Furnell, personal communication, 30 June 2020.
57. *Irish Times*, 14 Mar. 1997.
58. *The Examiner*, 13 Oct. 1997.
59. Simon Rees, *The Strange Success of Prepaid Mobile Phone Services* (1999) ULH MA thesis, http://telecomsblog.ie/wp-content/uploads/2015/06/Simon-Rees-Masters-dissertation-1999.pdf, 21 June 2020.
60. *Irish Times*, 24 Nov. 1997.
61. International Telecommunication Union, *Mobile-cellular subscriptions* (Geneva, 2019).
62. *The Examiner*, 3 Dec. 1997.
63. *Irish Times*, 23 Feb. 2001.
64. 'What's in a number?', www.siliconrepublic.com/gear/whats-in-a-number, 1 July 2020.
65. Commission for Communications Regulation, Quarterly Key Data Report (Dublin, Various).
66. Dónal Palcic and Eoin Reeves, *Privatisation in Ireland: Lessons from a European Economy* (London, 2014).
67. *Irish Independent*, 1 Aug. 2001.
68. Morten Tolstrup, *Indoor Radio Planning: A Practical Guide for GSM, DCS, UMTS and HSPA* (Hoboken NJ, 2008).
69. *The Examiner*, 12 Jan. 2000.
70. *Wall Street Journal*, 6 Feb. 2001.
71. 'Telefonica bids £18bn for UK's O2', http://news.bbc.co.uk/2/hi/business/4391754.stm, 29 June 2020.
72. *Irish Examiner*, 25 July 2005.
73. *Irish Independent*, 24 May 2003.
74. *Irish Times*, 4 Jan. 2000.
75. *Sunday Tribune*, 23 Jan. 2000.

76. 'Matti Makkonen: Finnish pioneer of texting tech dies', www.bbc.com/news/technology-33324708, 24 June 2020.
77. Finn Trosby, et al., *Short Message Service (SMS): The Creation of Personal Global Text Messaging* (Chichester, 2010).
78. 'Logica: History and key milestones', https://web.archive.org/web/20081006154352/www.logica.com/history+and+key+milestones/350233679, 24 June 2020.
79. Nevan Bermingham, personal communication, 30 May 2020.
80. Thierry Van de Velde, *Value-Added Services for Next Generation Networks* (Boca Raton FL, 2007).
81. *Xchange*, February 1997.
82. *Irish Independent*, 11 Dec. 1999.
83. Amy O'Connor, '11 legendary moments in Irish advertising history', www.dailyedge.ie/iconic-irish-ads-2149502-Jun2015, 3 July 2020.
84. 'Text Nation', *RTÉ News* (2004), www.rte.ie/archives/2018/1220/1018305-irelands-mobile-phone-love. 30 June 2020.
85. *Irish Times*, 24 May 2004.
86. *Irish Independent*, 30 Nov. 2000.
87. Ibid., 17 Jan. 2006.
88. *Irish Times*, 4 Aug. 2000.
89. Ibid., 12 Apr. 2002.
90. Ibid., 28 Mar. 2002.
91. Ibid., 22 Oct. 2002.
92. Ibid., 31 May 2005.
93. *Irish Examiner*, 26 July 2005.
94. *Irish Times*, 27 Mar. 2013.
95. Dan Farber, 'Jobs: Today Apple is going to reinvent the phone', www.zdnet.com/article/jobs-today-apple-is-going-to-reinvent-the-phone, 4 July 2020.
96. 'O2 and Apple to Launch iPhone 3G in the UK & Ireland on July 11th', press release, www.apple.com/ie/newsroom/2008/06/09O2-and-Apple-to-Launch-iPhone-3G-in-the-UK-Ireland-on-July-11, 30 June 2020.
97. *Irish Independent,* 22 Dec. 2009.
98. John Kennedy, 'ComReg reveals 4G auction results – 450m instant windfall for Irish Govt', www.siliconrepublic.com/comms/comreg-reveals-4g-auction-results-450m-instant-windfall-for-irish-govt, 5 July 2020.
99. *Sunday Independent*, 29 Sept. 2013.
100. *Irish Independent*, 27 Sept. 2019.
101. *Irish Times*, 27 Oct. 2007.
102. Ibid., 13 May 2010.
103. Ibid., 24 June 2013.
104. Ibid., 28 May 2014.

105. 'Worldwide mobile data pricing: The cost of 1GB of mobile data in 228 countries', www.cable.co.uk/mobiles/worldwide-data-pricing, 2 June 2020.
106. *Xchange*, Jan/Feb 1996.
107. *Irish Independent*, 3 May 2014.
108. 'Commission adopts regulatory proposals for a Connected Continent', press release, https://ec.europa.eu/commission/presscorner/detail/en/MEMO_13_779, 2 July 2020.
109. Michelle Cini and Marián Šuplata, 'Policy leadership in the European Commission: the regulation of EU mobile roaming charges' in *Journal of European Integration* 392 (2017) pp 143–56.
110. 'Roaming charges and open Internet: questions and answers', press release, https://ec.europa.eu/commission/presscorner/detail/en/MEMO_15_5275, 2 July 2020.
111. Deloitte Ireland, Analysis: *Global Mobile Consumer Survey 2019: The Irish Cut* (Dublin, 2019).
112. J. P. Morgan, 'E-commerce payments trends: Ireland', www.jpmorgan.com/europe/merchant-services/insights/reports/ireland, 3 June 2020.
113. *Irish Times*, 22 May 2017.
114. @adrianweckler (A. Weckler), 'Speedtest on Huawei Mate 20 X (5G)', Twitter, https://twitter.com/adrianweckler/status/1161217728532889601?s=20, 13 Aug. 2019.
115. Inez Hunt, *Lightning in His Hand: The Life Story of Nikola Tesla* (Hawthorne CA, 1981).

CHAPTER 12: ONLINE AND OFFLINE

1. John Beckett, personal communication, 2 Sept. 2020.
2. John Kennedy, 'John Beckett, entrepreneur who built the first Ryanair website (video)', www.siliconrepublic.com/companies/the-interview-john-beckett-entrepreneur-who-built-the-first-ryanair-website-video, 18 Aug. 2020.
3. Alan Ruddock, *Michael O'Leary: A Life in Full Flight* (Dublin, 2008).
4. 'History of Ryanair', https://corporate.ryanair.com/about-us/history-of-ryanair/, 15 Nov. 2020.
5. 'Ryanair Delivers Record 3rd Quarter Profits' (2001), press release, www.ryanair.com/doc/investor/2001/results_quarter3_2001.pdf, 16 Jan. 2021.
6. 'Ryanair Announces Record Annual Results' (2000), press release, www.ryanair.com/doc/investor/2000/results_end_2000.pdf, 18 Aug. 2020.
7. *Guardian*, 16 May 2005.
8. *Irish Times*, 13 Dec. 1999.
9. Ibid., 20 June 2000.
10. Ibid., 26 Mar. 2001.
11. 'Ireland's first computers 1956–69', https://techarchives.irish/irelands-first-computers-1956-69, 10 Apr. 2020.

12. *PagusT*, Feb. 1970.
13. 'David Kennedy, Aer Lingus systems manager 1965-69', https://techarchives.irish/irelands-first-computers-1956-69/david-kennedy/, 17 May 2020.
14. *Banks' Computers*, Hibernia, 5 Sept. 1975.
15. Andrew Wheen, *Dot-Dash to Dot.com: How Modern Telecommunications Evolved from the Telegraph to the Internet* (New York, 2011).
16. Dennis Jennings, personal communication, 16 Oct. 2020.
17. John Kennedy, 'Internet pioneer joins board of ICANN', www.siliconrepublic.com/comms/internet-pioneer-joins-board-of-icann, 23 Oct. 2020.
18. Liam Ferrie, personal communication, 15 July 2020.
19. *Irish Times*, 27 Feb. 1984.
20. Ibid., 21 Feb. 1990.
21. *New York Times*, 7 Mar. 2016.
22. *Irish Times*, 23 Jan. 1989.
23. *Oeust France*, 20 Aug. 2011.
24. Hugh Schofield, 'Minitel: The rise and fall of the France-wide web', www.bbc.com/news/magazine-18610692, 18 July 2020.
25. *Irish Independent*, 5 Sept. 1988.
26. Ibid., 22 Oct. 1991.
27. John Naughton, 'The evolution of the internet: From military experiment to General Purpose Technology', in *Journal of Cyber Policy* 11 (2016) pp 5–28.
28. Olivier Martin, *The 'Hidden' Prehistory of European Research Networking* (Bloomington IN, 2012).
29. *Irish Independent*, 27 Jan. 2018.
30. Michael Hanke, *Airline e-Commerce: Log on, Take off* (Abingdon, 2016).
31. Mary Robinson, *Cherishing the Irish Diaspora: Address by President Mary Robinson to the Houses of the Oireachtas* (1995).
32. History of the *Irish Times*, www.irishtimes.com/about-us/the-irish-times-trust, 26 Oct. 2020.
33. Helen McQuillan, 'Culture, Identity and Representation in an Information Age Town', in L. Stillman and G. Johanson (eds), *Constructing and Sharing Memory: Community Informatics, Identity and Empowerment* (Newcastle-upon-Tyne, 2009).
34. *Irish Independent,* 25 Sept. 1997.
35. *Irish Times*, 24 Sept. 1999.
36. *Clare Champion*, 6 Mar. 1998.
37. *Xchange*, February 1998.
38. *Irish Times*, 25 Sept. 1997.
39. Ibid., 19 Oct. 2000.
40. Ibid., 15 Sept. 1997.
41. Ibid., 8 Oct. 2001.
42. *Xchange*, February 1998.

43. Daphne Lavers, 'History of Cable Television', www.broadcasting-history.ca/industry-government/history-cable-television, 24 Aug. 2020.
44. Marlies Van der Wee, Sofie Verbrugge and Wolter Lemstra, 'Understanding the dynamics of broadband markets: A comparative case study of Flanders and the Netherlands', *23rd European Regional Conference of the International Telecommunications Society (ITS)* (Vienna, 2012)
45. OECD, 'The Development of Broadband Access in the OECD Countries', in *OECD Digital Economy Papers* 56 (2002)
46. Hans J. Kleinsteuber, 'New media technologies in Europe: The politics of satellite, HDTV and DAB', in *Irish Communication Review* 51 (1995)
47. Analysys Ltd. and Forfás, *Telecommunications and Enterprise: Building and Investing for our Future* (Dublin, 1996).
48. *Irish Times*, 1 May 1999.
49. Ibid., 27 Sept. 2002.
50. Ibid., 3 Mar. 1999.
51. John Kennedy, 'Irish telcos have worst online customer service on earth', www.siliconrepublic.com/companies/irish-telcos-have-worst-online-customer-service-on-earth, 8 Oct. 2020.
52. Thomas A. McDonnell, *The Economics of Broadband in Ireland: Country Endowments, Telecommunications Capital Stock, and Household Adoption Decisions* (2013) NUIG PhD thesis, http://hdl.handle.net/10379/3566, 19 July 2020.
53. 'Overview since privatisation', www.eolasmagazine.ie/an-overview-since-privatisation/, 18 July 2020.
54. Siobhan Creaton, *A Mobile Fortune* (London, 2010).
55. Matt Cooper, *Who Really Runs Ireland?: The Story of the Elite who Led Ireland from Bust to Boom – and Back Again* (Dublin, 2009).
56. *Irish Times*, 11 Nov. 2008.
57. Dáil Debates, 21 June 1923.
58. John Kennedy, 'Eircom sells off radio masts division for €155m', www.siliconrepublic.com/companies/eircom-sells-off-radio-masts-division-for-155m, 2 Oct. 2020.
59. *Irish Times*, 30 Mar. 2012.
60. Dónal Palcic and Eoin Reeves, *Privatisation in Ireland: Lessons from a European Economy* (London, 2014).
61. *Irish Times*, 19 Mar. 2001.
62. *Irish Independent*, 16 May 2002.
63. *Irish Times*, 25 Apr. 2003.
64. Department of Communications Climate Action and Environment, *Briefing for the Public Accounts Committee* (Dublin, 2019).
65. *Xchange*, October 1996.
66. Information Society Commission, *Ireland's Broadband Future* (Dublin, 2003).
67. 'Ryanair shares dive on profit warning', www.rte.ie/news/business/2004/0128/47912-ryanair, 4 July 2021.

68. *Irish Independent*, 3 Apr. 2008.
69. John Kennedy, 'Broadband urgency as Ireland hurtles towards a damaging digital divide', www.siliconrepublic.com/business/broadband-urgency-as-ireland-hurtles-towards-a-damaging-digital-divide, 1 Oct. 2020.
70. *Irish Times*, 7 June 2005.
71. 'Ryanair to charge for check-in luggage', www.rte.ie/news/business/2006/0125/72251-ryanair, 13 June 2020.
72. Nicky Ryan, '"You can only do a funeral once, and you have to do it right": The story behind RIP.ie', www.thejournal.ie/death-notices-ireland-rip-ie-3415982-Jun2017, 13 Oct. 2020.
73. Commission for Communications Regulation, *Quarterly Key Data Report* (Dublin, Various).
74. John Kennedy, 'Eircom goes live with up to 70 Mbps fibre network', www.siliconrepublic.com/comms/eircom-goes-live-with-up-to-70mbps-fibre-network, 31 Aug. 2020.
75. J. C. Herz, 'The bandwidth capital of the world', www.wired.com/2002/08/korea, 2 Oct. 2020.
76. Gordon Smith, 'Magnet to attract consumers with all-in-one service', www.siliconrepublic.com/life/magnet-to-attract-consumers-with-all-in-one-service, 29 Sept. 2020.
77. 'Europe lags behind in fibre to the home, warns FTTH Council president Chris Holden', www.gpondoctor.com/2011/11/24/europe-lags-behind-in-fibre-to-the-home-warns-ftth-council-president-chris-holden, 29 Sept. 2020.
78. Susan Ryan, 'Netflix launches in Ireland: But how does it compare?', https://jrnl.ie/323515, 13 Oct. 2020.
79. 'Joint receivers appointed to Xtra-Vision', www.rte.ie/news/business/2013/0429/387458-xtra-vision-receiver, 13 Oct. 2020.
80. John Kennedy, 'Eircom to trial fibre-to-the-home speeds up to 150 Mbps', www.siliconrepublic.com/enterprise/eircom-to-trial-fibre-to-the-home-speeds-up-to-150mbps, 13 Oct. 2020.
81. 'First customer goes live in BT/SIRO Network Interconnect agreement' (2017), press release, www.btireland.com/company/communications/newsroom/first-customer-goes-live-btsiro-network-interconnect-agreement, 13 Oct. 2020.
82. N. Czernich, et al., 'Broadband Infrastructure and Economic Growth', in *Econ. J. Economic Journal* 121552 (2011) pp 505–32.
83. John Kennedy, 'First signs of Ireland's digital divide emerge as Apple and Amazon jobs arrive', www.siliconrepublic.com/business/first-signs-of-irelands-digital-divide-emerge-as-apple-and-amazon-jobs-arrive, 1 Oct. 2020.
84. Government of Ireland, *Ireland: National Development Plan 2000–2006* (Dublin, 1999).
85. *Irish Times*, 17 June 1997.
86. Marine and Natural Resources Joint Committee on Communications, *Second Report: Provision of a national high speed broadband infrastructure – Vol. 1* (Dublin, 2004).
87. Waterways Ireland, *Annual Report and Accounts* (Enniskillen, 2007).

88. John Kennedy, 'Group Broadband Scheme labelled an "abject failure"', https://web.archive.org/web/20071117155941/www.siliconrepublic.com/news/news.nv?storyid=single7307, 4 Sept. 2020.
89. John Kennedy, 'Govt reveals €220m rural broadband scheme – 170 new jobs', www.siliconrepublic.com/comms/govt-reveals-220m-rural-broadband-scheme-170-new-jobs, 1 Oct. 2020.
90. *Irish Times,* 13 Sept. 2019.
91. Commission for Communications Regulation, *Improving connectivity in Ireland: Challenges, solutions and actions* (Dublin, 2018).
92. *Irish Times*, 26 Apr. 2014.
93. *Sunday Independent*, 10 Mar. 2013.
94. *Irish Times*, 15 Nov. 2019.
95. Ibid., 13 Nov. 2015.
96. Ibid., 31 Aug. 2020.
97. 'FCC gives SpaceX the go-ahead to drop Starlink satellite orbits by 500 kilometres or so', www.theregister.com/2021/04/28/fcc_spacex/, 4 July 2021.
98. 'Satellite internet service for Kerry's Black Valley moves a step closer', www.rte.ie/news/regional/2021/0329/1206850-black-valley-internet/, 4 July 2021.
99. *Irish Examiner*, 26 May 2021.
100. *Irish Times*, 31 Oct. 2000.
101. Ibid., 11 Sept. 2019.
102. Ibid., 6 July 1999.
103. 'The submarine cable map', www.submarinecablemap.com, 7 Nov. 2020.
104. Anton A. Huurdeman, *The Worldwide History of Telecommunications* (Hoboken NJ, 2003).
105. John Kennedy, 'US$15m UK–Ireland fibre network begins rollout today', www.siliconrepublic.com/comms/us15m-uk-ireland-fibre-network-begins-rollout-today, 25 Oct. 2020.

CHAPTER 13: CONCLUSION

1. Larry Press, 'Cuban Telecommunication Infrastructure and Investment', www.ascecuba.org/asce_proceedings/cuban-telecommunication-infrastructure-and-investment, 28 Nov. 2020.
2. Dónal Palcic and Eoin Reeves, *Privatisation in Ireland: Lessons from a European Economy* (London, 2014).
3. Dáil Debates,14 May 2019.
4. Adrian Weckler, 'A frank chat with eir's CEO', *The Big Tech Show* (2019) https://soundcloud.com/the-big-tech-show/a-frank-chat-with-eirs-ceo?utm_source=clipboard&utm_campaign=wtshare&utm_medium=widget&utm_content=https%253A%252F%252Fsoundcloud.com%252Fthe-big-tech-show%252Fa-frank-chat-with-eirs-ceo, 18 May 2019.

5. 'Global Speeds September 2020', www.speedtest.net/global-index, 13 Oct. 2020.
6. F. H. A. Aalen, Kevin Whelan, and Matthew Stout, *Atlas of the Irish Rural Landscape* (Cork, 1997).
7. An Foras Forbartha, *Urban Generated Housing in Rural Areas* (Dublin, 1976).
8. Commission for Communications Regulation, *Improving Connectivity in Ireland: Challenges, Solutions and Actions* (Dublin, 2018).
9. Department of Communications Climate Action and Environment, *Briefing for the Public Accounts Committee* (Dublin, 2019).
10. Rupert Wood, et al., *Full-fibre Access as Strategic Infrastructure: Strengthening Public Policy for Europe* (London, 2020).
11. Inez Hunt, *Lightning in his Hand: The Life Story of Nikola Tesla* (Hawthorne CA, 1981).
12. Massimo Guarnieri, 'A Question of Coherence [Historical]', in *IEEE Industrial Electronics Magazine* 103 (2016) pp 54–8.

BIBLIOGRAPHY

ARCHIVES

BT Archives, various papers
Committee on Finance Debates, 8 July 1927
Dáil Debates, various dates
Hansard, HC Debates, various dates
IET Archive, various papers
Irish Architectutal Archive, STC photo albums
Military Archives, Bureau of Military History, various witness statements
Military Service Pensions Collection, Kevin O'Reilly, MSP34REF50604
The National Archives of Ireland, Census of Ireland 1901
The National Archives of Ireland, various papers
The National Archives (UK), various papers
Oxford, Bodleian Libraries, Marconi papers
Oxford, Bodleian Libraries, Archive of Sir Matthew Nathan
Public Record Office of Northern Ireland, various papers
Seanad Debates, various dates
Waterville Cable Station Exhibition, Tech Amergin

NEWSPAPERS AND MAGAZINES

Anglo-Celt, various
Ballymena Observer, 1921
Ballymena Weekly Telegraph, 1953
Belfast Morning News, 1866
Belfast Newsletter, various
Belfast Telegraph, various

Catholic Telegraph, 1852
City Tribune, various
Clare Champion, 1998
Connacht Sentinel, various
Connacht Tribune, various
Daily Chronicle, 1915
Daily Express (Dublin), various
Derry Journal, various
Donegal Democrat, various
Donegal News, various
Dublin Evening Mail, various
Evening Echo, various
Evening Freeman, various
Evening Herald, various
Evening Packet and Correspondent, 1852
Fermanagh Herald, various
Frankfurter Allgemeine, 2016
Freeman's Journal, various
Guardian, various
Illustrated London News, 1852
Irish Farmers Journal, various
Irish Independent, various
Irish News, 1893
Irish Press, various
Irish Times, various
Kerry Evening Post, 1907
Kilkenny Moderator, 1922.
Kings County Chronicle, 1858
Larne Times, 1923
Liverpool Daily Post, 1857
London Evening Standard, 1858
London Gazette, 1918
Londonderry Sentinel, 1935
Longford Journal, 1909
Longford Leader, 1956
Magill, various
Meath Chronicle, various
Midland Counties Tribune, 1919
Le Monde, 1956
Munster Express, 1987

New York Times, various
Northern Whig, various
Oeust France, 2011
Offaly Independent, 1961
PagusT, various
Portadown Times, 1929
Saunders's News-Letter, various
Sligo Champion, various
Southern Star, various
Sunday Independent, various
Sunday Times, 2021
Sunday Tribune, various
Telecho, various
The Advocate or Irish Industrial Journal, 1853
The Examiner and predecessors, various
The Kerryman, various
The Times (London), 1854
Tuam Herald, 1974
Wall Street Journal, 2001
Warder and Dublin Weekly Mail, various
Waterford News and Star, 1971
Waterford Standard, various
Westmeath Independent, 1957
Wicklow People, 1956
Xchange, various

Theses

Biddlecombe, D., *Colonel Dan Bryan and the evolution of Irish Military Intelligence, 1919–1945* (1999) NUIM, MA thesis, http://mural.maynoothuniversity.ie/5176/1/Darragh_Biddlecombe_20140708161420.pdf.

Cullen, F. J. , *Local Government and the Management of Urban Space: A Comparative Study of Belfast and Dublin, 1830-1922* (2005) NUIM PhD thesis, www.jstor.org/stable/24338495.

Flynn, R., *The Development of Universal Telephone Service in Ireland 1880 - 1993* (1998) DCU, Doctor of Arts thesis, http://doras.dcu.ie/18734/1/Roddy_Fynn_20130509140820.pdf.

McDonnell, Thomas A., *The Economics of Broadband in Ireland: Country Endowments, Telecommunications Capital Stock, and Household Adoption Decisions* (2013) NUIG, PhD thesis, http://hdl.handle.net/10379/3566.

Rees, Simon, *The Strange Success of Prepaid Mobile Phone Services* (1999) ULH, MA thesis, http://telecomsblog.ie/wp-content/uploads/2015/06/Simon-Rees-Masters-dissertation-1999.pdf.

Scheibelhut, Andreas, *The Unprecedented Assimilation of Mobile Telephony in Ireland: A Phenomenon of the Celtic Tiger Era or a Result of Cultural Traits?* (2012) DIT, Masters dissertation, 10.21427/D7WP6F.

Stein, Jeremy Leon, *Ideology and the Telephone* (1996) UCL, PhD thesis, http://discovery.ucl.ac.uk/1381834/1/389060.pdf.

Press Releases

'Commission adopts regulatory proposals for a Connected Continent', press release, https://ec.europa.eu/commission/presscorner/detail/en/MEMO_13_779

'Get Things Done Vote No 1 Gildea Thomas', election leaflet (1998).

'First customer goes live in BT/SIRO Network Interconnect agreement' (2017), press release, https://www.btireland.com/company/communications/newsroom/first- customer-goes-live-btsiro-network-interconnect-agreement

'O2 and Apple to Launch iPhone 3G in the UK & Ireland on July 11th', press release, https://www.apple.com/ie/newsroom/2008/06/09O2-and-Apple-to-Launch-iPhone-3G-in-the-UK-Ireland-on-July-11/

'Roaming charges and open Internet: questions and answers', press release, https://ec.europa.eu/commission/presscorner/detail/en/MEMO_15_5275

'Ryanair Announces Record Annual Results' (2000), press release, https://www.ryanair.com/doc/investor/2000/results_end_2000.pdf

'Ryanair Delivers Record 3rd Quarter Profits' (2001), press release, https://www.ryanair.com/doc/investor/2001/results_quarter3_2001.pdf

Personal Communication

Allen, Tom
Beckett, John
Bermingham, Nevan
Bugler, Florence
Carey, Sarah
Ferrie, Liam
Furnell, Margaret
Goggin, Isolde
Jennings, Dennis
Jones, Ben
Kearney, Joe
Noble, Brian
O'Rourke, Mary
Ring, Colum

Sweeney, Vincent
Wall, Thomas F.
Ward, Michelle

Secondary Sources

Aalen, F. H. A., et al., *Atlas of the Irish Rural Landscape* (Cork: Cork University Press, 1997).

Abercrombie, P. et al., *Dublin of the Future* (Liverpool: University Press of Liverpool, Hodder and Stoughton, 1922).

Ahonen, Tomi T., 'Celebrating 30 Years of Mobile Phones, Thank You NTT of Japan', https://communities-dominate.blogs.com/brands/2009/11/celebrating-30-years-of-mobile-phones-thank-you-ntt-of-japan.html

Analysys Ltd. and Forfás, *Telecommunications and Enterprise: Building and Investing for our Future* (Dublin: Forfás, 1996).

'The Anglo-American Telegraph Company's Station, Valentia, Ireland.', in *The Telegraphist*, 1 Dec. 1885 (1885)

'At The Third Stroke: The story of the speaking clock in Britain', https://strowger-net.telefoniemuseum.nl/tel_hist_tim.html

Atkinson, J., *Telephony Vol. 1* (London: The New Era Publishing Company, 1948).

Aughey, E. G., 'Eurovision Song Contest 1971' in *The Post Office Electrical Engineers' Journal,* 1971, p. 198.

Baker, William John, *A History of the Marconi Company* (London: Routledge, 1970).

'Ballygomartin', http://www.dgsys.co.uk/btmicrowave/sites/86.php

'Banks' Computers', in *Hibernia*, 5 Sept. 1975

Barbrook, Richard, 'Broadcasting and National Identity in Ireland', http://www.imaginaryfutures.net/2007/01/20/broadcasting-and-national-identity-in-ireland-by-richard-barbrook/

Barry, Richard G., *Ireland as a Haven for International Banking and Financial Services* (Dublin: 1987).

Bart, David and Julia Bart, 'Sir William Thomson, on the 150th Anniversary of the Atlantic Cable', http://atlantic-cable.com/CablePioneers/Kelvin/

Barton, Brian, *The Belfast Blitz: The city in the War Years* (Belfast: Ulster Historical Foundation, 2015).

Barton, R. N., 'Brief Lives: Three British Telegraph Companies 1850–56' in *The International Journal for the History of Engineering & Technology,* 2, 2010, pp. 183–198.

BBC Engineering Division, *Visit to Telefis Eireann 5th and 6th March 1964* (London: BBC, 1964).

Beauchamp, Ken and Institution of Electrical Engineers, *History of telegraphy: IEE history of technology series*, vol. 26, (London: Institution of Electrical Engineers, 2001).

'Bell, Gray and the invention of the telephone', https://www.ericsson.com/en/about-us/history/communication/early-developments/bell-gray-and-the-invention-of-the-telephone

Bielenberg, Andy and Raymond Ryan, *An Economic History of Ireland Since Independence* (Abingdon: Taylor & Francis, 2013).

Bohan, Eddie, 'Ireland, The Coronation and Television 1953', http://ibhof.blogspot.com/2019/04/ireland-coronation-and-television-1953.html

—, *Rebel Radio: Ireland's First International Radio Station 1916* (Dublin: Kilmainham Tales Teo, 2016).

—, 'The Saint: A Shocking Irish Discovery', http://ibhof.blogspot.com/2013/06/the-saint-shocking-irish-discovery.html

Boyd, Gary A. and John McLaughlin, *Infrastructure and the Architectures of Modernity in Ireland 1916-2016* (London: Ashgate, 2015).

Boyd, Hugh Alexander 'Marconi and Ballycastle', http://antrimhistory.net/marconi-and-ballycastle/

Brady, Joseph, *Dublin, 1950–1970: Houses, Flats and High Rise* (Dublin: Four Courts Press, 2016).

Brennan, Cathal, 'The Postal Strike of 1922', http://www.theirishstory.com/2012/06/08/the-postal-strike-of-1922/#.XBlp5qecau4

—, 'Radio and the 1916 Rising', https://timeline.ie/radio-and-the-1916-rising/

Bresolor, Sara, 'Gordon Gekko's Cell Phone', *Slate*, https://slate.com/culture/2010/09/gordon-gekkos-cell-phone.html.

Bright, Charles, *The Story of the Atlantic Cable* (New York: D. Appleton & Co,, 1903).

Broadcasting Authority of Ireland, *A Report on Market Structure, Dynamics and Developments in Irish Media* (Dublin: 2017).

Brown, Howard R., 'British Telecom's ISDN Experience' in *IEEE Communications Magazine,* 12, 1987, pp. 70-73.

BT, *Annual Report and Accounts 1996* (London: 1996).

BT Archives, 'Private and municipal telephone companies', available at https://web.archive.org/web/20170924141940/https://www.btplc.com/Thegroup/BTsHistory/BTgrouparchives/Informationsheetsandtimelines/Info_sheet_Private_Telephone_Co_Aug_2008.pdf

—, 'The history of telecommunications 1912–1968', https://www.bt.com/about/bt/our-history/history-of-telecommunications/1912-to-1968

Bull, John, 'The Forecaster: The Man Who Decided D-Day', https://medium.com/p/the-weatherman-the-man-who-decided-d-day-d0eb5cad3f7e

Burnham, J. B., 'Why Ireland Boomed' in *The Independent Review,* Spring, 2003.

Burns, Bill, 'The Curious Story of the Tiffany Cables', http://atlantic-cable.com/Article/Lanello/index.htm

—, 'Messages Carried by the 1858 Atlantic Telegraph Cable', https://atlantic-cable.com/Article/1858Messages/index.htm

Burns, Russell W., *Communications: An International History of the Formative Years* (Stevenage: IET, 2004).

'Cappagh', http://www.dgsys.co.uk/btmicrowave/sites/90.php

Carey, Tim, *Dublin Since 1922* (Dublin: Hachette Books, 2016).

Caulfield, Max, *The Easter Rebellion* (London: New English Library, 1965).

Chalaby, Jean K., *Transnational Television in Europe: Reconfiguring Global Communications Networks* (London: I.B. Tauris, 2009).

Chapuis, R.J. and A.E. Joel, *100 Years of Telephone Switching*vol. pt. 1, (Amsterdam: IOS Press, 2003).

'Chronologie von Fernsprechwesen und Telefonbuch in Berlin', http://www.schwender.in-berlin.de/Fernsprechwesen.html

Chubb, Basil, *The Government and Politics of Ireland* (Abingdon: Taylor & Francis, 2014).

Cini, Michelle and Marián Šuplata, 'Policy leadership in the European Commission: the regulation of EU mobile roaming charges' in *Journal of European Integration,* 2, 2017, pp. 143-156.

Clark, Paddy, *Dublin Calling* (Dublin: Radio Telefís Éireann, 1986).

Clarke, Arthur C., *Voices Across the Sea* (New York: Harper & Row, 1974).

Clarke, Bob, *From Grub Street to Fleet Street: An Illustrated History of English Newspapers to 1899* (London: Ashgate, 2017).

Clarke, Desmond John and Richard Lovell Edgeworth, *The Ingenious Mr. Edgeworth* (London: Oldbourne, 1965).

Clements, William H., *Billy Pitt Had Them Built: the Napoleonic Towers in Ireland* (Stamford: The Holliwell Press, 2013).

Coakley, John and Michael Gallagher, *Politics in the Republic of Ireland* (Abingdon: Taylor & Francis, 2017).

Coggeshall, Ivan Stoddart et al., *An Annotated History of Submarine Cables and Overseas Company Radiotelegraphs, 1851–1934: With Special Reference to the Western Union Telegraph* (Norwich: School of Information Systems, UEA, 1993).

Commission for Communications Regulation, *Emergency Call Answering Services: Call Handling Fee review 2017* (Dublin: 2016).

—, *Improving Connectivity in Ireland: Challenges, Solutions and Actions* (Dublin: 2018).

—, *Quarterly Key Data Report* (Dublin: Various).

Condon, Gerald, 'Telecommunications', in Ronald C. Cox (ed.), *Engineering Ireland* (Cork: Collins Press, 2006).

'Confidential report from John W. Dulanty to Joseph P. Walshe', https://www.difp.ie/books/?volume=6&docid=3351

Connolly, Cornelia, 'The Transatlantic Cable – An Irish Perspective', https://corneliathinks.wordpress.com/2017/06/05/the-transatlantic-cable-an-irish-perspective/

—, 'The Transatlantic Cable Stations – An Irish Perspective', http://atlantic-cable.com/NF2001/CCPaper/

Cookson, Gillian, *The Cable: The Wire That Changed the World* (England: The History Press, 2012).

Cooney, D. A. Levistone, 'Switzer's of Grafton Street' in *Dublin Historical Record,* 2, 2002, pp 154–165.

Cooper, Matt, *Who Really Runs Ireland?: The Story of the Elite who Led Ireland from Bust to Boom — and Back Again* (Dublin: Penguin, 2009).

Corcoran, Donal P., *Freedom to Achieve Freedom: The Irish Free State 1922–1932* (Dublin: Gill & Macmillan, 2013).
Creaton, Siobhan, *A Mobile Fortune* (London: Aurum, 2010).
Crowley, John et al., *Atlas of the Irish Revolution* (Cork: Cork University Press, 2018).
Cruise O'Brien, Conor, 'An Open Television Area for Ireland', in *Fortnight*, 22 Feb. 1974 pp 7–8.
Curran, Noel, 'Telecom Eireann's uncertain future', in *Business and Finance*, 30 July 1992
Czernich, N. et al., 'Broadband Infrastructure and Economic Growth' in *Econ. J. Economic Journal,* 552, 2011, pp 505–532.
Dargan, M. J., *Report of Posts and Telegraphs Review Group: 1978–1979* (Dublin: Stationery Office, 1979).
Davidson, Peter A., *An Introduction to Electrodynamics* (Oxford: OUP 2019).
Davis, Jonathan, *Historical Dictionary of the Russian Revolution* (Lanham MD: Rowman & Littlefield Publishers, 2020).
De Cogan, Donard, 'Background to the 1858 telegraph cable', *Institution of Engineering and Technology Seminar on the Story of Transatlantic Communications 2008* (Manchester, 2008)
—, 'Dr. E.O.W. Whitehouse and the 1858 trans-Atlantic cable', http://atlantic-cable.com/Books/Whitehouse/DDC/index.htm
—, 'Harbour Grace and the "Direct": the amazing story of a telegraph cable', https://dandadec.files.wordpress.com/2013/07/direct-us-telegraph-company.pdf
—, 'Ireland, Telecommunications and International Politics' in *History Ireland,* 1993.
—, 'The Irish Submarine Cable Stations: A Technological History', https://dandadec.files.wordpress.com/2013/07/valentia-technology1.pdf.
—, *Thirty Six years in the Telegraphic Service, 1852 To 1888 :Being a Brief Autobiography of James Graves MSTE* (Norwich: Dosanda, 2016).
De Cogan, Donard and Bill Burns, *They Talk Along the Deep: A Global History of the Valentia Island Telegraph Cables* (Norwich: Dosanda, 2016).
Deloitte Ireland, *Analysis: Global Mobile Consumer Survey 2019: The Irish cut* (Dublin: Deloitte, 2019).
Department of Communications Climate Action and Environment, *Briefing for the Public Accounts Committee* (Dublin: 2019).
Devine, Francis, *Connecting Communities: A Pictorial History of the Communications Workers' Union* (Dublin: Communications Workers' Union, 2013).
Doyle, Oliver and Stephen Hirsch, *Railways in Ireland 1834-1984* (Dublin: Signal Press, 1983).
Doyle, Tom and Joseph Styles, 'A 126 km unrepeatered optical fibre submarine cable between Ireland and the UK' in *International Journal of Digital & Analog Cabled Systems,* 1, 1989, pp. 35-48.
Duggan, Agnes, 'Operating Reminiscences' in *National Telephone Journal,* 1909, pp. 96–97.
—, 'Operators in Training' in *National Telephone Journal,* 1907, pp 142–143.
—, 'Telephone Women' in *National Telephone Journal,* 1909, p. 99.
Eastwood, Eric, 'Marconi: His wireless telegraphy and the modern world' in *Journal of the Royal Society of Arts,* 1974, pp 599–611.

Edgeworth, Richard Lovell, 'An Essay on the Art of Conveying Secret and Swift Intelligence' in *The Transactions of the Royal Irish Academy,* 1797, pp 95–139.

Eircom, 'Eircom Reference Interconnect Offer Price List 2.17.', http://www.eircomwholesale.ie/dynamic/pdf/eircomRIOPriceList 2.17unmarked.pdf

Emmerson, Andrew, 'From Strowger to System X', http://strowger-net.telefoniemuseum.nl/tel_hist_edgelane.html

'Europe lags behind in fibre to the home, warns FTTH Council president Chris Holden', *GPONDoctor*, https://www.gpondoctor.com/2011/11/24/europe-lags-behind-in-fibre-to-the-home-warns-ftth-council-president-chris-holden/.

Fallon, Donal, 'Dublin's first public telephone', https://comeheretome.com/2012/04/06/dublins-first-public-telephone/

Farber, Dan, 'Jobs: Today Apple is going to reinvent the phone', *ZDNet*, https://www.zdnet.com/article/jobs-today-apple-is-going-to-reinvent-the-phone/.

Fari, Simone, *Victorian Telegraphy Before Nationalization* (Basingstoke: Palgrave Macmillan, 2015).

Farrell, Daniel, 'Robert Halpin: The Irishman who Revolutionized Global Communications', http://coastmonkey.ie/robert-halpin-irishman-global-communication/

Fava-Verde, Jean-Francois 'A tale of two telegraphs', *Science Museum Group Journal*, vol. 8, no. Autumn 2017, 2017, http://journal.sciencemuseum.ac.uk/browse/issue-08/a-tale-of-two-telegraphs/.

Federal Communications Commission, FCC Record: A Comprehensive Compilation of Decisions, Reports, Public Notices, and Other Documents of the Federal Communications Commission of the United Statesvol. v. 10, p. 13, (Washington DC: Federal Communications Commission, 1995).

Ferguson, Stephen, *GPO Staff in 1916: Business as Usual* (Cork: Mercier Press, 2012).

—, *The Post Office in Ireland: An Illustrated History* (Newbridge: Irish Academic Press, 2016).

'Fernamt Berlin', https://dewiki.de/Lexikon/Fernamt_Berlin

Ferriter, Diarmaid, *Ambiguous Republic: Ireland in the 1970s* (London: Profile Books, 2013).

—, *The Transformation of Ireland: 1900-2000* (London: Profile Books, 2004).

Field, H.M., *History of the Atlantic Telegraph* (New York: C. Scribner & Company, 1867).

Finn, Bernard S. and Daqing Yang, *Communications Under the Seas: The Evolving Cable Network and its Implications* (Cambridge MA and London: MIT, 2009).

Fleming, Gerard, interviewed on 'Storm Front in Mayo', RTÉ, 6 June 2019.

Flynn, Roderick and Paschal Preston, 'The long-run diffusion and techno-economic performance of national telephone networks: a case study of Ireland, 1922–1998' in *Telecommunications Policy,* 5, 1999, pp 437–457.

An Foras Forbartha, *Urban Generated Housing in Rural Areas* (Dublin, 1976).

Forfás, *Broadband Telecommunications Investment in Ireland* (Dublin: 1998).

Foster, Roy F., *Luck and the Irish: A Brief History of Change from 1970* (Oxford and New York: OUP, 2007).

Fox, Don, 'James Graves (1833 – 1911), his Descendants, the Electric Telegraph and Valentia Island', http://donaldpfox.blogspot.com/2018/01/james-graves-1833-1911-his-descendants.html

Friel, Brian, *Dancing at Lughnasa* (London and Boston: Faber and Faber, 1990).

Fuchs, Gerhard, 'ISDN: The telecommunications highway for Europe after 1992?' in *Telecommunications Policy,* 8, 1992, pp 635–645.

Garvin, Tom, *Preventing the Future: Why Was Ireland so Poor for so Long?* (Dublin: Gill & Macmillan, 2005).

'Global Speeds September 2020', https://www.speedtest.net/global-index

Gomersall, E., 'The impact of the 1916 rebellion on the refurbished GPO' in *The Post Office Electrical Engineers' Journal,* 1916.

Goodman, Joseph W., *Telecommunications Policy-Making in the European Union* (Cheltenham: Edward Elgar, 2006).

Gordon, John Steele, *A Thread Across the Ocean: The Heroic Story of the Transatlantic Cable* (New York: Perennial, 2003).

Green, Allan, 'A Meeting of Minds? Dr E. O. Wildman Whitehouse, Electrician to The Atlantic Telegraph Company, and his relationship with Professor William Thomson during the development and construction of the first transatlantic telegraph cable 1857/8', https://atlantic-cable.com/Books/Whitehouse/AG/amom.htm

Guarnieri, Massimo, 'The Conquest of the Atlantic' in *IEEE Industrial Electronics Magazine,* 1, 2014, pp 53–67.

—, 'A Question of Coherence [Historical]' in *IEEE Industrial Electronics Magazine*, 103, 2016, pp 54–58.

Hall, Eamonn G., *The Electronic Age* (Dublin: Oak Tree Press, 1993).

Hanke, Michael, *Airline e-Commerce: Log on. Take off* (Abingdon: Taylor & Francis, 2016).

Haroon, Muhammad et al., 'Perceptions and attitudes of hospital staff toward paging system and the use of mobile phones' in *International Journal of Technology Assessment in Health Care,* 4, 2010, pp 377–381.

Headrick, Daniel R. and Pascal Griset, 'Submarine Telegraph Cables: Business and Politics, 1838-1939' in *The Business History Review,* 3, 2001, pp 543–578.

Hertz, J C, 'The Bandwidth Capital of the World', *Wired*, https://www.wired.com/2002/08/korea/.

Hills, Jill, *The Struggle for Control of Global Communications: the Formative Century* (Chicago: University of Illinois 2002).

'History of Ryanair', https://corporate.ryanair.com/about-us/history-of-ryanair/

'History of The Irish Times', https://www.irishtimes.com/about-us/the-irish-times-trust

'History of Tinakilly House', https://tinakilly.ie/wp-content/uploads/2017/12/TK_History.pdf

Holmes, Andrew R., 'Professor James Thomson Sr. and Lord Kelvin: Religion, Science, and Liberal Unionism in Ulster and Scotland' in *Journal of British Studies,* 1, 2011, pp 100–124.

Homer, Peter, 'Malin Head Marconi Radio Station', http://www.malinhead.net/marconi/Marconi.htm

Houses of the Oireachtas - Joint Committee on Justice, Equality, Defence and Women's Rights, *Interim Report on the Report of the Independent Commission of Inquiry into the Dublin Bombings of 1972 and 1973* (Dublin: Houses of the Oireachtas, 2004)

Hugill, Peter J., *Global Communications Since 1844: Geopolitics and Technology* (Baltimore: Johns Hopkins University Press, 1999).

Hunt, Bruce J., 'Scientists, engineers and Wildman Whitehouse: Measurement and credibility in early cable telegraphy', http://atlantic-cable.com/Books/Whitehouse/BJH/index.htm

Hunt, Inez, *Lightning in His Hand: The Life Story of Nikola Tesla* (Hawthorne CA: Omni, 1981).

Huurdeman, Anton A., *The Worldwide History of Telecommunications* (Hoboken NJ: Wiley-Interscience, 2003).

Information Society Commission, *Ireland's Broadband Future*, (Dublin: 2003).

Infrapedia, https://www.infrapedia.com/app, 08/11/2020.

International Telecommunication Union, *Annual Report by the Secretary-General of the International Telecommunication Union 1957* (Geneva, 1958).

—, *General Telephone Statistics* (Geneva: ITU, 1958).

—, *Mobile-cellular subscriptions* (Geneva: ITU, 2019).

—, *Operational Bulletin No. 765 - 1.VI.2002* (Geneva: ITU, 2002).

—, *Yearbook of Statistics: Telecommunication Services* (Geneva: ITU, 2004).

Government of Ireland, *Ireland: National Development Plan 2000-2006*, (Dublin: Stationery Office, 1999).

'Ireland's first computers 1956-69', https://techarchives.irish/irelands-first-computers-1956-69/

Irish telephone directory - Eolaí telefóin na hÉireann, (Dublin: Department of Posts and Telegraphs, 1977).

Ivory, Gareth, 'The Provision of Irish Television in Northern Ireland: A Slow British–Irish Success Story' in *Irish Political Studies,* 1, 2014, pp. 134-153.

Jeffery, Keith, *The GPO and the Easter Rising* (Dublin and Portland OR: Irish Academic Press, 2006).

—, *The Sinn Fein Rebellion as they saw it* (Newbridge: Irish Academic Press, 1999).

Johnlepo, 'The price of a pint from 1928-2015 in todays money', http://publin.ie/2015/the-price-of-a-pint-from-1928-2015-in-todays-money/0

Johnston, Jamie, *Victorian Belfast* (Belfast: Ulster Historical Foundation, 1993).

Joint Committee on Communications, Marine and Natural Resources, *Second Report: Provision of a national high speed broadband infrastructure* (Dublin: 2004).

'Joint receivers appointed to Xtra-Vision', *RTÉ News*, https://www.rte.ie/news/business/2013/0429/387458-xtra-vision-receiver/.

Jordan, Eamonn and Eric Weitz, *The Palgrave Handbook of Contemporary Irish Theatre and Performance* (London: Palgrave Macmillan, 2018).

Joyce, James, *Ulysses* (New York: Garland, 1986).

Joyce, Shane, 'The Marconi Station', https://www.connemara.net/the-marconi-station/

Karlsson, Svenolof and Anders Lugn, 'The first GSM meeting', https://www.ericsson.com/en/about-us/history/changing-the-world/world-leadership/the-first-gsm-meeting

—, 'The launch of NMT', https://www.ericsson.com/en/about-us/history/changing-the-world/the-nordics-take-charge/the-launch-of-nmt

Kelleher, Terry, *The Essential Dublin* (Dublin: Gill and Macmillan, 1972).

Kennedy, David, https://techarchives.irish/irelands-first-computers-1956-69/david-kennedy/

Kennedy, John, 'Broadband urgency as Ireland hurtles towards a damaging digital divide', *Silicon Republic*, https://www.siliconrepublic.com/business/broadband-urgency-as-ireland-hurtles-towards-a-damaging-digital-divide.

—, 'ComReg reveals 4G auction results – 450m instant windfall for Irish Govt', *Silicon Republic*, https://www.siliconrepublic.com/comms/comreg-reveals-4g-auction-results-450m-instant-windfall-for-irish-govt.

—, 'Eircom goes live with up to 70Mbps fibre network', *Silicon Republic*, https://www.siliconrepublic.com/comms/eircom-goes-live-with-up-to-70mbps-fibre-network.

—, 'Eircom sells off radio masts division for 155m', *Silicon Republic*, https://www.siliconrepublic.com/companies/eircom-sells-off-radio-masts-division-for-155m.

—, 'Eircom to trial fibre-to-the-home speeds up to 150Mbps', *Silicon Republic*, https://www.siliconrepublic.com/enterprise/eircom-to-trial-fibre-to-the-home-speeds-up-to-150mbps.

—, 'First signs of Ireland's digital divide emerge as Apple and Amazon jobs arrive', *Silicon Republic*, https://www.siliconrepublic.com/business/first-signs-of-irelands-digital-divide-emerge-as-apple-and-amazon-jobs-arrive.

—, 'Govt reveals €220m rural broadband scheme – 170 new jobs', *Silicon Republic*, https://www.siliconrepublic.com/comms/govt-reveals-220m-rural-broadband-scheme-170-new-jobs.

—, 'Group Broadband Scheme labelled an "abject failure" ', *Silicon Republic*, http://www.siliconrepublic.com/news/news.nv?storyid=single7307 Retrieved August 28, 2020, from https://web.archive.org/web/20071117155941/http://www.siliconrepublic.com/news/news.nv?storyid=single7307.

—, 'Internet pioneer joins board of ICANN', *Silicon Republic*, https://www.siliconrepublic.com/comms/internet-pioneer-joins-board-of-icann.

—, 'Irish telcos have worst online customer service on earth', *Silicon Republic*, https://www.siliconrepublic.com/companies/irish-telcos-have-worst-online-customer-service-on-earth.

—, 'John Beckett, entrepreneur who built the first Ryanair website (video)', *Silicon Republic*, 2015, https://www.siliconrepublic.com/companies/the-interview-john-beckett-entrepreneur-who-built-the-first-ryanair-website-video.

—, 'US$15m UK-Ireland fibre network begins rollout today', *Silicon Republic*, https://www.siliconrepublic.com/comms/us15m-uk-ireland-fibre-network-begins-rollout-today.

Kennedy, Michael, *Guarding Neutral Ireland: The Coast Watching Service and Military Intelligence, 1939–1945* (Dublin: Four Courts Press, 2008).

Kern, Stephen, *The Culture of Time and Space, 1880-1918: With a New Preface* (Cambridge MA and London: Harvard University Press, 2003).

Kieve, Jeffrey L., *The Electric Telegraph: A Social and Economic History* (England: David and Charles, 1973).

Kirwan, A. J., 'R.L. Edgeworth and optical telegraphy in Ireland, c. 1790-1805' in Proceedings of the Royal Irish Academy, Section C: Archaeology, Celtic Studies, History, Linguistics and Literature, 2017, pp 209–235.

Kirwan, M., 'Foynes Aeradio in the years 1936 to 1945', http://www.limerickcity.ie/media/olj%202013%20p007%20to%20010.pdf

Kleberg, L., 'Silence And Surveillance: A History Of Culture And Communication', *Baltic Worlds*, http://balticworlds.com/a-history-of-culture-and-communication/.

Kleinsteuber, Hans J., 'New Media Technologies in Europe: The Politics of Satellite, HDTV and DAB' in *Irish Communication Review,* 1, 1995.

L. M. Ericsson Holdings, *Ireland's Communication Company* (Dublin: Ericsson, 1985).

Laakso, Seija-Riitta, *Across the Oceans: Development of Overseas Business Information Transmission, 1815-1875* (Helsinki: Helsinki University Printing House, 2007).

Lambe, A, 'This Kerry peninsula shows how the Irish have communicated for thousands of years', https://www.thejournal.ie/rock-art-ireland-2975807-Sep2016/.

Lavelle, Des, interviewed on 'Valentia Lecture and Fireside Chat', https://valentiacable.com/events/valentia-lecture-2020.

Launiainen, Petri, *A Brief History of Everything Wireless: How Invisible Waves Have Changed the World* (Switzerland: Springer Nature, 2018).

Lavers, Daphne, 'History of Cable Television', https://www.broadcasting-history.ca/industry-government/history-cable-television

Leabhar Seolta Telefona Saorstat Éireann (Dublin: Department of Posts and Telegraphs, 1929).

Lee, Joseph, *Ireland 1912-1985: Politics and Society* (Cambridge: Cambridge University Press, 2006).

Litton, A.J., 'The Growth and Development of the Irish Telephone System' in *Journal of the Statistical and Social Inquiry Society of Ireland,* 1962, pp 79–115.

'Logica: History and key milestones', https://web.archive.org/web/20081006154352/http://www.logica.com/history+and+key+milestones/350233679

Mac Sharry, Ray et al., *The Making of the Celtic Tiger: The Inside Story of Ireland's Boom Economy* (Cork: Mercier Press, 2001).

Madrigal, Alexis C., 'The 1947 Paper That First Described a Cell-Phone Network', *The Atlantic*, 2011, https://www.theatlantic.com/technology/archive/2011/09/the-1947-paper-that-first-described-a-cell-phone-network/245222/

Malanowski, Gregory, *The Race for Wireless: How Radio Was Invented (or Discovered?)* (Bloomington IN: AuthorHouse, 2011).

Maltby, Arthur and Jean Maltby, *Ireland in the 19th Century: A Breviate of Official Publications* (Elkins Park PA: Franklin Book, 1994).

Manning, Maurice et al., *Electricity Supply in Ireland: The History of the ESB* (Dublin: Gill and Macmillan, 1985).

'Marconigrams [advertisment]', in *Marconigraph*, November 1912, p. iv.

Marine Licence Vetting Committee, Report of the Marine Licence Vetting Committee (MLVC) on Foreshore Licence Application by Emerald Bridge Fibres Ltd for Installation of a Subsea Telecommunciations Cable from Portmarnock, Dublin to Port Darfach, Wales. (Dublin: 2012).

Martin, Olivier, *The "Hidden" Prehistory of European Research Networking* (Bloomington IN: Trafford Publishing, 2012).

Masini, Giancarlo, *Marconi* (New York: Marsilio Publishers, 1998).

'Matti Makkonen: Finnish pioneer of texting tech dies', *BBC News*, https://www.bbc.com/news/technology-33324708.

McAuliffe, Bridget et al., *Kerry 1916: Histories and Legacies of the Easter Rising: A Centenary Record* (Listowel: Red Hen Publishing, 2016).

McCartney, Mark and Andrew Whitaker, *Physicists of Ireland: Passion and Precision* (Bristol and Philadelphia: Institute of Physics Pub., 2003).

McCurdy, Augustine, 'Marconi and Rathlin', http://www.culturenorthernireland.org/article/1316/marconi-and-rathlin

McMahon, Paul, *British Sspies and Irish Rebels: British Intelligence and Ireland, 1916-1945* (England: Boydell, 2011).

McNally, Michael and Peter Dennis, *Easter Rising 1916: birth of the Irish Republic* (Oxford and New York: Osprey Pub., 2007).

McQuillan, Helen, 'Culture, Identity and Representation in an Information Age Town', in L. Stillman and G. Johanson (eds), *Constructing and Sharing Memory: Community Informatics, Identity and Empowerment* (Newcastle-upon-Tyne: Cambridge Scholars Publisher, 2009).

McVeigh, Daniel P, 'Stephen Mitchell Yeates ', http://oceanofk.org/telephone/html/part10.html

Memorandum of Understanding on the Implementation of a Pan European 900 MHz Cellular Mobile Telecommunications Service by 1991 (Copenhagen: 1987).

Mitchell, Ken, 'Robert Halpin – the Wicklow master mariner who connected the world', https://web.archive.org/web/20180329231854/www.engineersjournal.ie/2017/12/05/robert-halpin-transatlantic-cable-wicklow/

Mitchell, Kenneth 'Marconi and Ireland – the small country that played a big role in the radio age', www.engineersireland.ie/Engineers-Journal/News/marconi-and-ireland-the-small-country-that-played-a-big-role-in-the-radio-age

Mollan, Charles, *It's Part of What We Are - Volume 1* (Dublin: RDS, 2007).

Moran, Gerard, 'James Hack Tuke and his schemes for assisted emigration from the west of Ireland' in *History Ireland,* 3, 2013.

Morash, Christopher, *A History of the Media in Ireland* (Cambridge and New York: Cambridge University Press, 2010).

Morgan, J.P. 'E-commerce payments trends: Ireland', https://www.jpmorgan.com/europe/merchant-services/insights/reports/ireland

'Moriarty Tribunal report - The main points', *RTÉ News*, https://www.rte.ie/news/2011/0322/298955-moriartymainpoints/.

Morse, Samuel Finley Breese, *Foreign Conspiracy Against the Liberties of the United States: The Numbers of Brutus, Originally Published in the New-York Observer* (New York: Leavitt, Lord, 1835).

'Motorola to close plant in Cork', *RTÉ News*, https://www.rte.ie/news/2007/0308/86574-motorola/.

Muldowney, Mary, 'Working on the Railway in Dublin, 1900-1925 - Transcript', http://www.dublincity.ie/story/working-railway-dublin-1900-1925-transcript

Mulrane, John, 'Early Irish Dial Telephones' in *Telecommunications Heritage Journal,* 103, 2018.

Murray, K A., 'The Armagh Collision, 1889' in *Journal of the Irish Railway Record Society,* 54, 1971, pp 10–28.

Nagel, Brian, 'Shannon Airadio ' in *Monitoring Times,* April 1988, 1988.

Naughton, John, 'The evolution of the Internet: from military experiment to General Purpose Technology' in *Journal of Cyber Policy,* 1, 2016, pp. 5-28.

NESC, *The Importance of Infrastructure to Industrial Development in Ireland vol. 59* (Dublin, NESC: 1981).

'New digital service starts on trial basis', https://www.rte.ie/news/2010/1029/137461-dtt/

'New telephone fault handling system for Telecom Eireann, ' in *Telecommunication Journal,* 9, 1984.

Noam, Eli M., *Telecommunications in Europe* (New York and Oxford: OUP, 1992).

'Northern Atlantic Communications: History', https://www.iaa.ie/air-traffic-management/north-atlantic-communications/north-atlantic-communications—history

'Nurses in the Crimea', http://www.nationalarchives.gov.uk/womeninuniform/crimea_intro.htm

Ó Bréartúin, Mícheál, *Sreangscéal: Forbairt na Teileagrafaíochta Leictrí* (Baile Átha Cliath: Coiscéim, 2005).

Ó Broin, León, *Just like yesterday: an autobiography* (Dublin: Gill and Macmillan, 1986).

Ó Drisceoil, Donal, *Censorship in Ireland, 1939-1945: neutrality, politics and society* (Cork: Cork University Press, 1996).

Ó Gráda, Cormac, *A Rocky Road: The Irish Economy since the 1920s* (Manchester: University Press, 1997).

O'Connor, Amy, '11 legendary moments in Irish advertising history', *The Daily Edge*, https://www.dailyedge.ie/iconic-irish-ads-2149502-Jun2015/.

O'Connor, Jimmy, 'Aspects of Galway Postal History 1638-1984' in *Journal of the Galway Archaeological and Historical Society,* 1992, pp 119–194.

O'Halloran, Louis, interviewed on 'Irish Life and Lore', https://www.irishlifeandlore.com/product/louis-ohalloran-b-1935-former-lecturer-department-of-electrical-engineering/.

O'Halpin, Eunan, *Defending Ireland: The Irish state and its enemies since 1922* (Oxford: OUP, 2002).

—, 'Intelligence and security in Ireland, 1922-45' in *Intelligence and National Security,* 1, 1990, pp 50–83.

—, *Spying on Ireland: British Intelligence and Irish Neutrality during the Second World War* (Oxford: OUP, 2008).

O'Kelly, Brendan, *The Esat Story* (Dublin: Gill & Macmillan, 2001).

O'Regan, Gerard, *A Brief History of Computing* (Berlin: Springer, 2008).

O'Rourke, Mary, *Just Mary: A Memoir* (Dublin: Gill & Macmillan, 2013).

O'Sullivan, John L., *From Morse to Mobile* (Cork: Ballyheada Press, 2000).

O'Keeffe, Helene, 'Generational Change in Rising Memories', http://theirishrevolution.ie/generational-change-rising-memories/#.W-WOJpP7SM9

OECD, 'The Development of Broadband Access in the OECD Countries' in *OECD Digital Economy Papers,* 2002.

Ofcom, *Communications Market Report: Northern Ireland* (Belfast: 2017).

'Orphan of the Storm', in *Hibernia*, 25 May 1973

'An overview of the Shannon Scheme', https://esbarchives.ie/2016/02/13/shannon-scheme/

'Overview since privatisation', *Eolas Magazine*, https://www.eolasmagazine.ie/an-overview-since-privatisation/.

Palcic, Dónal and Eoin Reeves, *Privatisation in Ireland: Lessons from a European Economy* (London: Palgrave Macmillan, 2014).

Pedler, Garth, *Rail Operations Viewed from South Devon* (Leicester: Troubador, 2017).

Perry, Charles R., 'Frank Ives Scudamore and the Post Office Telegraphs' in *Albion: A Quarterly Journal Concerned with British Studies,* 4, 1980, pp 350–367.

—, *The Victorian Post Office: The Growth of a Bureaucracy* (London: Royal Historical Society, 1992).

Pine, Richard, *2RN and the Origins of Irish Radio* (Dublin: Four Courts Press, 2002).

'Plastics Historical Society - William Montgomerie ', http://plastiquarian.com/?page_id=14213

Press, Larry, 'Cuban Telecommunication Infrastructure and Investment', https://www.asce-cuba.org/asce_proceedings/cuban-telecommunication-infrastructure-and-investment/

Programme For Economic Expansion (Dublin: The Stationery Office, 1958).

Raboy, Marc, *Marconi: The Man Who Networked the World* (Oxford: OUP, 2016).

'Radio Signal from Kerry to Canada, 1965', https://www.rte.ie/archives/2015/0518/701871-hello-canada-hello-canada/

Radley, W. G., 'Telephony and Telegraphy' in *Journal of the Institution of Electrical Engineers,* 507, 1939, pp 359–367.

Ramirez, Ainissa, 'A Wire Across the Ocean', https://www.americanscientist.org/article/a-wire-across-the-ocean

'Record reductions for DQ, ' in *BT Journal,* 2, 1981, pp 4–6.

'Registration of the United Kingdom, ' in *Journal of the Statistical Society of London,* 1870, pp 153–157.

Research and Technology Survey Team, Science and Irish economic development: report of the Research and Technology Survey Team appointed by the Minister for Industry and Commerce in November 1963 (in association with OECD) (Dublin: Stationery Office, 1966).

Richardson, Neil, *According to their Lights* (Cork: Collins Press, 2015).

Roberts, Steven, 'Anglo-Irish Cables', http://atlantic-cable.com//Cables/Anglo-Irish/index.htm

—, 'Distant Writing – 8 Non-Competitors', http://distantwriting.co.uk/noncompetitors.html

—, 'Distant Writing – 5 Competitors and Allies', http://distantwriting.co.uk/competitorsallies.html

—, *Distant Writing: A History of the Telegraph Companies in Britain between 1838 and 1868,* http://distantwriting.co.uk/Documents/Distant%20Writing%202012.pdf.

Robertson, J. H., *The Story of the Telephone: A History of the Telecommunications Industry of Britain* (London: Pitman, 1948).

Robinson, Andrew, 'Marconi forged today's interconnected world of communica-tion', in *New Scientist*, 10 Aug. 2016, www.newscientist.com/article/mg23130862-900-marconi-forged-todays-interconnected-world-of-communication

Robinson, Mary, 'Cherishing the Irish Diaspora, Address By President Mary Robinson To The Houses Of The Oireachtas' (1995).

Rowland, Wade. *Spirit of the Web: The Age of Information from Telegraph to Internet.* (Vancouver, BC: Langara College, 2007).

'"Rozmowy kontrolowane" - satyra na stan wojenny', *Polskie Radio Historia*, https://www.polskieradio.pl/39/156/Artykul/2420638,Rozmowy-kontrolowane-satyra-na-stan-wojenny.

Ruddock, Alan, *Michael O'Leary: A Life in Full Flight* (Dublin: Penguin Ireland, 2008).

Russell, William Howard, *The Atlantic Telegraph* (London: Dawson Ltd., 1898).

Ryan, Nicky '"You can only do a funeral once, and you have to do it right': The story behind RIP.ie", *TheJournal.ie*, https://www.thejournal.ie/death-notices-ireland-rip-ie-3415982-Jun2017/.

Ryan, Susan, 'Netflix launches in Ireland: but how does it compare?', *TheJournal.ie*, https://www.thejournal.ie/netflix-launches-in-ireland-but-how-does-it-compare-323515-Jan2012/.

'Ryanair to charge for check-in luggage', *RTÉ News*, https://www.rte.ie/news/business/2006/0125/72251-ryanair/.

'Satellite internet service for Kerry's Black Valley moves a step closer', *RTÉ News*, https://www.rte.ie/news/regional/2021/0329/1206850-black-valley-internet/.

Savage, Robert J., *Irish Television: The Political and Social Origins* (Cork: Cork University Press, 1996).

Schiffer, Michael B., *Power Struggles: Scientific Authority and the Creation of Practical Electricity Before Edison* (Cambridge MA and London: MIT Press, 2011).

Schofield, Hugh, 'Minitel: The rise and fall of the France-wide web', *BBC News*, https://www.bbc.com/news/magazine-18610692.

Sexton, Michael, *Marconi: The Irish Connection* (Dublin: Four Courts, 2005).

Share, Bernard, *In Time of Civil War: The Conflict on the Irish Railways, 1922-1923* (Cork: Collins Press, 2006).

Shiel, Michael J., *The Quiet Revolution: The Electrification of Rural Ireland, 1946-1976* (Dublin: O'Brien Press, 2005).

Shiels, Maggie 'A chat with the man behind mobiles', *BBC News Online*, 2003, http://news.bbc.co.uk/2/hi/uk_news/2963619.stm.

'The Sinn Fein Rising, ' in *The Telephone and Telegraph Journal,* 21, 1916, p. 177.

'Sinn Fein volunteers have attacked the castle and have possession of the GPO', https://twitter.com/1916Live/status/724200305810903040.

Sloan, G., 'The British state and the Irish rebellion of 1916: an intelligence failure or an failure of response' in *Intelligence and National Security,* 4, 2013, p 453–494.

Smith, Gordon, 'Magnet to attract consumers with all-in-one service', *Silicon Republic*, https://www.siliconrepublic.com/life/magnet-to-attract-consumers-with-all-in-one-service.

Smurfit, Michael, *A Life Worth Living: The Autobiography* (Cork: Oak Tree Press, 2014).

Speed, Richard, 'FCC Gives SpaceX the Go-Ahead to Drop Starlink Satellite Orbits by 500 Kilometres or So', *The Register*, 28 Apr. 2021, https://www.theregister.com/2021/04/28/fcc_spacex/.

Standage, Tom, *The Victorian Internet: The Remarkable Story of the Telegraph and the Nineteenth Century's On-Line Pioneers* (New York: Walker and Co., 1998).

'Standing Stones', http://www.dgsys.co.uk/btmicrowave/sites/89.php

Stanton, Aaron 'The Periphery At The Center: Valentia Island And The Transatlantic Telegraph' in *Brown Journal of History,* Spring 2008, 2008, pp 127–149.

Sterne, John, *Adventures in Code: The Story of the Irish Software Industry* (Dublin: Liffey Press, 2004).

Technical Centre of the European Broadcasting Union, 'Eurovision links' in *EBU Review. A, Technical,* Issues 89–94, 1965.

'Telecom faces financial crunch', in *Business and Finance* 30 July 1992

'Telefonica bids £18bn for UK's O2', *BBC News*, http://news.bbc.co.uk/2/hi/business/4391754.stm.

TeleGeography, 'The Submarine Cable Map', https://www.submarinecablemap.com

'Telegram from Winston Churchill to Eamon de Valera 8 Dec. 1941', https://www.difp.ie/docs/1941/-Now-or-Never.-A-Nation-once-again-/3577.htm

Telephone Directory: Scotland/ North Western/ Northern/ Ireland (London: The National Telephone Co., 1904).

Telephone Directory Vol 5 (Belfast, Dublin and Cork) (London: The Post Office Telephone Service, 1921).

Thanei, Luca, 'Please connect me! Automating our telephone exchanges', https://blog.national-museum.ch/en/2019/12/please-connect-me-automating-our-telephone-exchanges/

Thom's Official Directory of the United Kingdom of Great Britain and Ireland (Dublin: Thom, 1898).

Thom's Official Directory of the United Kingdom of Great Britain and Ireland (Dublin: Thom, 1881).

Thompson, Silvanus P., *Philipp Reis: Inventor of the Telephone, A Biographical Sketch* (London: Spon, 1883).

Thompson, Silvanus Phillips, *The Life of William Thomson, Baron Kelvin of Largs*vol. II, (Cambridge: Cambridge University Press, 1910).

'Thursday Television', in *Radio Times*, 15 July 1955 (1955)

'Timeline of telephony in Sweden', https://timelines.issarice.com/wiki/Timeline_of_ telephony_in_Sweden

Tolstrup, Morten, *Indoor Radio Planning: A Practical Guide for GSM, DCS, UMTS and HSPA* (Hoboken NJ: Wiley, 2008).

Trosby, Finn et al., *Short Message Service (SMS): The Creation of Personal Global Text Messaging* (Chichester: Wiley, 2010).

'UAX5 General Features', http://www.uax.me.uk/uax5/uax5general.html

'Valentia Cable Station 1885', in *The Telegraphist* 1 Dec. 1885.

Van der Wee, Marlies et al., 'Understanding the dynamics of broadband markets: A comparative case study of Flanders and the Netherlands', *23rd European Regional Conference of the International Telecommunications Society (ITS)* (Vienna, 2012).

Varney, Alan Leon, 'NetValley: A New Home for the Mind?', http://www.netvalley.com/archives/mirrors/telegraph__radio_timeline-3.htm

Vaughan, W. E. and A.J. [eds] Fitzpatrick, *Irish Historical Statistics: Population, 1821–1971* (Dublin: Royal Irish Academy, 1978).

Van de Velde, Thierry, *Value-Added Services for Next Generation Networks* (Boca Raton FL: CRC Press, 2007).

'Volledige Lijst der Europeesche Omroepzenders ', http://www.radioheritage.net/europe/images/lists/frequentieplan1933.pdf

Wall, Thomas F., 'Railways & Telecommunications – 1' in *Journal of the Irish Railway Record Society,* 142, 2000, pp 479–485.

—, 'Railways & Telecommunications – 4' in *Journal of the Irish Railway Record Society,* 147, 2002, pp 194–203.

—, *Some notes towards a history of telecommunications with particular reference to Ireland* (2005), Dublin City Archive.

Ward, Keith, 'A short history of telecommunications transmission in the UK' in *Journal Of The Communications Network,* January–March, 2006, pp 30–41.

Warren, Stanley, 'Montrose House and the Jameson Family in Dublin and Wexford: A Personal Reminiscence' in *The Past: The Organ of the Uí Cinsealaigh Historical Society,* 28, 2007, pp 87–97.

Waterways Ireland, *Annual Report and Accounts* (Enniskillen: 2007).

Weckler, Adrian (@adrianweckler) 'Speedtest on Huawei Mate 20 X (5G).' *Twitter,* 13 Aug. 2019.

'What's in a number?', *Silicon Republic*, https://www.siliconrepublic.com/gear/whats- in-a-number.

Wheen, Andrew, *Dot-Dash to Dot.com: How Modern Telecommunications Evolved from the Telegraph to the Internet* (New York: Springer, 2011).

Whelan, Bernadette, *Ireland and the Marshall Plan, 1947-57* (Dublin: Four Courts Press, 2000).

Wills, Clair, *That Neutral Island: A Cultural History of Ireland during the Second World War* (London: Faber, 2007).

Winseck, Dwayne R. and Robert M. Pike, 'Communication and Empire: Media, Markets, and Globalization, 1860-1930' in *Business History Review,* 2008, pp 138–178.

Wood, Ian S., *Ireland during the Second World War* (London: Caxton Editions, 2002).

Wood, Rupert, et al., *Full-fibre Access as Strategic Infrastructure: Strengthening Public Policy for Europe* (London: Analysys Mason, 2020).

Woods, Damian, *The Fateful Day: A Commemorative Book of the Armagh Railway Disaster June 12th 1889* (Armagh: Armagh District Council, 1989).

'Worldwide mobile data pricing: The cost of 1GB of mobile data in 228 countries', https://www.cable.co.uk/mobiles/worldwide-data-pricing/

Young, Peter, *Power of Speech: A History of Standard Telephones and Cables, 1883–1983* (London and Boston: Allen & Unwin, 1983).

'Zugfunk 1918 – 1926 – 1940', http://www.oebl.de/A-Netz/Rest/Zugfunk/Zug1926.html

Broadcast Media

'Could You Hold A Minute Please?', *Week In*, RTÉ, 25 Feb. 1980.

'The End Of The Speaking Clock', *The Ray D'Arcy Show*, RTÉ Radio 1, 31 July 2018.

Marks, Jonathan, 'MN.20.March.1987 - Ireland Calling on Shortwave', *The Media Network Vintage Vault,* podcast, 2 July 2010.

'Minister Launches New Technology: The Payphone,' *RTÉ News*, RTÉ, 3 Mar. 1981, www.rte.ie/archives/2016/0302/772141-introduction-of-payphone

'Mountshannon Goes Automatic', *RTÉ News*, RTÉ, 28 May 1987, www.rte.ie/archives/collections/news/21249852-mountshannon-goes-automatic.

'President First On New Television Service *1961'*, RTÉ, 31 Dec. 1961.

'Text Nation,' *RTÉ News*, RTÉ, 8 Jan. 2004.

Weckler, Adrian, 'A frank chat with eir's CEO', *The Big Tech Show,* podcast, 10 May 2019.

INDEX